LASER INTERFEROMETER SPACE ANTENNA

THE SECOND INTERNATIONAL LISA SYMPOSIUM

ON THE DETECTION AND OBSERVATION OF

GRAVITATIONAL WAVES IN SPACE

was sponsored by

NASA - Office of Space Science
Jet Propulsion Laboratory,
California Institute of Technology

LASER INTERFEROMETER SPACE ANTENNA

Second International LISA Symposium on the Detection and Observation of Gravitational Waves in Space

Pasadena, California July 1998

EDITOR
William M. Folkner
Jet Propulsion Laboratory,
California Institute of Technology

American Institute of Physics

AIP CONFERENCE PROCEEDINGS 456

Woodbury, New York

Editor:

William M. Folkner
Jet Propulsion Laboratory,
California Institute of Technology
Mail Stop 238-600
4800 Oak Grove Drive
Pasadena, CA 91109

E-mail: William.Folkner@jpl.nasa.gov

L.C. Catalog Card No. 98-88783
ISBN 1-56396-848-7
ISSN 0094-243X
DOE CONF- 980761

Printed in the United States of America

PDA Stack
1/7/97

CONTENTS

THE LISA MISSION

SOURCES OF GRAVITATIONAL WAVES

ANALYSIS TECHNIQUES FOR GRAVITATIONAL-WAVE DETECTORS

TECHNOLOGY FOR GRAVITATIONAL-WAVE DETECTORS IN SPACE

GROUND AND OTHER GRAVITATIONAL-WAVE DETECTORS

Preface

Since shortly after the pioneering efforts to detect gravitational radiation by Joseph Weber in the late 1960's, concepts for gravitational-wave detectors in space have been explored. By the mid 1980's, with the experience in the development of ground-based laser interferometer detectors, the concepts for space-based gravitational-wave detectors were settling towards a constellation of spacecraft to form an interferometer detector with arm lengths of order one million kilometers. Such space-based interferometers will be able to detect gravitational waves with periods from one second to thousands of seconds. These 'low frequency' gravitational waves can never be observed by ground-based detectors due to the interference of ground vibrations and local changes in the Earth's gravitational field. Thus the space-based detectors will complement present and planned ground-based detectors, and observe gravitational waves from different types of astrophysical sources.

In the early 1990's major efforts to develop large ground-based detectors were underway, and at the same time the technology needed to implement a spaced-based detector was significantly advanced. In 1993 the Laser Interferometer Space Antenna (LISA) project was proposed by an international consortium of scientists. In 1995 LISA was recommended as the third Cornerstone mission of the then 'Horizons 2000' program of the European Space Agency, and in 1997 LISA was included in the Mission Roadmap of the Structure and Evolution of the Universe theme of the National Aeronautics and Space Administration.

The First International LISA Symposium was held in Oxford, United Kingdom in July 1996 to address the field of space detection of gravitational waves. The symposium covered the topics of astrophysical sources of low-frequency gravitational waves, technological and engineering issues associated with a space-based detection of gravitational waves, and the relationship to ground-based detectors in operation and under construction.

The Second International LISA Symposium was held in Pasadena, California, July 1998 to present the significant advances made between 1996 and 1998 in our knowledge of the astrophysical sources of gravitational waves, and in the continually maturing technology and mission studies for space-based detectors. Papers included in this volume include new estimates of the characteristics and numbers of sources of low-frequency gravitational radiation, including those associated with massive black holes in the centers of galaxies; evaluations of the ability of a space-based gravitational-wave detector to measure the properties of astrophysical sources; refined mission design and engineering studies associated with the LISA project; technology advances enabling space-based gravitational-wave detection; and an update on the status of ground-based gravitational-wave detectors, many of which are rapidly nearing completion. These advances led to an enthusiastic meeting with very good prospects for the successful development and exciting discoveries from a space-based gravitational-wave mission within the next decade.

The program was arranged by the Science Program Committee, consisting of P.L. Bender, C. Cutler, K.V. Danzmann, J. Hough, Y.R. Jafry, G.M. Keiser, T.A. Prince, R. Reinhard, D.O. Richstone, M.C W. Sandford, B.F. Schutz, D.H. Shoemaker, S.L. Shapiro, R.T. Stebbins, J.-Y. Vinet, and S. Vitale. The Symposium was made possible by the sponsorship of the Jet Propulsion Laboratory, California Institute of Technology and the National Aeronautics and Space Administration.

W. M. Folkner

THE LISA MISSION

LISA and Ground-Based Detectors for Gravitational Waves: An Overview

Karsten Danzmann* for the LISA Study Team

**Institut für Atom- und Molekülphysik, Universität Hannover, and Max-Planck Institut für Quantenoptik, Außenstelle Hannover, Callinstr. 38, D-30167 Hannover, Germany*

Abstract. The gravitational wave spectrum covers many decades in frequency. Sources in the audio-frequency regime above 1 Hz are accessible to ground-based detectors while sources in the low-frequency regime can only be observed from space because of the unshieldable background of local gravitational noise on the ground and because ground-based interferometers are limited in length to a few kilometres. Laser interferometry is a promising technique to observe the minute distance changes caused by gravitational waves, but the actual implementation is very different on ground and in space. An overview of LISA and other detectors will be given.

I THE NATURE OF GRAVITATIONAL WAVES

In Newton's theory of gravity the gravitational interaction between two bodies is instantaneous, but according to Special Relativity this should be impossible, because the speed of light represents the limiting speed for all interactions. If a body changes its shape the resulting change in the force field will make its way outward at the speed of light. It is interesting to note that already in 1805, Laplace, in his famous *Traité de Mécanique Céleste* stated that, if Gravitation propagates with finite speed, the force in a binary star system should not point along the line connecting the stars, and the angular momentum of the system must slowly decrease with time. Today we would say that this happens because the binary star is losing energy and angular momentum by emitting gravitational waves. It was no less than 188 years later in 1993 that Hulse and Taylor were awarded the Nobel prize in physics for the indirect proof of the existence of Gravitational Waves using exactly this kind of observation on the binary pulsar PSR 1913+16. A direct detection of gravitational waves has not been achieved up to this day.

Einstein's paper on gravitational waves was published in 1916, and that was about all that was heard on the subject for over fourty years. It was not until the late 1950s that some relativity theorists, H. Bondi in particular, proved rigorously that gravitational radiation was in fact a physically observable phenomenon, that gravitational waves carry energy and that, as a result, a system that emits gravitational waves should lose energy.

General Relativity replaces the Newtonian picture of Gravitation by a geometric one that is very intuitive if we are willing to accept the fact that space and time do not have an independent existence but rather are in intense interaction with the physical world. Massive bodies produce "indentations" in the fabric of spacetime, and other bodies move in this curved spacetime taking the shortest path, much like a system of billard balls on a springy surface. In fact, the Einstein field equations relate mass (energy) and curvature in just the same way that Hooke's law relates force and spring deformation, or phrased somewhat poignantly: spacetime is an elastic medium.

If a mass distribution moves in an asymmetric way, then the spacetime indentations travel outwards as ripples in spacetime called gravitational waves. Gravitational waves are fundamentally different from the familiar electromagnetic waves. While electromagnetic waves, created by the acceleration of electric charges,

CP456, *Laser Interferometer Space Antenna*
edited by William M. Folkner
 1-56396-848-7/98/$15.00

propagate IN the framework of space and time, gravitational waves, created by the acceleration of masses, are waves of the spacetime fabric ITSELF.

Unlike charge, which exists in two polarities, mass always come with the same sign. This is why the lowest order asymmetry producing *electro-magnetic* radiation is the dipole moment of the charge distribution, whereas for *gravitational* waves it is a change in the quadrupole moment of the mass distribution. Hence those gravitational effects which are spherically symmetric will not give rise to gravitational radiation. A perfectly symmetric collapse of a supernova will produce no waves, a non-spherical one will emit gravitational radiation. A binary system will always radiate.

Gravitational waves distort spacetime, in other words they change the distances between free macroscopic bodies. A gravitational wave passing through the Solar System creates a time-varying strain in space that periodically changes the distances between all bodies in the Solar System in a direction that is perpendicular to the direction of wave propagation. These could be the distances between spacecraft and the Earth, as in the case of ULYSSES or CASSINI (attempts were and will be made to measure these distance fluctuations) or the distances between shielded proof masses inside spacecraft that are separated by a large distance, as in the case of LISA. The main problem is that the relative length change due to the passage of a gravitational wave is exceedingly small. For example, the periodic change in distance between two proof masses, separated by a sufficiently large distance, due to a typical white dwarf binary at a distance of 50 pc is only 10^{-10} m. This is not to mean that gravitational waves are weak in the sense that they carry little energy. On the contrary, a supernova in a not too distant galaxy will drench every square meter here on earth with kilowatts of gravitational radiation intensity. The resulting length changes, though, are very small because spacetime is an extremely stiff elastic medium so that it takes extremely large energies to produce even minute distortions.

II SOURCES OF GRAVITATIONAL WAVES

The two main categories of gravitational waves sources for LISA are the galactic binaries and the massive black holes (MBHs) expected to exist in the centres of most galaxies.

Because the masses involved in typical binary star systems are small (a few solar masses), the observation of binaries is limited to our Galaxy. Galactic sources that can be detected by LISA include a wide variety of binaries, such as pairs of close white dwarfs, pairs of neutron stars, neutron star and black hole ($5 - -20\,M_\odot$) binaries, pairs of contacting normal stars, normal star and white dwarf (cataclysmic) binaries, and possibly also pairs of black holes. It is likely that there are so many white dwarf binaries in our Galaxy that they cannot be resolved at frequencies below 10^{-3} Hz, leading to a confusion-limited background. Some galactic binaries are so well studied, especially the X-ray binary 4U1820-30, that it is one of the most reliable sources. If LISA would not detect the gravitational waves from known binaries with the intensity and polarisation predicted by General Relativity, it will shake the very foundations of gravitational physics.

The main objective of the LISA mission, however, is to learn about the formation, growth, space density and surroundings of massive black holes (MBHs). There is now compelling indirect evidence for the existence of MBHs with masses of 10^6 to $10^8\,M_\odot$ in the centres of most galaxies, including our own. The most powerful sources are the mergers of MBHs in distant galaxies, with amplitude signal-to-noise ratios of several thousand for $10^6\,M_\odot$ black holes. Observations of signals from these sources would test General Relativity and particularly black-hole theory to unprecedented accuracy. Not much is currently known about black holes with masses ranging from about $100\,M_\odot$ to $10^6\,M_\odot$. LISA can provide unique new information throughout this mass range.

III COMPLEMENTARITY OF DETECTION ON THE GROUND AND IN SPACE

Astronomical observations of electromagnetic waves cover a range of 20 orders of magnitude in frequency, from ULF radio waves to high-energy gamma-rays. Almost all of these frequencies (except for visible and radio) cannot be detected from the Earth, and therefore it is necessary to place detectors optimised for a particular frequency range (e.g. radio, infrared, ultraviolet, X-ray, gamma-ray) in space.

The situation is similar for gravitational waves. The range of frequencies spanned by ground- and space-based detectors, as shown schematically in Figure 1, is comparable to the range from high frequency radio waves up to X-rays. Ground-based detectors will never be sensitive below about 1 Hz, because of terrestrial gravity-gradient noise. A space-based detector is free from such noise and can be made very large, thereby

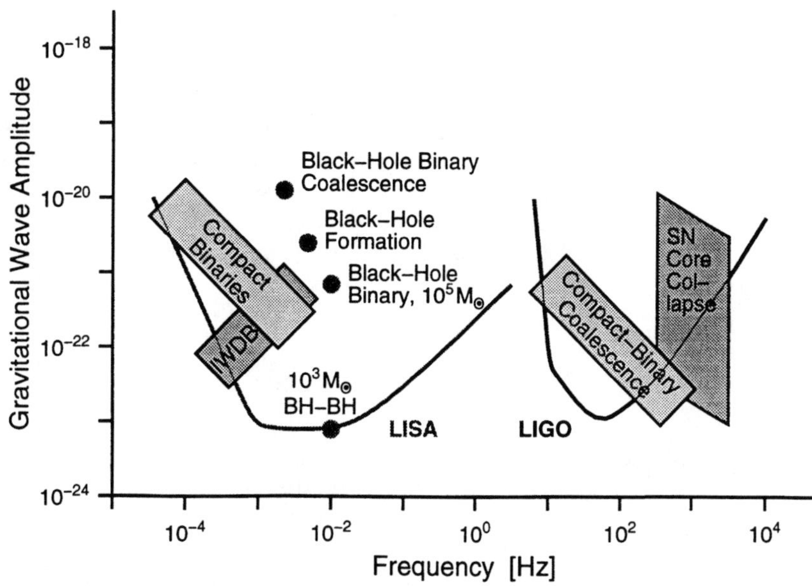

FIGURE 1. Comparison of frequency range of sources for ground-based and space-based gravitational wave detectors. Only a few typical sources are indicated, ranging in frequency from the kHz region of supernovae and final mergers of binary stars down to mHz events due to formation and coalescence of supermassive black holes, compact binaries and interacting white dwarf binaries. The sources shown are in two clearly separated regimes: events in the range from, say, 10 Hz to several kHz (and only these will be detectable with terrestrial antennas), and a low-frequency regime, 10^{-4} to 10^{-1} Hz, accessible only with a space project. Sensitivities of LISA for periodic sources, and of (the "Advanced") LIGO for burst sources, are indicated.

opening the range from 10^{-4} Hz to 1 Hz, where both the most certain and the most exciting gravitational-wave sources radiate most of their power.

The importance of low frequencies is a consequence of Newton's laws. For systems involving solar-mass objects, lower frequencies imply larger orbital radii, and the range down to 10^{-4} Hz includes sources with the typical dimensions of many galactic neutron star binaries, cataclysmic binaries, and some known binaries. These are the most certain sources. For highly relativistic systems, where the orbital velocities approach the speed of light, lower frequencies imply larger masses ($M \propto 1/f$), and the range down to 10^{-4} Hz reaches masses of $10^7\, M_\odot$, typical of the black holes that are believed to exist in the centres of many, if not most, galaxies. Their formation and coalescences could be seen anywhere in the Universe and are among the most exciting of possible sources. Detecting them would test the strong-field limit of gravitational theory and illuminate galaxy formation and quasar models.

For ground-based detectors, on the other hand, their higher frequency range implies that even stellar-mass systems can last only for short durations, so these detectors will mainly search for sporadic short-lived catastrophic events (supernovae, coalescing neutron-star binaries). Normally, several detectors are required for directional information. If such events are not detected in the expected way, this will upset the astrophysical models assumed for such systems, but not necessarily contradict gravitation theory.

By contrast, if a space-based interferometer does not detect the gravitational waves from known binaries with the intensity and polarisation predicted by General Relativity, it will undermine the very foundations of gravitational physics. Furthermore, even some highly relativistic events, such as massive black hole coalescences with masses below $10^5\, M_\odot$, last roughly a year or longer. This allows a single space-based detector to provide directional information as it orbits the Sun during the observation.

Both ground- and space-based detectors will also search for a cosmological background of gravitational waves. Since both kinds of detectors have similar energy sensitivities, their different observing frequencies are ideally complementary: observations can supply crucial spectral information.

The space-based interferometer proposal has the full support of the ground-based detector community. Just as it is important to make observations at radio, optical, X-ray, and all other electromagnetic wavelengths, so too is it important to cover different gravitational-wave frequency ranges. Ground-based and space-based observations will therefore complement each other in an essential way.

IV GROUND-BASED DETECTORS

The highest frequencies expected for the emission of strong gravitational waves are around 10 kHz because a gravitational wave source cannot emit strongly at periods shorter than the light travel time across its gravitational radius. At frequencies below 1 Hz, observations on the ground are impossible because of an unshieldable background due to Newtonian gravity gradients on the earth. These two frequencies define the limits of the high-frequency band of gravitational radiation, mainly populated by signals from neutron star and stellar mass black hole binaries. This band is the domain of ground-based detectors: laser interferometers and resonant-mass detectors.

A Resonant-mass detectors

The history of attempts to detect gravitational waves began in the 1960s with the famous bar experiments of Joseph Weber [5]. A resonant-mass antenna is, in principle, a simple object. It consists of a solid body that during the passage of a gravitational wave gets excited similarly to being struck with a hammer, and then rings like a bell.

The solid body traditionally used to be a cylinder, that is why resonant-mass detectors are usually called bar detectors. But in the future we may see very promising designs in the shape of a sphere or sphere-like object like a truncated icosahedron. The resonant mass is usually made from an aluminum alloy and has a mass of several tons. Occasionally, other materials are used, e.g. silicon, sapphire or niobium.

The first bar detectors were operated at room temperature, but the present generation of bars is operating below liquid-helium temperature. The next generation, which is already under construction (NAUTILUS in Frascati, AURIGA in Legnaro), will operate at a temperature around 50 mK.

Resonant-mass detectors are equipped with transducers that monitor the complex amplitudes of one or several of the bar's vibrational modes. A passing gravitational wave changes these amplitudes due to its frequency content near the normal mode frequencies. Present-day resonant mass antennas are fairly narrowband devices, with bandwidths of only a few Hz around centre-frequencies in the kHz range. With improved transducer designs in the future, we may see the bandwidth improve to 100 Hz or better.

The sensitivities of bar antennas have steadily improved since the first experiments of Joe Weber. Currently, we see a network of antennas at Rome, Louisiana State, and Perth operating at an rms noise level equivalent to $h_{\rm rms} \approx 6 \times 10^{-19}$. In the first decade of the next millennium, planned sphere-like detectors operating near the standard quantum limit may reach burst sensitivities below 10^{-21} in the kHz range [6].

B Laser Interferometers

Although the seeds of the idea can be found in early papers by Pirani [7] and Gertsenshtein and Pustovoit [8], it was really in the early 1970s that the idea emerged that laser interferometers might have a better chance of detecting gravitational waves, mainly promoted by Weiss [9] and Forward [10].

A Michelson interferometer measures the phase difference between two light fields having propagated up and down two perpendicular directions, i.e. essentially the length difference between the two arms (see Figure IV B). This is the quantity that would be changed by a properly oriented gravitational wave. The phase difference measured can be increased by increasing the armlength, or, equivalently, the interaction time of the light with the gravitational wave, up to an optimum for an interaction time equal to half a gravitational wave period. For a gravitational wave frequency of 100 Hz this corresponds to five milliseconds or an armlength of 750 km. On the ground it is clearly impractical to build such large interferometers, but there are ways to increase the interaction time without increasing the physical armlength beyond reasonable limits. Several variants have been developed, all of them relying on storing and enhancing the laser light, or the gravitational-wave induced sidebands, or both. The technology and techniques for such interferometers have now been under development for nearly 30 years. Table 3 gives an impression of the wide international scope of the interferometer efforts. After pioneering work at MIT, other groups at Munich/Garching, at Glasgow, then Caltech, Paris/Orsay, Pisa, and later in Japan, also entered the scene. Their prototypes range from a few meters up to 30, 40, and even 100 m.

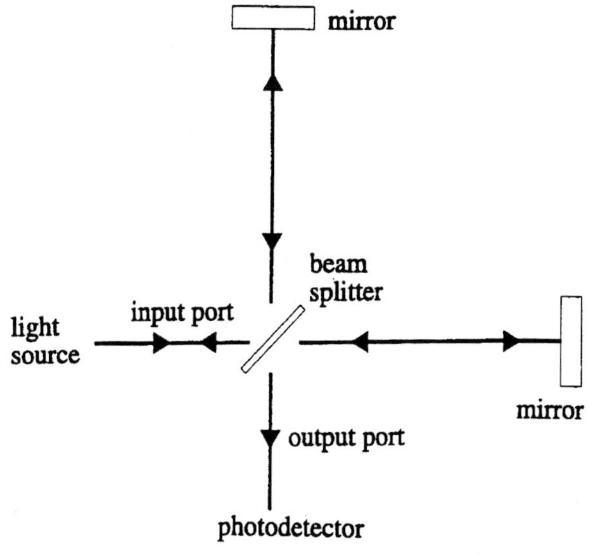

FIGURE 2. Schematic of a two-arm Michelson interferometer. Interference of the two returning beams on the photodetector gives a measure of their relative phase. Any variation in the mirror distances caused by a passing gravitational wave will modulate this phase signal. By having two arms fed from the same light source, the phase noise inherent to the non-ideal source is the *same* in each arm. In essence, the interferometry amounts to a differencing of the phases from the light returning along each arm, so the common-mode noise in the light is cancelled. This is the main reason for having two arms in a Michelson interferometer.

Country: *Institute:*	**USA** MIT	**USA** Caltech	**GER** MPQ	**GBR** Glasgow	**FRA** CNRS	**ITA** INFN	**JPN** ISAS	**JPN** NAO
Prototypes:								
Start:	1972	1980	1975	1977	1983	1986	1986	1991
Laser:	Ar^+	Ar^+	Ar^+	Ar^+	(Ar^+)	(Ar^+)	Ar^+	YAG
Arm length ℓ:	40 m		30 m	10 m	0.5 m		100 m	20 m
Strain sensitivity $\tilde{h}$ $[Hz^{-\frac{1}{2}}]$:	$1 \cdot 10^{-20}$ 1995		$11 \cdot 10^{-20}$ 1986	$6 \cdot 10^{-20}$ 1992			$8 \cdot 10^{-20}$ 1996	$2 \cdot 10^{-18}$ 1996
Large Interferometric Detectors:								
Planning (start):	1982	1984	1985	1986	1986	1986	1987	1994
Arm length ℓ:	4 km 2 km	4 km	600 m		3 km		300 m	
Site *(State)*	Hanford (WA)	Livingston (LA)	Hannover GER		Pisa ITA		Mitaka JPN	
Cost (10^6 US$):	292		7		90		15	
Project name:	**LIGO**		**GEO 600**		**VIRGO**		**TAMA 300**	

TABLE 1. Funded ground-based interferometric gravitational wave detectors: List of prototypes (upper part) and long-baseline projects (lower part).

Today, these prototype interferometers are routinely operating at a displacement noise level of the order $10^{-19}\,\mathrm{m}/\sqrt{\mathrm{Hz}}$ over a frequency range from 200 Hz to 1000 Hz, corresponding to an rms gravitational-wave amplitude noise level of $h_{\mathrm{rms}} \approx 10^{-19}$.

Plans for kilometer-size interferometers have been developed for the last 15 years. All of these large-scale projects will use low-noise Nd:YAG lasers (wavelength $1.064\,\mu$m), pumped with laser diodes, just as is intended for LISA, which will greatly benefit from their efforts for achieving extreme stability and high overall efficiency.

The US project LIGO calls for *two* facilities at two widely separated sites [11]. Both will house a 4 km interferometer, Hanford an additional 2 km interferometer. At both sites ground-work and construction have been finished, vacuum tests (of the "world's largest vacuum chamber") successfully completed, and installation of optics is to begin in the latter half of 1998.

In the French-Italian project VIRGO, being built near Pisa, an elaborate seismic isolation system will allow this project to measure down to a frequency of 10 Hz or even below [12]. Construction is in full progress.

A British-German collaboration has de-scoped the project of a 3 km antenna to a length of only 600 m: GEO 600 [13]. It will employ advanced optical techniques to make up for the shorter arms. Ground work and construction at the site near Hannover are completed, the vacuum system tested.

In Japan, after a merger of efforts at ISAS and other institutions, construction and vacuum verification of a common 300 m project called TAMA 300 [14] is completed, and first optical tests in single arms have been performed.

Not included in Table 3 is the (not yet funded) Australian project of a 500 m detector to be built near Perth. The site would allow later extension to 3 km arms.

LIGO, VIRGO, GEO 600 and TAMA 300 are scheduled to be completed by the end of this century. Observations may begin in 2000 or 2001, although the sensitivity of the first stage detectors may be only marginally sufficient to detect gravitational waves. However, step-by-step improvements will be made, until the network finally reaches the advanced detector sensitivity sometime between 2005 and 2010. At that point, one can be confident that signals will be observed from sources such as supernovae, compact binary coalescences and pulsars, unless something is fundamentally wrong with our current estimates of their strength and distribution.

V PULSAR TIMING

Man-made gravitational wave detectors operate by detecting the effect of gravitational waves on the apparatus. It is also possible to detect gravitational waves by observing their effect on electromagnetic waves as they travel to us from astronomical objects. Such methods of detection are like "one-arm interferometers" – the second arm is not needed if there is another way to provide a reference clock stable enough to sense the changes in propagation time produced by gravitational waves.

Pulsar timing makes use of the fact that the pulsar is a very steady clock. If we have a clock on the Earth that is as stable as the pulsar, then irregularities in the arrival times of pulses that are larger than expected from the two "clocks" can be attributed to external disturbances, and in particular possibly to gravitational waves. Since the physics near a pulsar is poorly known, it might be difficult to prove that observed irregularities are caused by gravitational waves. But where irregularities are absent, this provides an upper limit to the gravitational wave field. This is how such observations have been used so far.

All pulsars slow down, and a few have shown systematic changes in the slowing down rate. Therefore, it is safer to use random irregularities in the pulsar rate as the detection criterion, rather than systematic changes. Such random irregularities set limits on random gravitational waves: the stochastic background.

The arrival times of individual pulses from most pulsars can be very irregular. Pulsar periods are stable only when averaged over considerable times. The longer the averaging period, the smaller are the effects of this intrinsic irregularity. Therefore, pulsar timing is used to set limits on random gravitational waves whose period is of the same order as the total time the pulsar has been observed, from its discovery to the present epoch. Millisecond pulsars seem to be the most stable over these long periods, and a number of them are being used for these observations.

The best limits come from the first discovered millisecond pulsar, PSR 1937+21. At a frequency of approximately 1 per 10 years the pulsar sets an upper limit on the energy density of the gravitational wave background of $\Omega_{GW} < 10^{-7}$ [15]. This is in an ultra-low frequency range that is 10^5 times lower than the LISA band and 10^{10} times lower than the ground-based band. If one believes a theoretical prediction of the spectrum of a cosmic gravitational wave background, then one can extrapolate this limit to the other bands. But this may be naive, and it is probably wiser to regard observations in the higher-frequency bands as independent searches for a background.

More-recently discovered millisecond pulsars are also being monitored and will soon allow these limits to be strengthened. If irregularities are seen in all of them at the same level, and if these are independent of the radio frequency used for the observations, then that will be strong evidence that gravitational waves are indeed responsible.

These observations have the potential of being extended to higher frequencies by directly cross-correlating the data of two pulsars. In this way one might detect a correlated component caused by gravitational waves passing the Earth at the moment of reception of the radio signals from the two pulsars. Higher frequencies are accessible because the higher intrinsic timing noise is reduced by the cross-correlation. Again, seeing the effect in many pairs of pulsars independently of the radio frequency would be strong evidence for gravitational waves.

VI SPACECRAFT TRACKING

Precise, multi-frequency transponding of microwave signals from interplanetary probes, such as the ULYSSES, GALILEO and CASSINI spacecraft, can set upper limits on low-frequency gravitational waves. These appear as irregularities in the time-of-communication residuals after the orbit of the spacecraft has been fitted. The irregularities have a particular signature. Searches for gravitational waves have produced only upper limits so far, but this is not surprising: their sensitivity is far short of predicted wave amplitudes. This technique is inexpensive and well worth pursuing, but will be limited for the forseeable future by some combination of measurement noise, the stability of the frequency standards, and the uncorrected parts of the fluctuations in propagation delays due to the interplanetary plasma and the Earth's atmosphere. Consequently, it is unlikely that this method will realise an *rms* strain sensitivity much better than 10^{-17}, which is six orders of magnitude worse than that of a space-based interferometer.

VII THE LISA MISSION

The LISA mission comprises three identical spacecraft located 5×10^6 km apart forming an equilateral triangle. LISA is basically a giant Michelson interferometer placed in space, with a third arm added to give independent information on the two gravitational wave polarizations, and for redundancy. The distance between the spacecraft – the interferometer arm length – determines the frequency range in which LISA can make observations; it was carefully chosen to allow for the observation of most of the interesting sources of gravitational radiation. The centre of the triangular formation is in the ecliptic plane, 1 AU from the Sun and 20° behind the Earth. The plane of the triangle is inclined at 60° with respect to the ecliptic. These particular heliocentric orbits for the three spacecraft were chosen such that the triangular formation is maintained throughout the year with the triangle appearing to rotate about the centre of the formation once per year.

While LISA can be described as a big Michelson interferometer, the actual implementation in space is very different from a laser interferometer on the ground and is much more reminiscent of the technique called spacecraft tracking, but here realized with infrared laser light instead of radio waves. The laser light going out from the center spacecraft to the other corners is not directly reflected back because very little light intensity would be left over that way. Instead, in complete analogy with an RF transponder scheme, the laser on the distant spacecraft is phase-locked to the incoming light providing a return beam with full intensity again. After being transponded back from the far spacecraft to the center spacecraft, the light is superposed with the on-board laser light serving as a local oscillator in a heterodyne detection. This gives information on the length of one arm modulo the laser frequency. The other arm is treated the same way, giving information on the length of the other arm modulo the same laser frequency. The difference between these two signals will thus give the difference between the two arm lengths (i.e. the gravitational wave signal). The sum will give information on laser frequency fluctuations.

Each spacecraft contains two optical assemblies. The two assemblies on one spacecraft are each pointing towards an identical assembly on each of the other two spacecraft to form a Michelson interferometer. A 1 W infrared laser beam is transmitted to the corresponding remote spacecraft via a 30-cm aperture $f/1$ Cassegrain telescope. The same telescope is used to focus the very weak beam (a few pW) coming from the distant spacecraft and to direct the light to a sensitive photodetector where it is superimposed with a fraction of the original local light. At the heart of each assembly is a vacuum enclosure containing a free-flying polished platinum-gold cube, 4 cm in size, referred to as the proof mass, which serves as an optical reference ("mirror") for the light beams. A passing gravitational wave will change the length of the optical path between the proof masses of one arm of the interferometer relative to the other arm. The distance fluctuations are measured to sub-Ångstrom precision which, when combined with the large separation between the spacecraft, allows LISA to detect gravitational-wave strains down to a level of order $\Delta\ell/\ell = 10^{-23}$ in one year of observation, with a signal-to-noise ratio of 5.

The spacecraft mainly serve to shield the proof masses from the adverse effects due to the solar radiation pressure, and the spacecraft position does not directly enter into the measurement. It is nevertheless necessary to keep all spacecraft moderately accurately ($10^{-8}\,\mathrm{m}/\sqrt{\mathrm{Hz}}$ in the measurement band) centered on their respective proof masses to reduce spurious local noise forces. This is achieved by a "drag-free" control system, consisting of an accelerometer (or inertial sensor) and a system of electrical thrusters.

Capacitive sensing in three dimensions is used to measure the displacements of the proof masses relative to the spacecraft. These position signals are used in a feedback loop to command micro-Newton ion-emitting proportional thrusters to enable the spacecraft to follow its proof masses precisely. The thrusters are also used to control the attitude of the spacecraft relative to the incoming optical wavefronts, using signals derived from quadrant photodiodes. As the three-spacecraft constellation orbits the Sun in the course of one year, the observed gravitational waves are Doppler-shifted by the orbital motion. For periodic waves with sufficient signal-to-noise ratio, this allows the direction of the source to be determined (to arc minute or degree precision, depending on source strength).

Each of the three LISA spacecraft has a launch mass of about 400 kg (plus margin) including the payload, ion drive, all propellants and the spacecraft adapter. The ion drives are used for the transfer from the Earth orbit to the final position in interplanetary orbit. All three spacecraft can be launched by a single Delta II 7925H. Each spacecraft carries a 30 cm steerable antenna used for transmitting the science and engineering data, stored on board for two days, at a rate of 7 kBps in the Ka-band to the 34-m network of the DSN. Nominal mission lifetime is two years.

LISA is envisaged as a NASA/ESA collaborative project, with NASA providing the launch vehicle, the Ka-band telecommunications system on board the spacecraft, mission and science operations and about 50 % of the payload, ESA providing the three spacecraft including the ion drives, and European institutes, funded nationally, providing the other 50 % of the payload. The collaborative NASA/ESA LISA mission is aimed at a launch in the 2008–2010 time frame.

REFERENCES

1. C.W. Misner, K.S. Thorne, and J.A. Wheeler, *Gravitation* (Freeman & Co., San Francisco, 1973).
2. P.R. Saulson, *Fundamentals of Interferometric Gravitational Wave Detectors* (World Scientific, Singapore, 1994).
3. B.F. Schutz, *A First Course in General Relativity* (Cambridge University Press, Cambridge, 1985).
4. K.S. Thorne, *Gravitational Radiation*, in: S.W. Hawking and W. Israel, eds., *300 Years of Gravitation* (Cambridge University Press, Cambridge, 1987) 330–458.
5. J. Weber, Phys. Rev. **117** (1960) 306.
6. M. Bassan, Class. Quant. Grav. Supplement A **39** (1994) 11.
7. F.A.E. Pirani, Acta Physica Polonica **15** (1956) 389.
8. M.E. Gertsenshtein and V.I. Pustovoit, JETP **16** (1963) 433.
9. R. Weiss, Quarterly Progress Report of RLE, MIT **105** (1971) 54.
10. G.E. Moss, L.R. Miller, and R.L. Forward, Appl. Opt. **10** (1971) 2495.
11. A. Abramovici *et al.*, Science **256** (1992) 325.
12. G. Bradaschia *et al.*, Nucl. Instrum. and Methods A **289** (1990) 518.
13. K. Danzmann *et al.*, *GEO 600 – A 300 m laser-interferometric gravitational wave antenna*, Proc. 1st Edoardo Amaldi Conference, Frascati, June 1994; and also: J. Hough *et al.*, Proc. MG7, Stanford, July 1994.
14. K. Tsubono and TAMA collaboration, TAMA *Project* in: K. Tsubono, M.-K. Fujimoto, K. Kuroda, (Eds.), *Gravitational Wave Detection, Proc.* TAMA *Intern. Workshop, Nov. 1996*, p. 183–191, Universal Academy Press (Tokyo, 1997).
15. V.M. Kaspi, J.H. Taylor, M.F. Ryba, Astrophys. J. **428** (1994) 713.

The LISA Mission Design

W. M. Folkner* for the LISA Team

* *Jet Propulsion Laboratory, California Institute of Technology, Pasadena, CA 91109 USA*

Abstract: The Laser Interferometer Space Antenna (LISA) will be capable of detecting gravitational waves with frequencies from 0.1 mHz to 1 Hz by using laser interferometers to monitor changes in the distances between test masses in spacecraft separated by five million km. LISA will detect strains as low as 10^{-23} with a one year observation time and a signal-to-noise ratio of five. The sensitivity will be sufficient to detect gravitational waves from sources connected with massive black holes in the centers in many galaxies, and from many binary systems within the Milky Way galaxy. Under the concept presented, LISA will be formed by three spacecraft at the vertices of an equilateral triangle. The orbits are chosen so that the triangle formation trails the Earth by 20 degrees. Each spacecraft will contain two independent payloads containing a test mass, laser and 30 cm diameter telescope for the transmission and reception of laser signals. Two independent Michelson interferometers will be formed allowing both polarizations of gravitational waves to be detected.

INTRODUCTION

The goal of LISA (Laser Interferometer Space Antenna) is to detect and study low-frequency astrophysical gravitational radiation. The data will be used for research in astrophysics, cosmology, and fundamental physics. LISA is designed to detect the gravitational radiation from regions of the universe that are strongly relativistic, e.g., in the vicinity of black holes. The types of exciting astrophysical sources potentially visible to LISA include extra-galactic massive black hole binaries at cosmological distances, binary systems composed of a compact star and a massive black hole, galactic neutron star-black hole binaries, and background radiation from the Big Bang. LISA will also observe galactic binary systems, which are known to exist.

The effect of a gravitational wave passing through a system of free test masses is to create a strain in space that changes distances between the masses. The LISA mission will comprise three spacecraft located 5×10^6 km apart forming an equilateral triangle. LISA will detect gravitational wave strains down to a level of order 10^{-23} in one year of observation time by measuring the fluctuations in separation between shielded test masses located within each spacecraft. The test masses will be shielded from extraneous disturbances (e.g., solar pressure) by the spacecraft in which it is accommodated so that changes in separation will be due to gravitational forces only. Laser interferometry will be used to measure the separation between the test masses with an accuracy of 10 picometer/$\sqrt{\text{Hz}}$ over the frequency range of 10^{-4} to 10^{-1} Hz.

Figure 1 shows the sensitivity to gravitational waves for the LISA mission and some of the expected signals from galactic binaries. The solid line indicates the LISA sensitivity limit based on measurements of the distance between spacecraft separated by 5 million kilometers with 20 picometer/$\sqrt{\text{Hz}}$ accuracy. The sensitivity from 3×10^{-3} Hz to 3×10^{-2} Hz is limited by the shot noise in the laser interferometer system used to make the distance measurements. The sensitivity for frequencies higher than 3×10^{-2} Hz is also limited by shot noise but with degraded response because the gravitational wavelength is less than the distance between the spacecraft. At frequencies below 3×10^{-3} Hz the sensitivity is limited by noise forces on the test masses within each spacecraft used as reference points for the distance measurements. Source strengths and frequencies for several known binary star systems are shown, as are those for several known interacting white-dwarf binaries (e. g. AmCVn). Within the LISA sensitivity range, thousands of close-white-dwarf binary(CWDB) systems are expected to be observed. At lower levels there may be so many binary systems that their signals cannot be distinguished, which may lead to a confusion noise limit indicated by the dashed line.

CP456, *Laser Interferometer Space Antenna*
edited by William M. Folkner
 1-56396-848-7/98/$15.00

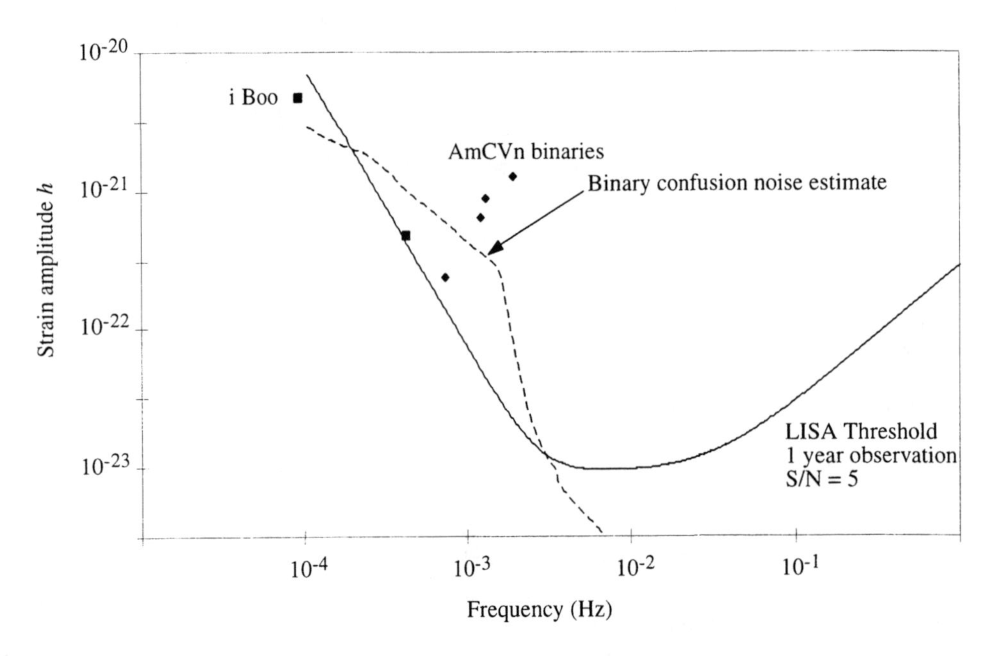

FIGURE 1. LISA sensitivity

MISSION DESIGN OPTIONS

The mission design for LISA has evolved through several configurations. Earliest mission concepts focused on forming a single Michelson interferometer with a central main spacecraft transmitting light to, and receiving light from, two distant spacecraft functioning as the mirrors at the ends of the arms of the interferometers (1). A later option studied was to split the central spacecraft into two, so that there would be four identical spacecraft with two spacecraft close together forming the central part of the Michelson interferometer (2). From this concept the idea of using six spacecraft of identical design arose with two spacecraft at each vertex of a triangle formation (3). The advantage with six spacecraft is that two independent Michelson interferometers can be formed using two of the vertices as centers of separate Michelson interferometers. (The information from the third possible Michelson interferometer would be linearly related to the other two.) This has the advantage of providing measurements of both possible polarizations of gravitational waves. Also, if one of the six spacecraft failed, the remaining spacecraft could still be used to form a single Michelson interferometer and continue observations.

A key issue in the mission cost is the number of spacecraft and the degree of reliability needed for each spacecraft. The mission design presented here is based on the philosophy that each spacecraft should be robust, with no single-point failure modes. Then the mission cost is minimized by having three spacecraft of identical design with backup spacecraft and payload systems. The three spacecraft each transmit and receive laser signals from the other two spacecraft with independent instruments. With all systems operational, two independent interferometers would be formed to observe both gravitational-wave polarizations. If any payload element leads to failure of one instrument, the mission would degrade gracefully into a single interferometer.

ORBIT SELECTION

The LISA laser interferometry measurements will be more difficult to make if the distances between spacecraft are not nearly equal. Thus the preferred orbits are chosen to minimize changes in the distance between spacecraft. The nominal orbits are shown in the Fig. 2a. Each spacecraft will be in an Earth-like orbit with a period of one year going around the Sun. The spacecraft orbits will be slightly elliptical and slightly tilted with respect to each other and with respect to the plane of the Earth's orbit (the ecliptic). By careful choice of the tilts of the orbits, the three spacecraft will maintain a triangular configuration even though each will be separately orbiting about the Sun. Figure 2b highlights the motion of one of the spacecraft and indicates how the distance between spacecraft remains the same and how the triangular formation changes orientation over one year. The change in orientation of the triangle formation is helpful in determining the direction of the sources of observed gravitational waves.

Changes in the spacecraft separation will be caused mainly by the gravitational pull of the Earth. The location of the center of the formation 20 degrees behind the Earth represents a compromise between the desire to reduce the gravitational pull by the Earth and the desire to be closer to the Earth to reduce the amount of propellant needed to reach the operational configuration and to ease the requirements on the telecommunications system. With these orbits, the angle between the two distant spacecraft, as seen from any one spacecraft, will change slowly through the year, by ±1°. This will require the angle between the two instruments on each spacecraft to be adjustable.

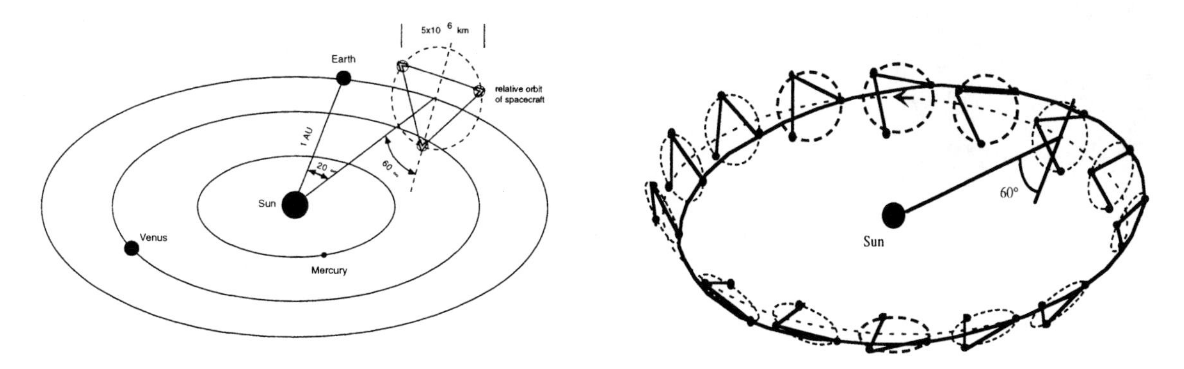

FIGURE 2.a) Initial orbit configuration for the three LISA spacecraft. b) Evolution of orbit configuration over one year. The figures are not to scale. The separation between spacecraft is actually 1/30 of the Sun-Earth distance.

SPACECRAFT DESCRIPTION

The spacecraft design is based on a short structural cylinder 1.8 m in diameter and 0.48 m high. Figure 3a shows an artist's concept of the spacecraft design. The cylindrical shape is efficient for attaching multiple spacecraft together on a single launch vehicle. The cylinder is as short as possible while still accommodating the 30 cm telescopes of the instrument, and as large in diameter as allowed by the launch vehicle's shroud, to maximize resistance to vibration during launch. The cylindrical structure supports a Y-shaped tubular structure which contains the two instruments. The spacecraft equipment will be mounted on the inside wall of the structural cylinder. A sun-shield will extend out from the top of the structural cylinder. During science operations the sun will be 30 degrees from the normal to the top of the cylinder and the sun shield will keep sunlight off the cylinder wall. The main solar panels for the spacecraft will be mounted on this sun shield. A cover across the top of the cylinder (not shown in the artist's concept) will prevent sunlight from striking the Y-shaped structure. The Y-shaped structure is gold-coated and suspended by fiberglass bands from the spacecraft cylinder to thermally isolate it from the spacecraft. The optical assemblies in turn are thermally isolated from the payload thermal shield. The spacecraft cylinder and payload thermal shield are made of a graphite-epoxy composite chosen for its low coefficient of thermal expansion. Two 30 cm diameter X-band radio antennas (not shown) will be mounted to the outside of the spacecraft for communication to the Earth.

INSTRUMENTATION

Figure 3b shows a cross-section of the two instruments. Each instrument contains a 30 cm diameter telescope (f/1 Cassegrain) for transmission and reception of laser signals to another spacecraft. Each instrument also has an optical bench, machined from a block of Ultra-Low Expansion glass with dimensions 20 x 35 x 4 cm, which contains injection, detection and beam shaping optics. An inertial sensor is mounted to the center of each optical bench, containing a test mass shielded from non-gravitational disturbances and a capacitor plate arrangement for measuring the position of the spacecraft with respect to the test mass. The interferometer measures changes in the distance between test masses in the different spacecraft. The spacecraft is kept centered on the test masses, based on the capacitive measurements and using small ion thrusters, to keep motions of the spacecraft from disturbing the test masses. The two instruments are supported at the front by pointing actuators near the telescope primary and by flexures at the other end to allow the angle between them to be changed as the spacecraft configuration changes.

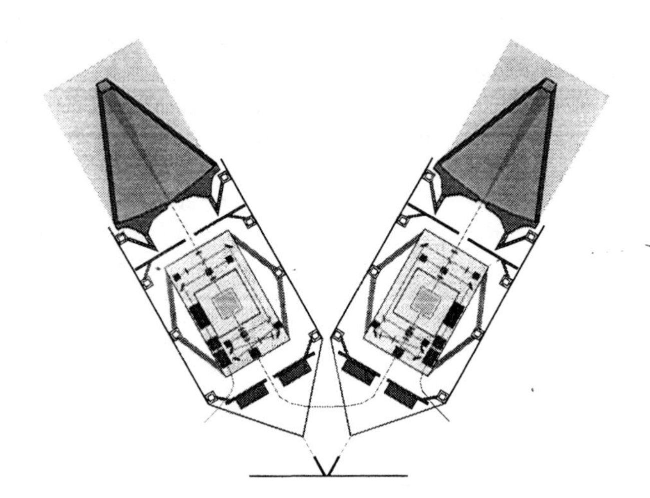

FIGURE 3.a) Artist's concept of the LISA spacecraft. Not shown is a cover over the top of the cylinder that prevents sunlight from striking the Y-shaped payload enclosure. b) Cross section of the two optical assemblies comprising the main part of the payload on each LISA spacecraft. The two assemblies are mounted from flexures at the back (bottom of figure) and from pointing actuators (not shown) at the front, near the primary mirrors.

LAUNCH CONFIGURATION

The three LISA spacecraft are designed to be launched on a single Delta-II 7925H rocket. Figure 4a shows the launch configuration. At launch, each spacecraft will be attached to a propulsion module that will, after launch, be used to guide the three spacecraft into their individual orbits. The three LISA spacecraft and propulsion modules will be stacked vertically inside the launch vehicle's payload envelope. The launch configuration has a propulsion module on the top of the launch stack.

The three spacecraft will be injected into an Earth-escape orbit by the launch vehicle. The Earth-escape orbit will cause the three spacecraft will slowly drift behind the Earth. After launch and injection to the Earth-escape trajectory, the three spacecraft with their propulsion modules will be separated and individually targeted to their desired operational orbits.

The thrust necessary to reach the final orbit configuration will be achieved through solar-electric propulsion, where electrical energy from converted sunlight is used to ionize and accelerate atoms in the direction of the desired thrust. This system requires much less mass at launch than a traditional chemical propulsion system and allows the mission to be launched on a smaller (and less expensive) launch vehicle.

TRANSFER PHASE

Because the solar-electric engines will only be needed to slightly change the tilt of the orbit and then to stop the slow drift with respect to the Earth induced by the launch vehicle, the engines can be smaller than those developed for delivering spacecraft to other planets. The LISA mission can use engines developed for keeping large communications satellites in proper geostationary orbit, some of which are already in use. The electrical power needed for the solar-electric engine is more than can be provided by the solar panels for the science part of the mission. The electrical power will be generated by two circular solar arrays that will be stored within the propulsion module at launch and then deployed to power the engine. Figure 4b is an artist's representation of one LISA spacecraft attached to its propulsion module with the solar panels deployed.

After reaching the final orbits, about 13 months after launch, the propulsion modules will be separated from the spacecraft to avoid having excess mass, propellant, moving parts, and/or solar panels near the test masses within the spacecraft. After reaching the final orbits, the spacecraft positions will evolve under gravitational forces only. Micronewton ion thrusters will be used to keep the spacecraft centered about the shielded test masses within each spacecraft.

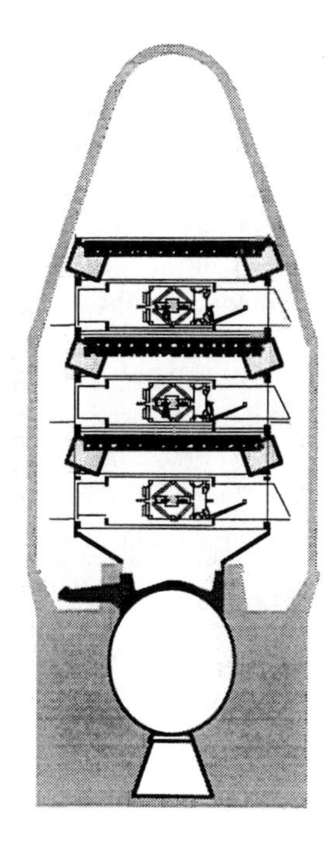

FIGURE 4.a) Artist's concept of the LISA spacecraft attached to the solar-electric propulsion module. The ion-engine is mounted at an angle to the wall of the main cylinder of the propulsion module in order to thrust through the combined center of mass. The ion engine requires power from two deployable solar arrays which are gimbaled to allow for tracking the Sun. Not shown is a cover over the top of the cylinder that prevents sunlight from striking the Y-shaped payload enclosure. b) Launch configuration for the three LISA spacecraft, each with attached propulsion module, within the 2.9 m (9.5-foot) fairing for the Delta-II 7925H. The propulsion module indicated includes two xenon-ion thrusters with two deployable solar panels in the stowed position. The spacecraft assembly is attached to the upper stage by a custom launch adapter.

MISSION OPERATIONS

The LISA mission collects data at a fairly low rate, making about ten measurements per second of the difference in distance between pairs of spacecraft. The data can be compressed using on-board algorithms to give a science data rate of 100 bits per second. Another 100 bits per second may be used for spacecraft and instrument monitoring purposes. The data can be stored on board indefinitely, although it is desirable to transmit the data to Earth at least weekly to monitor spacecraft and instrument performance. Nominally data from each spacecraft will be transmitted to the Earth through a 30 cm diameter antenna on the spacecraft to a 34 m antenna of the NASA Deep Space Network, using a data rate of 7000 bits per second with each spacecraft transmitting for 2.5 hours every other day over the mission lifetime.

CONCLUSION

The three-spacecraft mission design was refined through discussion with spacecraft design engineers of the Jet Propulsion Laboratory's Advanced Concepts Project Design Team. A short mission study was carried out to determine the feasibility of the mission design and to form preliminary subsystems designs. The results from the study are given in (4). More information about the current LISA mission design and science objectives can be found in (5) and at the web site; **http://lisa.jpl.nasa.gov**.

ACKNOWLEDGMENTS

The authors would like to thank the engineers of the JPL Advanced Project Design Team for their work in support of the mission design and to the LISA science team for their help in the mission and instrument definition. The work described in this paper was, in part, carried out by the Jet Propulsion Laboratory, California Institute of Technology, under a contract with the National Aeronautics and Space Administration.

REFERENCES

1. Faller, J.E., P.L. Bender, J.L. Hall, D. Hils and M.A. Vincent, "Space antenna for gravitational wave astronomy", in *Proceedings of the Colloquium on Kilometric Optical Arrays in Space*, Cargese, Corsica, 23-25 October 1984.
2. *LISA: Proposal for a Laser Interferometric Gravitational Wave Detector in Space*, Publication MPQ 177, Max-Planck Institute for Quantum Optics, Garching, Germany, May 1993.
3. *LISA pre-phase A report.*, Publication MPQ-208. Max-Planck Institute for Quantum Optics, Garching, Germany, February 1996.
4. Folkner, W. M., P. L. Bender, and R. T. Stebbins, LISA Mission Concept Study, Publication 97-16, Jet Propulsion Laboratory, California Institute of Technology, Pasadena, CA, March 1998.
5. *LISA Pre-Phase A report.*, 2nd Edition, Publication MPQ-233, Max-Planck Institute for Quantum Optics, Garching, Germany, July 1998.

LISA Operations and Sensitivity

R. T. Stebbins for the LISA Team

JILA/University of Colorado, Boulder, CO 80309 USA

Abstract: The three LISA spacecraft must monitor changes in their relative separation to about 10 pm. The various operating modes during the mission lifetime will be summarized, and the science mode will be described in detail. The science operations involve the optical system, the attitude and position control system, the on-board processing for signal extraction, telemetry and ground operations. The instrumental origins of the anticipated LISA sensitivity curve will also be described.

INTRODUCTION

To achieve the scientific goals of the mission, the LISA science instrument has to detect the strain in spacetime caused by gravitational waves of astrophysical interest. This amounts to monitoring the 5×10^6 km spacecraft separations for relative displacement changes of about 10 picometers (pm). This high strain sensitivity is achieved by interferometric measurement between free-falling test masses. However, the scientific performance of the mission relies on the combined operation of the optical system, the spacecraft attitude and position control system and a signal extraction process. These and other systems contribute to the LISA science operations. The current mission design has been described in the previous paper (1), and in greater detail in the references therein. In this paper, the science operations are explained, and other operating modes are briefly summarized

The LISA sensitivity curve is the result of two noise types and the instrumental response. At frequencies below about 2 mHz, several effects which look like spurious accelerations dominate the sensitivity. At frequencies between 2 and 30 mHz, measurement noise dominates. Above 30 mHz, the sensitivity of the instrument is controlled by the effect of the measurement arms being longer than the wavelength of the gravitational wave. The error budget for spurious accelerations and measurement noise on which the LISA sensitivity curve is based is given.

OPTICAL OPERATION

The operation of the LISA optical system and the topology of the interferometer can best be described by reference to Figs. 1, 2 and 3. Figure 1 shows the complete instrument payload found on each of the three spacecraft. The payload consists of two optical assemblies, each of which constitutes the end of an interferometer arm. An optical assembly, shown in Fig. 2, includes a supporting structure, a telescope for transmitting and receiving beams from the optical assembly at the far end of the associated arm and an optical bench that carries the interferometer components. The details of the optical bench are shown in Fig. 3.

Before describing the operation of the optical system, it's useful to point out three major design features. First, to limit diffraction losses from the very long arms, a laser transponder system is used in which a beam is sent down an arm, and the local laser at the far end generates a return beam offset-locked to it. In this manner a full power beam with the same phase information is returned. Secondly, the original laser beam in each arm is interfered with the return beam in that arm (i.e., a beam phase-locked to itself a round-trip travel time earlier) and separately with the laser of the other optical assembly in the same spacecraft. These combinations permit correction for laser phase noise as well as detection of the gravitational wave signals. Thirdly, at all but one end of one arm, the local laser is offset-locked to either the incoming beam from the long arm or the one from the other optical assembly. In this way, all of the lasers

CP456, *Laser Interferometer Space Antenna*
edited by William M. Folkner
 1-56396-848-7/98/$15.00

are forced to have the same phase noise, although time delayed in some cases. This interferometry locking scheme is described elsewhere (2, 3), as is the phase-locking (4).

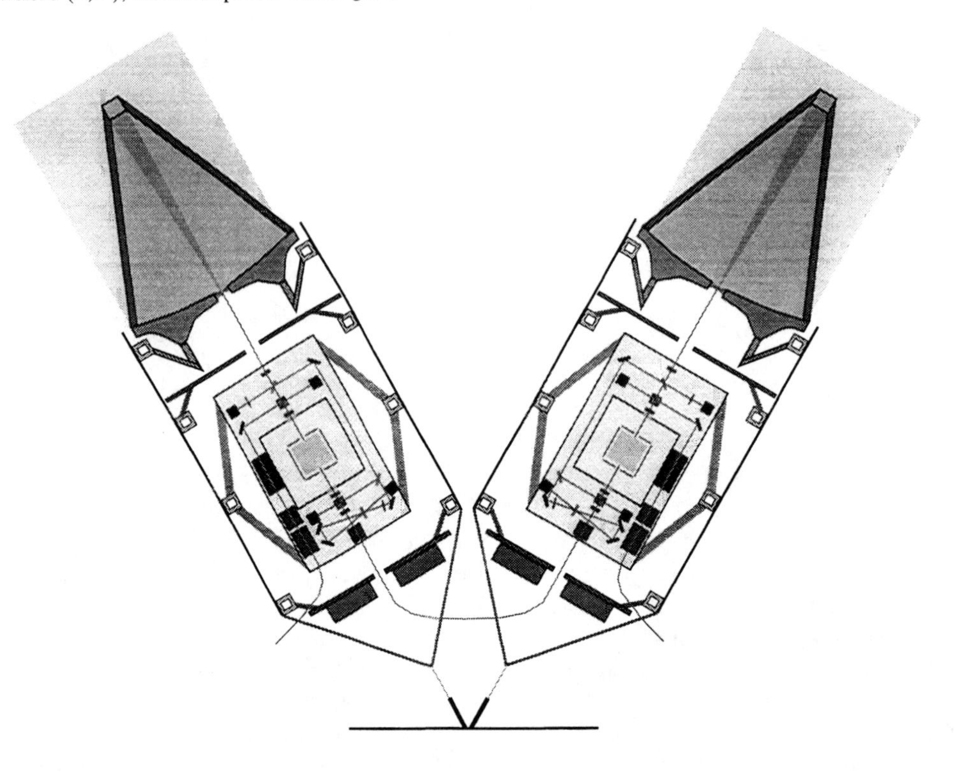

FIGURE 1. The geometry of the two optical assemblies that make up the LISA scientific payload.

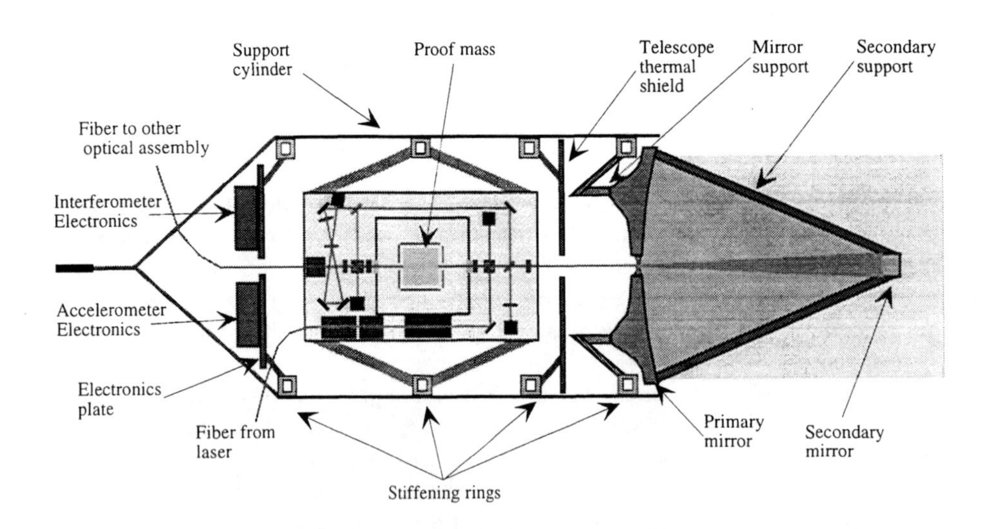

FIGURE 2. A LISA optical assembly.

The optical system in an arm operates as follows: Polarized light from Nd:YAG non-planar ring oscillator laser operating at 1.06 μ is brought onto the optical bench via a fiber on a motorized positioner. The beam passes through an electro-optic modulator that imposes a phase modulation on the beam for optical heterodyning and other purposes. The beam is then steered to a polarizing beamsplitter whence the main beam is directed through a quarter wave plate to the telescope for expansion and transmission to the distance spacecraft. A small amount of the laser beam leaks through the polarizing beamsplitter and interferes with the received beam on the main quadrant photodiode creating the primary science signal. The received beam is concentrated by the telescope, polarized so as to pass through the main polarizing beamsplitter, and reflected off the proof mass, rotating the polarization in the process so that on encountering the main beamsplitter a second time, the beam is reflected into the main quadrant photodiode. In this manner, a fringe signal is generated by interfering the received beam with the outgoing beam in the same arm.

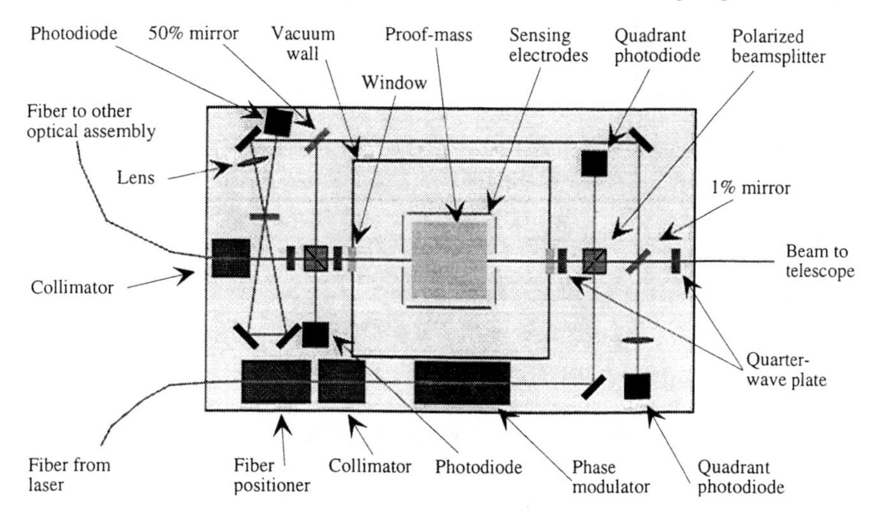

FIGURE 3. The optics bench.

To separate the phase modulation associated with a gravitational wave from that caused by laser phase noise, phase information from a second interferometric arm needs to be compared with the phase information from the first. Laser phase noise is common to both arms; spacetime strains cause phase changes with opposite signs. The comparison between the two arms is accomplished by diverting a small fraction of the outgoing beam with a 1% turning mirror and routing it to an aft polarizing beamsplitter, where, in the same manner as with the main polarizing beamsplitter, the beam is reflected off the back side of the proof mass and sent via a fiber as a phase reference signal to the other optical assembly in the same spacecraft. The phase comparison is made at the aft polarizing beamsplitter by interfering the phase reference signals exchanged between the two optical assemblies. Additionally, half of the light sent to the aft beamsplitter is directed to a triangular stabilization cavity. This cavity is used to stabilize the master laser; it can be used in all of the other lasers to stabilize them at frequencies higher than that associated with the round trip light travel time in the arms.

There are several other functions that are essential to the optical operation which will only be listed here. First, as mentioned above, the master laser frequency is controlled by reflection locking to a reference cavity (5). At a later date, locking to a molecular reference may be more desirable for better low frequency stability. Second, the remaining laser phase noise is measured using information from the fringe signals as described in (2). The phase noise of the onboard ultrastable oscillators used in the fringe timing measurement described below is also measured and corrected using an analogous algorithm. The science signal is corrected for both noises onboard.

Third, a hierarchical control system is required to maintain the outgoing beam pointing. Pointing information is acquired from the proof mass, a quadrant diode on the optical bench and a pointing dither sensed by the distant spacecraft. Fine pointing adjustment is done with the secondary mirror of the telescope. Each optical assembly is pointed by a PZT-based forcer at the telescope end acting on a flexure at the aft end, and the spacecraft has to be pointed by information from both optical assemblies.

Fourth, the laser amplitude will be controlled with a reference diode (5). And finally, the temperature of the laser pump diodes and the Nd:YAG crystal must also be actively stabilized.

ATTITUDE AND DISPLACEMENT CONTROL

The effectiveness of the drag-free system for attitude and displacement control determines the low frequency performance of LISA. A drag-free system works by sensing the relative position between the spacecraft and an enclosed, but freely falling, proof mass within that spacecraft and commanding the spacecraft thrusters to hold the relative position fixed. The drag-free system must shield the proof mass from the solar radiation pressure and the solar wind, limit proof mass disturbances caused by relative spacecraft motion and control other parasitic forces, e.g., thermal disturbances and electrostatic charging.

In LISA, the relative motion between the proof mass and its housing fixed to the optical bench are sensed capacitively in all 6 degrees of freedom (6). As mentioned in the previous section, there is also pointing information from local quad cells and from a pointing dither sensed by the distant spacecraft. In operation, these inputs are fed to a control law which generates commands for the micronewton thrusters on the spacecraft, for electrostatic trim forces and torques applied to the proof masses (outside of the measurement band) and for telescope and optical assembly pointing.

Controlling the electrostatic charging of the proof mass and its housing is an important operational feature. Charging driven by high energy particle impacts are sensed with the inertial sensor capacitor plates and neutralized by electrons produced by UV light delivered by fiber from a mercury lamp.

SIGNAL EXTRACTION

Information is extracted from the fringe signal as a time series of phase measurements. The fringe rate from two arms could be as low as 1 MHz with occasional orbital maintenance, but the rate from the third arm may be as high as 15 MHz. The two beams interfering to make the fringe signal may differ by as much as 6 orders of magnitude in power. The phase measurements need to be made with a resolution of at least 1×10^{-4} rad/$\sqrt{}$Hz from 0.001 to 1 Hz.

Several phase timing strategies appear feasible. In the baseline plan described in (2), the fringe signal is beat against a reference frequency selected from a comb of frequencies to obtain an intermediate frequency between 75 and 125 kHz. The result is filtered with a tracking filter to remove phase noise above 100 Hz, and then conditioned to produce zero crossing pulses for counting and timing. A promising alternate timing strategy (7) beats down the fringe signal with a programmable digital frequency synthesizer to a nearly constant IF, applies a tracking filter to the IF signal, conditions for zero crossing pulses and then times the result. The product is then fed back to the synthesizer to correct the IF. The conditioning and timing steps could be replaced by rapid digitization and fitting. In any event, the end product is a time series of phase measurements on a 0.5 sec grid, low-passed around 1 Hz.

As mentioned above, the signal from one arm is used to estimate the laser phase noise. With a knowledge of the arm lengths, initially from spacecraft tracking, the difference of the arm lengths can be corrected for the laser phase noise. This onboard correction is essential to compress the data for telemetry.

TELEMETRY OPERATIONS

The LISA mission generates science data at approximately 100 bps, and expects to generate another 100 bps of engineering data. The spacecraft use two X-band downlink modes, a 7 kbps rate for normal science data and a 10 bps rate for monitoring and emergencies. With 70 Mbits onboard storage, science downlink lasting 10.5 hr takes place every second day with a 34 m DSN antenna. Uplink commanding is carried out at 2 kbps. The high gain antennas on the spacecraft require periodic repointing.

GROUND OPERATIONS

A detailed plan for ground operations has not been worked out by any of the groups involved in mission definition. However, a general outline can be made. An operations control center must interface with the DSN and process the received telemetry to separate and reformat engineering and science data. At the operations center, engineering data is analyzed to monitor spacecraft health. Mission planning and command generation is performed. The science data and pertinent engineering data is forwarded to a science operations center where it is archived. At the science operations center, the health of the payloads is monitored, the science data is examined and validated. And finally the data is forwarded to the science team for reduction and analysis (8).

OTHER OPERATIONAL MODES

There are several other operational modes, corresponding to different phases of the mission, that can only be summarized here. The science spacecraft, paired with their propulsion modules, are stacked in the launch vehicle shroud (1). When the launch vehicle is spent, the stack must be separated from the upper stage of the Atlas, despun with a special attachment (PAM-D) for the purpose, and then each sciencecraft/propulsion module pair has to be separated from the stack. The paired units then unfold their stowed solar arrays and begin a ~13 month cruise phase in an elliptical orbit. During cruise, there is constant thrusting with the ion thrusters, the inertial sensor is vented and the spacecraft can thoroughly outgas. At the end of cruise, the three spacecraft will have been inserted in recircularized orbits with appropriate inclinations and nodal longitudes. Then propulsion modules are jettisoned, and a period of DSN tracking and final orbit touch up completes the ferry phase of the mission.

Next, the mission enters an alignment and pointing acquisition phase. The acquisition steps are: lasers are turned on, each spacecraft is oriented by star trackers, coaligned with the optical assemblies, locating the laser beam from the other two spacecraft. Next, the proof mass is uncaged and drag-free control is established. Outgoing beams are defocused using the fiber positioners so that received beams can be acquired with the quad cells after only a moderate search using the proof mass as an inertial reference. Finally, the outgoing beam is re-focused, the pointing control loops are locked, the optics are adjusted as necessary, and the lasers and clocks are slaved to their respective masters.

It is anticipated that occasional (weekly or monthly) orbital correction or optical re-alignment will punctuate intervals of science operations. There also must be operational modes for standby/safe mode/recovery. Very massive solar flares are capable of interrupting science operations with particle events. These will normally be anticipated a couple of days in advance, and may be cause for a standby mode. And, of course, there are the usual calamities of command/procedural errors, equipment failures, software errors and upset events for which there needs to be safe mode/recovery operations.

SENSITIVITY

The LISA sensitivity curve is an estimate of the signal strength that is expected to be detected with a signal-to-noise ratio (SNR) of 5 from a source with "average" direction and polarization. It is produced (9) by combining a noise budget at lower frequencies and an expected instrumental response curve above 30 mHz and multiplying the results by 5 for SNR and $\sqrt{5}$ for averaging over all directions and polarizations. A 1 yr. integration is taken as the standard for comparison. The noise budget is derived by considering the sources of noise for a specific design, at some level of definition, estimating their magnitude and then allocating a noise amount and a number of such sources to account for both known and unknown noise sources. In general, there are two types of noise: those with a frequency dependence like spurious accelerations which dominate from 0.1 to a couple of mHz, and those with a flat frequency character which dominate from a couple of mHz to about 30 mHz. The declining instrumental response to waves shorter than the arm length dominates from 30 mHz to 1 Hz.

Table 1 gives the noise allocation for sources of acceleration noise, per inertial sensor. To get the effect on a two-arm interferometer, these numbers have to be doubled for a roundtrip optical path, and doubled again in the difference of optical path lengths. Thermal distortions of the spacecraft and payload can give rise to local gravitational forces. Dielectric losses are the dissipation mechanism by which the electrostatic restoring forces on the proof mass introduce fluctuations. Residual motions of the spacecraft introduce gravity noise. Thermal gradients across the test mass cavity give rise to an imbalance of thermal radiation on the test mass. The residual charge on the proof-mass gives rise to electrical forces and a Lorentz force from the interaction with the interplanetary magnetic field. The residual gas impacts on the proof mass from variations in outgassing can generate balanced forces. Thermal fluctuations, expected to have a nearly f^{-2} spectrum, in the telescope will result in optical path changes that look like acceleration noise. And finally, there will be a residual coupling of the fluctuating interplanetary field with the proof mass.

Table 2 gives the noise allocation for sources of optical-path noise, per interferometer of four optical assemblies. Detector shot noise is the result of the selected optical power and anticipated efficiency of the system. There will be some residual noise in both the master clock and the laser phase after correction from the data. Likewise, residual laser beam pointing errors, phase measurement errors and frequency locking errors will also contribute optical-path noise.

TABLE 1. Acceleration Noise Budget in units of 10^{-15} m/s^2/$\sqrt{}$Hz at 10^{-4} Hz.

Error Source	Error	Number
Thermal distortion of spacecraft	1	1
Thermal distortion of payload	0.5	1
Thermal noise due to dielectric losses	1	1
Gravity noise due to spacecraft displacement	0.5	1
Temperature difference variations across cavity	1	1
Electrical force on charged proof mass	1	1
Lorentz force on charged proof mass from fluctuating interplanetary field	1	1
Residual gas impacts on proof mass	1	1
Telescope thermal expansion	0.5	1
Magnetic force on proof mass from fluctuating interplanetary field	0.5	1
Other substantial effects	0.5	4
Total Effect of Accelerations (for one inertial sensor)	3	
Effect in Optical Path Difference (second time derivative of path difference)	12	

TABLE 2. Optical-Path Noise Budget in units of 10^{-12} m/ $\sqrt{}$Hz

Error Source	Error	Number
Detector shot noise, 1 W laser, 30 cm optics	11	4
Master clock noise	10	1
Residual laser phase noise after correction	10	1
Laser beam-pointing instability	10	1
Laser phase measurement and offset lock	5	4
Scattered light effects	5	4
Other substantial effects	3	32
Total Path Difference	40	

Figure 4 shows the LISA sensitivity curve based on the acceleration and optical-path noise budgets above and the instrumental response above 30 mHz. Selected galactic sources of gravitational waves are shown for comparison.

SUMMARY

A feasible plan for science operations has been developed in some detail. Deeper study is in progress in several areas. Other operations modes have been considered, at least superficially. The likely performance of LISA has been examined in considerable detail, and a spectral sensitivity has been estimated.

ACKNOWLEDGMENTS

The author would like to acknowledge the work of the JPL Advanced Project Design Team and the LISA science team for their work in the mission and instrument design and definition. The work described in this paper was carried out in part with support from the Jet Propulsion Laboratory, California Institute of Technology, under contracts #960833, 961287 and 961533.

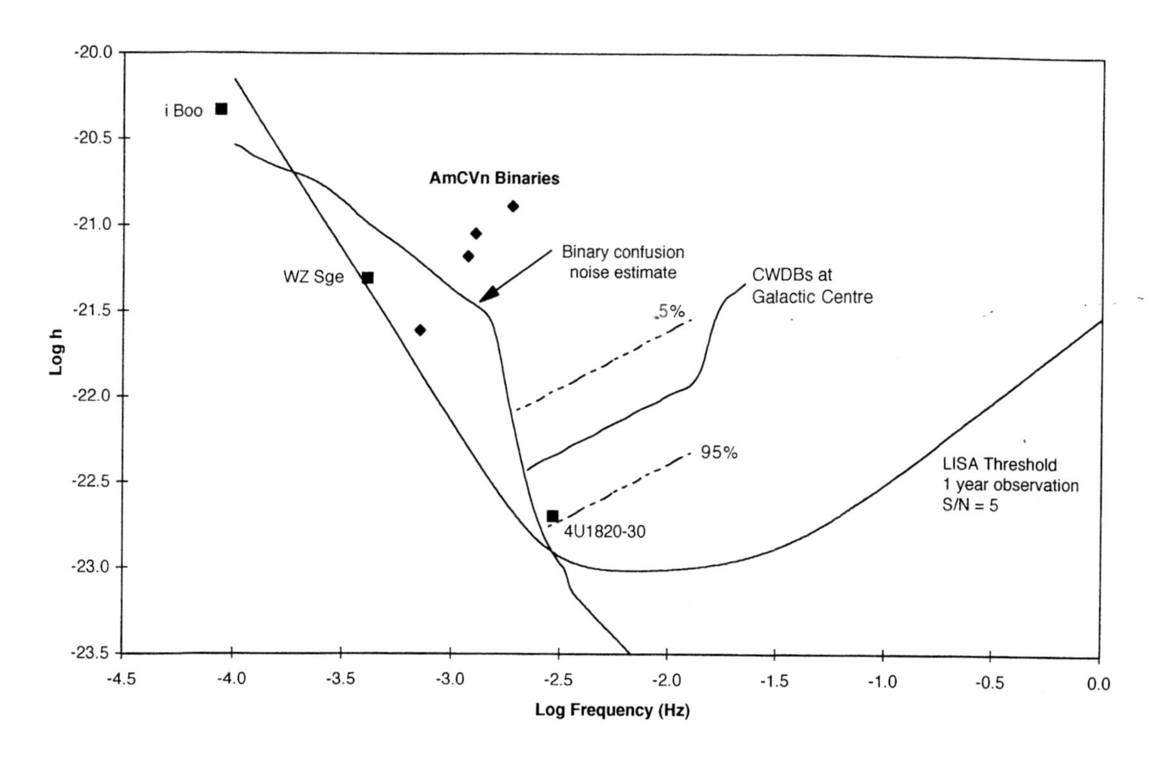

FIGURE 4. LISA sensitivity and selected galactic gravitational wave sources.

REFERENCES

1. Folkner, W. M., "The LISA mission design," in this proceedings, 1998.
2. Stebbins, R. T. , Bender, P. L., Folkner, W. M. and The LISA Science Team, *Classical and Quantum Gravity* **13**, A285-A289 (1996).
3. Robertson, D., "Interferometry for LISA - principles," in this proceedings, 1998.
4. McNamara, P. W., Ward, H. and Hough, J., "Phase locking for LISA: experimental status," in this proceedings, 1998.
5. Peterseim, M., Brozek, O. S., Danzmann, K., Tuennermann, A. and Welling, H., "Laser frequency and power stabilization for LISA," in this proceedings, 1998.
6. Josselin, V., Rodrigues, M. and Touboul, P., "Inertial sensor for the gravity-wave missions," in this proceedings, 1998.
7. Giaime, J., "Experimental demonstration of aspects of interferometry," in this proceedings, 1998.
8. Stebbins, R. T. , Bender, P. L., and Folkner, W. M., *Classical and Quantum Gravity* **14**, 1499-1505 (1997).
9. LISA Pre-Phase A report., 2nd Edition, Publication MPQ-233, Max-Planck Institute for Quantum Optics, Garching, Germany, July 1998.

Structural Design of the LISA Payload

Martin S. Whalley, Raymond F. Turner, Michael C. W. Sandford

Space Science Department, CCLRC Rutherford Appleton Laboratory,
Chilton, Didcot, Oxfordshire OX11 0QX, U.K.

Abstract. A CAD representation of the LISA payload has been developed and a finite element analysis conducted. The design closely follows the configurations developed by the science working teams, with independent pointing adjustment of the two telescopes. However, the accommodation of certain systems boxes needs further thought, with perhaps a reduction in unit sizes or transfer to a position on the Spacecraft main structure. Initial modal analysis has given an eigenfrequency for the payload of 50 Hz in the launch configuration. This is encouraging for a later design stage, when analysis of the combined Spacecraft and Payload is conducted, and consideration given to the dynamic characteristics of the multi-spacecraft stack. The principle adopted in the design is to restrain the payload for launch (with the pointing adjustment actuators carrying no load) and to remove the restraints after the spacecraft has been deployed. The finite element model has been linked with thermal modelling data, enabling sensitivity to temperatures to be assessed.

INTRODUCTION

An accommodation study for the LISA payload has been conducted using configuration information developed by the science working teams. The resulting 3D-CAD model formed the basis for creating two finite element models; a Mechanical FEM (Finite Element Model) used for modal analysis, and a Thermal FEM used to assess the thermal response of the payload (3). The Mechanical FEM was also used in conjunction with the results from the Thermal FEM to analyse thermally induced distortions in the structure. This paper describes the 3D-CAD model, the Mechanical FEM, and the results of the modal and thermal distortion analyses.

ACCOMMODATION STUDY

The information contained within the reference documents [1,2] was used to create a 3D-CAD model of the LISA payload and a representation of the surrounding spacecraft. This model was later updated, following the modal analysis conducted on the FEM. The resulting configuration is illustrated in Fig. 1. The payload sunshield, which covers the circular aperture in the spacecraft ring above the payload is not shown, and was not modelled.

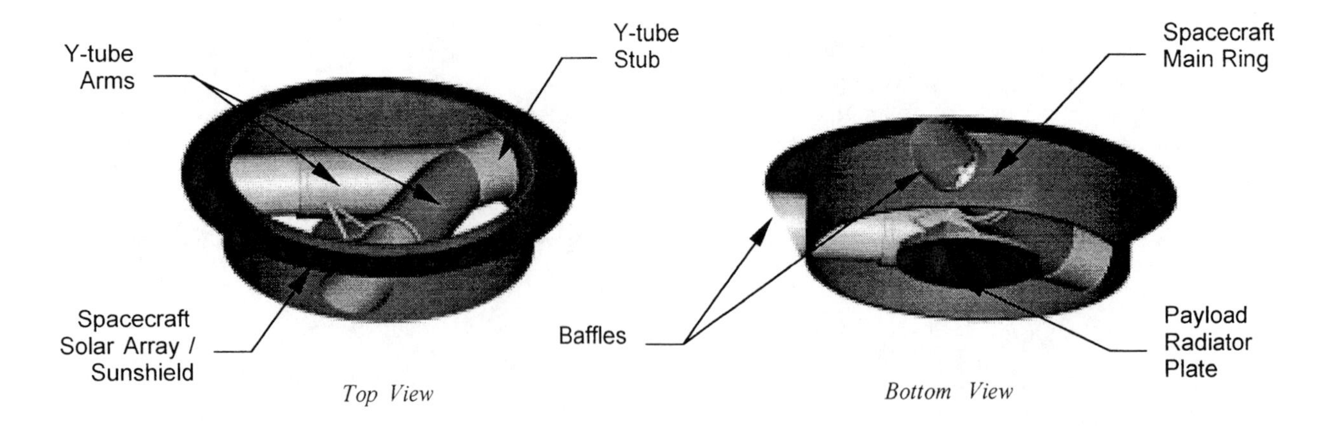

FIGURE 1. LISA Spacecraft

CP456, *Laser Interferometer Space Antenna*
edited by William M. Folkner

The LISA payload includes Y-shaped thermal shield, consisting of the Y-tube Arms and Stub, plus two Baffles. Two optical assemblies are located within the payload thermal shield as indicated in Fig. 2.

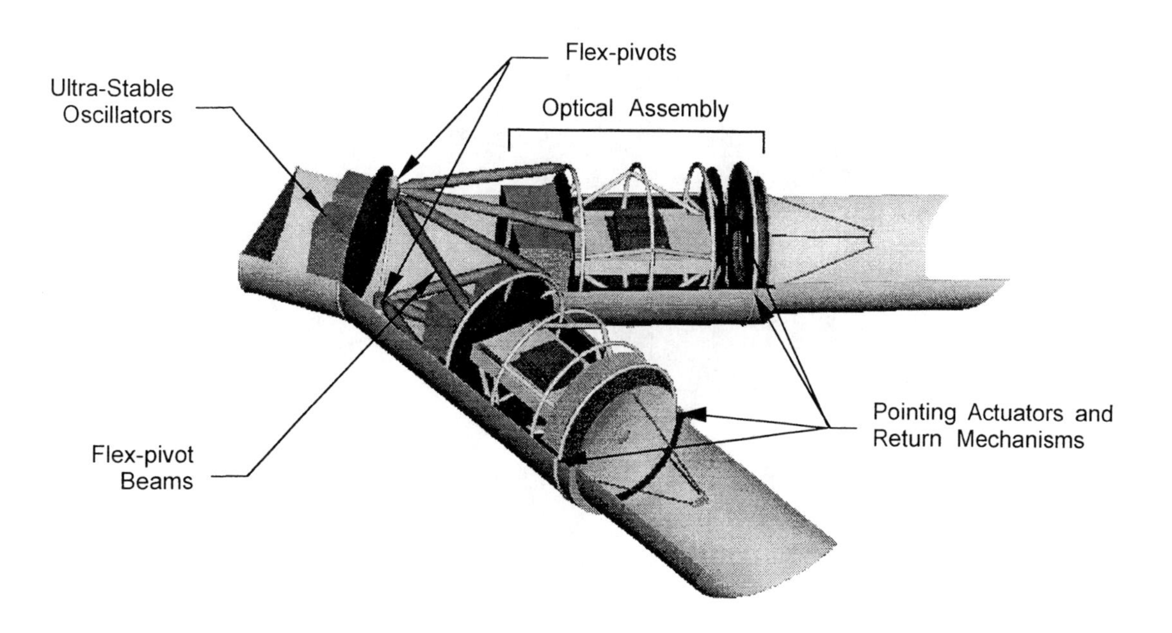

FIGURE 2. LISA Payload (Section View)

All of the optical components are mounted together within rigid subassemblies (the "*Optical Assemblies"*) to minimise changes to critical alignment dimensions. These two subassemblies are enclosed within the arms of the rigid Y-shaped payload thermal shield (as shown in Fig. 2). They are attached by flex-pivot assemblies and pointing actuators, which allow the alignment of the optical assemblies to be adjusted. These adjustments are made by moving each optical assembly *within* the payload thermal shield; the shield itself does not move with respect to the spacecraft during alignment. One optical assembly is shown below:

Finally, the lasers plus their control electronics and the UV discharge unit are mounted on the radiator plate, which is suspended from five struts attached to the Y-tube Arms, as shown in Fig. 4.

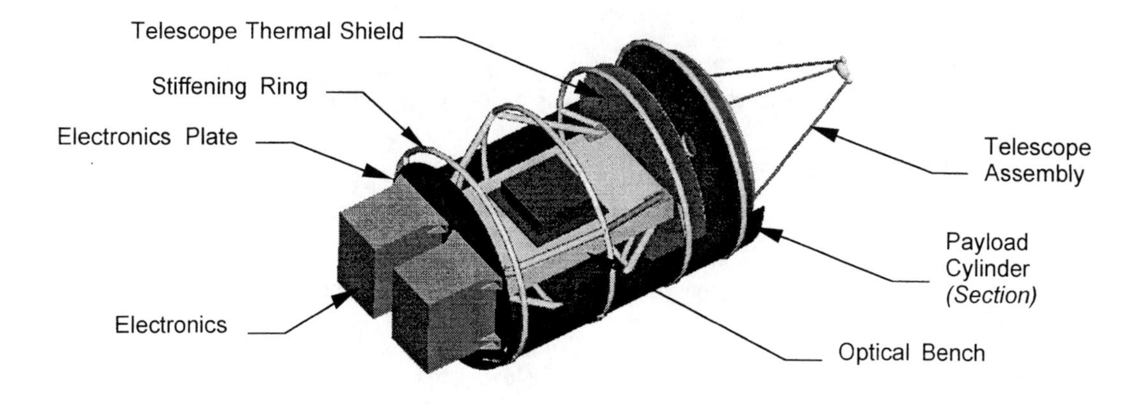

FIGURE 3. Optical Assembly (Section View)

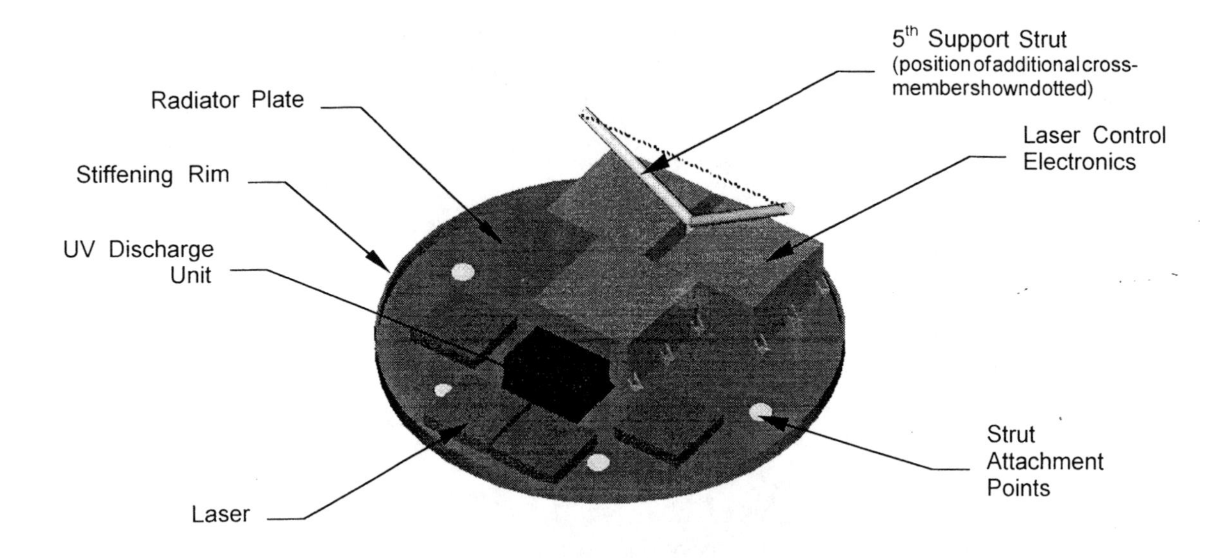

FIGURE 4. Radiator Plate

Accommodation Issues

Few problems were encountered in accommodating the components as required. The problems that did arise are described below.

The current LISA payload configuration calls for two electronics boxes plus an Ultra-Stable Oscillator (USO) unit to be mounted on the electronics plate at the rear of each optical assembly. However, the limited space available rendered this impossible. The solution was to reduce the height of the electronics boxes from the stated 100x150x150 mm to 100x150x130 mm, enlarge the diameter of the electronics plate, and mount the USOs for each optical assembly together on their own support plate, in the Stub of the Y-tube (see Figs. 2 and 3).

The presence of the Y-tube directly above the radiator plate imposes a height restriction on the components mounted on it. Hence there is only sufficient room for three laser control boxes, not the four originally specified. However, it has been suggested that three units would provide sufficient redundancy (and hence the fourth unit is superfluous). Thus the accommodation problem is resolved by mounting the components as shown in Fig. 4, with the lasers beneath the Y-tube Arms and their control electronics "between" the Arms.

The dimensions and location of the Spacecraft Solar Array is dependent on the position of the payload within the Spacecraft Ring. In order to achieve complete shading of the Spacecraft (from the Array position down to the bottom of the Spacecraft Ring) the outer diameter of the Solar Array/Sunshield was increased to 2290 mm.

Details of the primary and secondary mirror supports are limited, hence detailed modelling was not possible. However, a conceptual design was modelled, with the secondary mirror mounted from the primary mirror on a "spider" of three carbon-fibre rods. The primary mirror was attached to a bulkhead on the first stiffening ring (see Fig. 3) via three blade mounts, to accommodate the differential expansions of support structure and mirror.

Launch-Locks and Spacecraft Attachment Concept

The Stub of the Y-tube is permanently connected to the Spacecraft via three stressed fibreglass bands around its circumference. Blade-mounts attach the Baffles to the spacecraft, at the point where they meet the Spacecraft Ring. These blade-mounts allow longitudinal expansion of the Y-tube arms (due to thermal effects), whilst preventing motion in other directions.

Results from the FEM modal analysis indicate that additional load paths are required to produce a structure capable of surviving the vibrations generated during launch. These load paths take the form of launch-locks, which are released before scientific operations begin.

The first set of launch-locks reinforce the attachment of the payload to the spacecraft structure. They connect the baffles to the spacecraft ring during launch. The second set of launch-locks attach the rear of the optical assembles directly to the inside of the payload thermal shield. They are positioned at the top and bottom of the fourth stiffening ring as indicated in Fig. 5.

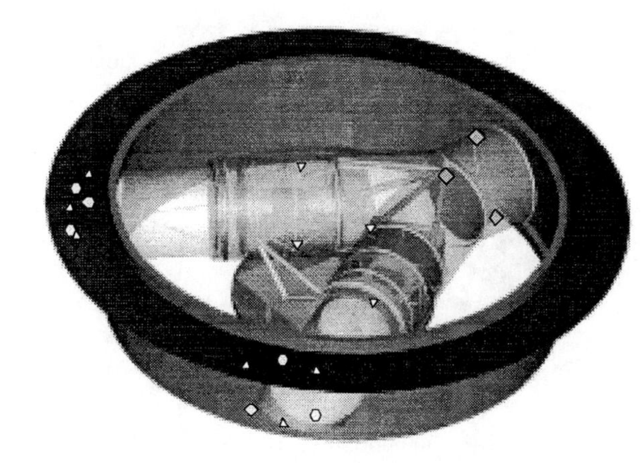

Attachment Type

◇ Stressed Fibreglass Bands

○ Blade-mounts

▵ Launch-locks (Set 1)

▿ Launch-locks (Set 2)

FIGURE 5. Attachment of Payload to Spacecraft

MODAL ANALYSIS

A finite element model of the payload was created using the 3D-CAD model for reference. Two configurations were investigated, with differing objectives:

- ***Launch Configuration*** - *launch-locks in place.*
  The structural design was iterated in order to raise the fundamental frequency of the payload to approximately 60 Hz.
- ***Operational Configuration*** - *launch-locks removed.*
  The design resulting from the Launch Configuration analysis was examined with the launch-locks removed to verify that all modal frequencies were outside the bandwidth of scientific interest (0.1 to 1×10^{-4} Hz).

The model encompassed all components of the payload, with certain simplifications made to reduce the size of the model. The spacecraftitself was not modelled; it was assumed to be a rigid structure, hence the interfaces were simply modelled as nodal restraints.

The approach adopted was to model the individual components and subassemblies separately, modifying their structures to stiffen them as required, before "assembling" them into a complete payload model. The modifications required described in the following section.

Structural Modifications

The Radiator Plate requires a fifth support strut and a rim around the edge (see Fig. 4). The fifth strut (in the shape of a "Y") connects the front of the radiator plate to the sides of the Y-tube Arms, and a cross-member (shown dotted in Fig. 4) spans the gap between them.

A preliminary design describes the Optical Bench as being supported from two points on the third stiffening ring. However, analysis showed this configuration to have a low fundamental frequency. On the advice of the LISA study team, two further support points were added. Thus the Optical Bench is suspended by eight support tubes, as illustrated in Fig. 6.

The electronics plate is attached to the fourth stiffening ring by six support tubes rather than the four suggested in preliminary designs.

Stiffening rings have been added to the Y-tube at points where the Optical Assemblies and Ultra-Stable Oscillator Plate are attached, and to the Baffles where they attach to the Spacecraft Ring.

Modal Analysis Results

A summary of the modal analysis results for the Launch Configuration is given in the Table 1. The first two modes are associated with vibrations of the telescope mounts, which were not modelled in detail. The motion of the Radiator Plate drives the next three modes in different directions.

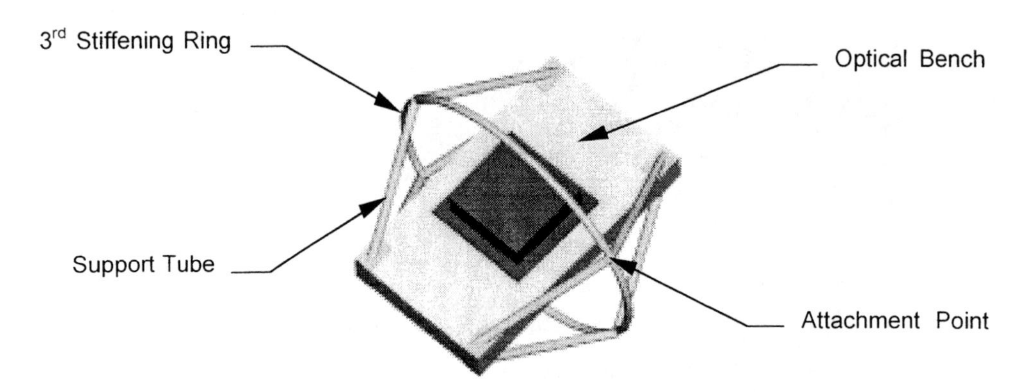

FIGURE 6. Optical Bench Support Structure

A summary of the modal analysis results for the Operational Configuration (launch-locks removed) is given in Table 2. The first two modes are associated with the motion of the Optical Assemblies within the Y-tube arms; this motion is prevented by the launch-locks in the Launch Configuration. The motion of the Radiator Plate again drives the next three modes. The frequencies and motions differ from the Launch Configuration because of the removal of the launch-locks.

TABLE 1. Modal response of LISA Payload in Launch Configuration

Mode	Frequency(Hz)	Description
1	55.1	Vibration of Telescope Bulkheads and USO Plate
2	55.9	Vibration of Telescope Bulkheads
3	66.7	Radiator Plate "swings" side-to-side
4	69.2	Radiator Plate "swings" fore-and-aft
5	71.0	Radiator Plate oscillates vertically

TABLE 2. Modal response of LISA Payload in Operational Configuration

Mode	Frequency(Hz)	Description
1	40.5	Vibration of Optical Assemblies (both in same direction)
2	48.2	Vibration of Optical Assemblies (in opposite directions)
3	50.7	Radiator Plate "swings" side-to-side
4	52.9	Radiator Plate oscillates vertically, driving other components in same direction.
5	55.6	Radiator Plate oscillates vertically, driving other components in opposite direction.

Modal Analysis - Conclusions

The results of the modal analyses for the two configurations show that:

- The addition of launch-locks raises the fundamental frequency of the Launch Configuration to almost 60 Hz, and
- The Operational Configuration has no modes within the 0.1 to 1×10^{-4} bandwidth.

Further detailed design and modelling of the telescope assemblies and their attachment to the Optical Assemblies are required before the first and second modal frequencies of the Launch Configuration can be increased.

Further improvements to the payload structure may be obtained by reassessing the design of the radiator plate (the cause of modal vibrations just above 60 Hz), and in particular its attachment to the underside of the payload. Moving the lasers and associated components to a radiator connected to the Spacecraft structure would increase the modal frequencies of the payload, and reduce the loads induced in the payload structure. It would also increase their separation from the test masses, and hence reduce their influence on the gravitational potentials within the instrument. This is currently dominated by the close proximity of the lasers and control units to the test masses.

THERMAL DISTORTION ANALYSIS

The distortions brought about by temperature distributions throughout the payload were investigated using the results from the thermal analysis and the mechanical FEM. In order to demonstrate the analysis process, comparisons were made between the undistorted structure and the structure subject to a nominal steady-state temperature case in orbit configuration (launch-locks retracted). The analysis process is described below:

1. The mechanical FEM was created with all components at the same temperature (an arbitrary 295K), so initially the mechanical FEM existed in an undistorted state.
2. Nodes connected to elements of the same physical and material properties were grouped together into "Zones" so that masses could be attributed to them. This was done simply by dividing the mass of a Zone equally between its nodes, since the elements were of approximately the same size. Lumped masses used to model individual components (e.g. electronics boxes) were treated separately. This information was used to investigate the gravitational field within the payload (3).
3. The "steady-state" temperatures calculated in the thermal analysis (3) were "mapped" onto corresponding nodes in the mechanical FEM. Due to the difference in mesh density of the two models, this mapping procedure left many nodes without temperature constraints.
4. The temperatures of these remaining nodes were calculated by heat transfer analysis within the mechanical FE analysis software. Due to the restrictions of this software, only conductive heat transfer was accounted for at this stage.
5. The mechanical FEM was analysed again, using the temperatures at each node to generate distortions in all the elements throughout the model.
6. The masses and co-ordinates of each node, both in undistorted and distorted positions, were then used to calculate the effects of thermal distortion on the gravitational field within the payload (3).

Thermal Distortion Results

The results of the thermal distortion analysis are best illustrated graphically. Figure 7 shows the temperatures throughout the payload, as predicted by the mechanical FE analysis software, and the distorted shape of the payload. Note that the material properties for the Y-tube and baffles used in the analysis give them a negative thermal expansion coefficient. Thus the baffles increase in radius because they are colder than the nominal 295 K. The "lobes" that appear in the baffles are the result of the blade-mounts attaching them to the Spacecraft restricting this radial increase.

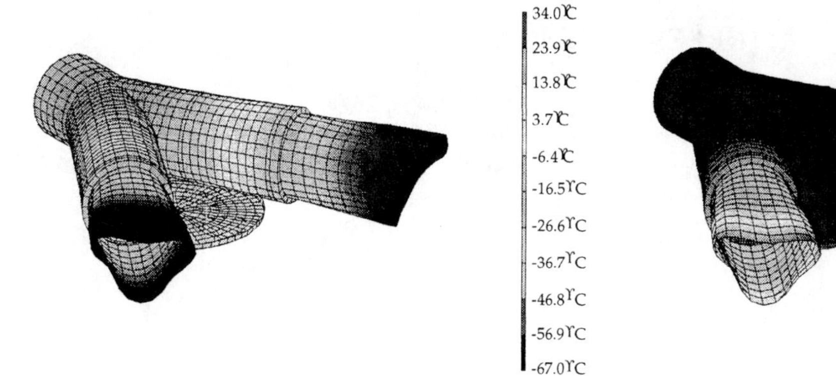

FIGURE 7. Temperatures

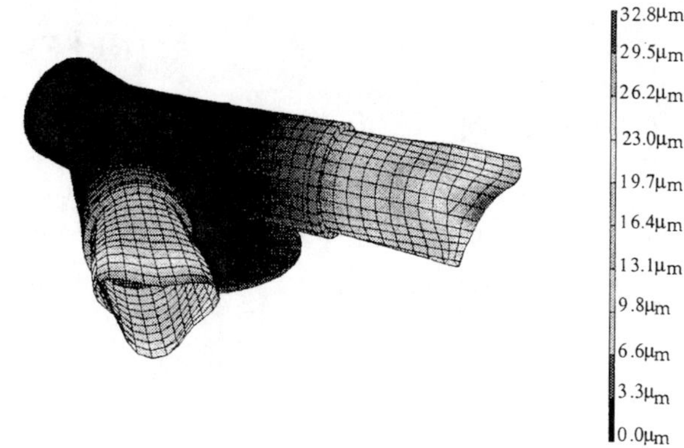

FIGURE 8. Distortions

Figure 8 shows the distorted shape of the payload, with colours illustrating the magnitudes of displacement. Note that the distortions in the structures are shown greatly magnified, to make them easily visible.

Figure 9 illustrates the displacements within the optical assembly. The colours represent displacements in microns. Note that the displacements are measured relative to the undistorted payload model, rather than displacements of one component in the optical assembly relative to another. The figure also appears to show the telescope thermal shield penetrating the optical bench. This is due to the magnification factor used to display the displacements and would not occur in reality.

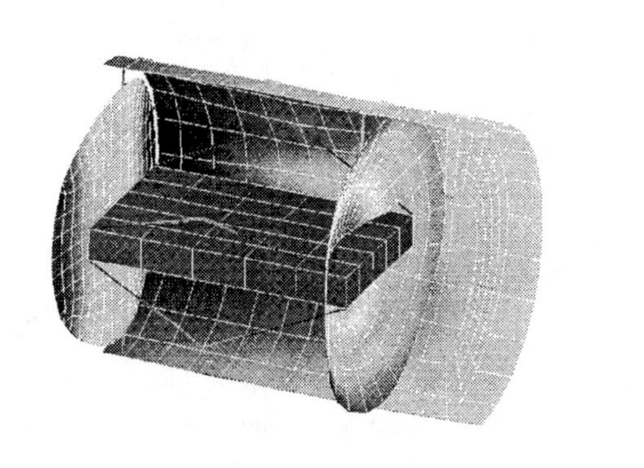

FIGURE 9. Optical Assembly Displacements

Comments on Thermal Distortion Analysis

This work has demonstrated that the thermo-mechanical interactions of the payload can be assessed using this process. The results produced to date compare two arbitrary steady-state cases, hence the data produced is not indicative of the true distortions to be experienced during payload operations. The accuracy of the results will be influenced by the resolutions of the thermal and mechanical FEMs, the assumptions made during the temperature mapping process (Steps 3 & 4), and the material properties attributed to components.

Further iterations of this process, coupled with the preliminary results of the gravitational analysis [3] and using more realistic temperature cases, will identify the components with dominant effectson the gravitational potential at the test masses. This will enable the FEM meshes to be selectively refined in order to assess their movements in more detail.

REFERENCES

1. LISA Science Study Team, "LISA Payload Definition Document" Version 0.3, 23/04/97
2. LISA Advanced Projects Design Team Final Report JPL D-14350 24/02/97
3. Peskett S., Kent B., Whalley M., Sandford M. C. W., "Thermal design and the gravitational influence of thermally induced mechanical changes on the LISA payload", this proceedings.

Thermal design and the gravitational influence of thermally induced mechanical changes on the LISA payload

Simon Peskett, Barry Kent, Martin Whalley, Michael Sandford

Space Science Department, CCLRC, Rutherford Appleton Laboratory,
Chilton Didcot, Oxfordshire OX11 OQX, United Kingdom.

Abstract: The current LISA concept has three spacecraft each with two telescope arms protected by a 'Y'-shaped' tube. Thermal mathematical models were established using ESARAD and ESATAN using the latest configuration drawings and design details contained within v0.3 of the Payload Definition Document. The thermal design was refined and the sensitivity of the optics bench and other payload components to disturbances in solar irradiation and power dissipation was determined and compared with defined requirements. The predicted steady state temperatures were used in a finite element model using I-DEAS to determine thermally induced distortions. This 4500 node model was used as input data to a spread sheet evaluation of the gravitational influence of the payload components at a point arbitrarily chosen to be the centre of the one of the payload optics benches. The resultant acceleration due to all of the nodes, initially all at uniform temperature, was calculated as a straight forward vector summation of the influence of each node. This assumes that each could be regarded as an independent point in space defined by a point mass and location as given in the finite element model. The thermally induced distortions were then applied to the model and the resultant acceleration recalculated.

INTRODUCTION

The LISA mission concept comprises three spacecraft at the corners of an equilateral triangle with sides of 5 million kilometers. The centre of this triangle is at 1AU and lags the Earth by 20°. The three spacecraft rotate around the triangle centre once a year and the plane of rotation makes an angle of 60° with respect to the ecliptic plane. Each LISA spacecraft is identical and full details of the payload configuration are found in (1).

PAYLOAD THERMAL DESIGN

The major science requirement on the payload thermal control subsystem is one of temperature stability, with the optical bench fluctuations due to solar intensity variations and other sources of disturbances kept below 10^{-6} K/Hz$^{1/2}$ at 10^{-3} Hz. It has also been stated that the optical bench part of the payload shall be maintained at 20°C±10°C with minimal temperature gradients. It is also evident that electronics boxes need to be maintained within their as yet unspecified operational temperature limits and in this respect 'sensible' target temperatures are assumed in this study. It has been suggested that multi-lary insulation (MLI) should be avoided and this has been done for the current thermal study.

It is clear that many thermal and system level trade-offs need to be performed before an optimised thermal design may be established. A thermal design, based upon the configuration and structural studies (1) together with the various structural and thermal design aspects discussed in (2), was developed which meets the specified thermal requirements stated above as a first iteration in these trade-off studies.

There is a solar shield attached to the service module having an optimised α_s/ε ratio yielding a temperature in the range -5.0°C to 15°C for the mission. The payload Y-shaped thermal shield is assumed to be goldized on the external surface of each arm between the aperture and the electronics plate. It is goldized on the internal surface of each arm in

CP456, *Laser Interferometer Space Antenna*
edited by William M. Folkner
 1-56396-848-7/98/$15.00

the area immediately surrounding the internal optical bench support cylinder. The internal surface forward of the primary mirror is assumed to be black painted to reduce scattered light. All other surfaces comprising the Y-shaped tube (i.e. aft of the two electronics plates) are assumed to be black in order to help radiate heat away from the various electronics boxes (i.e. those on the electronics plates and the USO boxes at the 'stub'). The electronics boxes contained within the Y-shaped tube are also assumed to be black painted. The external and internal surfaces of the optical bench support cylinders are goldized to radiatively isolate them from the Y-shaped tube but the optical bench and sensor assemblies have been assumed to have their natural surface properties. Conductive isolation is used throughout the payload and at the interfaces with the spacecraft. The types of isolation assumed are those discussed in (1) and (2). To summarise, 'pyroceram' cylinders are used to support the optical benches, the electronics plates and the telescope thermal shields off the internal support cylinders. Glass fibre reinforced bands are assumed for mounting the internal support cylinders off the Y-shaped tube and carbon fibre brackets for mounting the laser radiator off the Y shaped tube and the primary mirror off the support cylinder. The laser radiator was assumed to have a radiative coupling of $0.113m^2$ to space, equivalent to an actual black painted radiator diameter of about 0.4m.

PAYLOAD THERMAL MATHEMATICAL MODELS

Geometrical mathematical models have been established for the calculation of radiative couplings using ESARAD v3.2.6 and thermal mathematical models have been established using ESATAN v8.2.3. These models include major service module surfaces (the solar array, top sunshield, ring, and bottom cover) in order that radiation exchanges with these surfaces are calculated and the sensitivity of the payload to changes in spacecraft temperature and solar intensity determined. The thermal mesh permits the calculation of axial and circumferential gradients within the arms of the Y-shaped tube and internal support cylinders. Significant thermal components of the payload (the proof mass, sensor, titanium housing, optics bench, telescope shield, electronics plate, electronics boxes, primary mirror, secondary mirror, spider) are each represented by one node. However a separate, detailed model representing the optical bench with 28 nodes permitted the two dimensional prediction of temperature gradients within this component. The carbon fibre reinforced plastic used for the Y- shaped tube and the support cylinder was assumed to have a nominal conductivity of 2.5W/mK. All gold coated surfaces were assumed to have an emissivity of 0.05 (a conservative value to allow for certain irregularities in the coatings).

ANALYSIS RESULTS

A nominal steady state calculation case was established for the "Overall" model , with boundary conditions as given in Table 1. Table 2 summarises the temperature results. The detailed optical bench model predicts a maximum temperature in the region of the EOM of 22.9 °C and a minimum temperature of 20.1 °C. The temperature sensitivity of various components compared with these Nominal Case temperatures is given in Table 3 for various changes in the thermal boundary conditions.

TABLE 1. **Nominal Case Boundary Conditions**

Solar Constant	1370W
Sunshield α_s/ε (BOL value)	0.265
Spacecraft ring/base temperature	20°C
Optics bench dissipations	2 x
USO electronics dissipations	3.0W
Analogue electronics (on plates)	2 x
Digital electronics (on plates)	2 x
Radiator plate dissipation	41.4W

TABLE 2. Nominal Case Temperature Distribution

Location	T (°C)	Location	T (°C)
Optics bench	21.0	Support cylinder stiffening ring 4 (front	2.2
Proof mass	20.4	Support cylinder middle	14.3
Sensor	20.4	USO box plate	24.7
Primary mirror	-19.7	USO box	26.6
Secondary mirror	-45.3	End plate of Y-shaped tube	19.8
Telescope thermal shield	4.4	Y-shaped tube apex (surrounding USO	19.8
Electronics plate	32.1	Y-shaped tube between USO plate and	23.0
Analogue electronics box on plate	33.6	Y-shaped tube between electronics plate and	11.8
Digital electronics box on plate	33.8	Y-shaped tube in front of primary mirror,	-33.2
Support cylinder stiffening ring 1	18.6	Y-shaped tube in front of primary mirror,	-49.0
Support cylinder stiffening ring 2	16.0	Y-shaped tube in front of primary mirror,	-65.4
Support cylinder stiffening ring 3	9.7	Laser electronics radiator	12.2

TABLE 3. Component Temperature Changes (°C) compared with Nominal Case

	Proof Mass	Optics Bench	Payload Electronics	Primary Mirror	Support Cylinder
Nominal case (temperature °C)	*20.4*	*21.0*	*33.6*	*-19.7*	*14.3*
SVM Base/Ring increased by 10°C	+4.5	+4.5	+4.8	+4.2	+4.9
Solar Constant increased by $50 W/m^2$	+0.5	+0.5	+0.5	+0.5	+0.6
Shield α_s increased by 0.052 (to EOL) value)	+3.1	+3.1	+3.2	+2.7	+3.4
Electronics power increased by 1W	+1.4	+1.4	+2.5	+0.9	+1.4
Optical bench power increased by 0.5W	+9.4	+10.0	+1.3	+2.8	+4.9
CFRP conductivity doubled	-1.3	-1.3	-0.8	+4.2	-2.2

For the transient state analysis numerical convergence criteria were set to sufficiently small values and double precision values were used throughout the ESATAN analysis so as to allow the detection of the very small temperature changes important for LISA. Frequency response simulations were made at 10^{-4},10^{-3} and 10^{-2} Hz and sets of 'transfer functions' relating the rms temperature of various payload components to rms power, power density or temperature have been determined.

The power spectral density for observed solar insolation variations is given in (3) as

$$\delta L = 1.3 \times 10^{-4} f^{-1/3} L_0 \ (W/m^2)/Hz^{1/2}$$

where L_0 has been taken as $1350 W/m^2$. Thus at a frequency of 10^{-4} Hz the variation of solar insolation is 3.8 $(W/m^2)/Hz^{1/2}$, at 10^{-3} Hz it is 1.8 $(W/m^2)/Hz^{1/2}$ and at 10^{-2} Hz it is 0.8 $(W/m^2)/Hz^{1/2}$. The optical bench transfer functions for solar insolation were calculated to be 5.2×10^{-5} $K/(W/m^2)$ at 10^{-4} Hz, 2.4×10^{-7} $K/(W/m^2)$ at 10^{-3} Hz, and $<1.0 \times 10^{-9}$ $K/(W/m^2)$ at 10^{-2} Hz and therefore the optical bench temperature fluctuations at these frequencies are

$2.0x10^{-4}$ K/Hz$^{1/2}$, $4.3x10^{-7}$ K/Hz$^{1/2}$ and <$8.0x10^{-10}$ K/Hz$^{1/2}$ respectively. The requirement of $1.0x10^{-6}$ K/Hz$^{1/2}$ at 10^{-3} Hz is met, but only by a factor 2. The study of (3) established temperature fluctuations of $6.7x10^{-9}$ K/Hz$^{1/2}$ at 10^{-3} Hz which is very much better although the reasons for the discrepancy are not known. Another study (4) established $2.6x10^{-7}$ K/Hz$^{1/2}$ at 10^{-3} Hz which is similar to the results of the current study, but again the details of that analysis are not known.

To conform with the stability budget the payload electronics dissipation variations on the electronics plate would have to be less than $1.0x10^{-3}$ W/Hz$^{1/2}$ and variations in optical bench power dissipation would have to be less than $5.2x10^{-6}$ W/Hz$^{1/2}$. Spacecraft temperature variations would have to be less than $2.4x10^{-3}$ K/ Hz$^{1/2}$. It should be noted that power variations in the laser electronics will also be a source of disturbance and so the above stated disturbance budgets are likely to be reduced.

DISCUSSION

To further reduce the effect of solar insolation fluctuations on the optical bench the use of MLI or a thermally decoupled sunshield on the sun-facing side of the service module should be considered. Another possibility would be to goldize the optical bench to further improve its radiative isolation, although in this case its steady state temperature would increase significantly. Analysis has indicated that a 10°C rise in operational temperature would result for the same dissipations, but that fluctuations caused by solar insolation variations will be reduced to $9.0x10^{-8}$ K/Hz$^{1/2}$ at 10^{-3} Hz.

With the assumed power dissipations it was seen that the electronics boxes are somewhat warm. The design studied considered radiative losses only form the box to the Y-shaped tube and from there to the spacecraft. It is conceivable that an additional radiator could be accommodated in plane with the laser radiator and with heat straps to the payload electronics boxes. This approach may also benefit the stability of the optical bench since the whole of the Y-shaped tube could be goldized. It is clearly desirable to minimise the total power dissipation within the Y-shaped tube enclosure.

PAYLOAD SELF GRAVITY EFFECTS

The test masses at the centre of each optical bench are in the gravitational field of every other component of the spacecraft, they are thus subject to an acceleration resulting from the vector sum of the gravitational influence of all the spacecraft material parts. This section describes a study in which this influence for payload items has been modeled by means of a simple spread sheet program the format of which lends itself to indivudualy modeling each component included in the vector sum, and enables calculation of the overall gravitational effect of LISA payload component masses as measured at an arbitrary location within the payload.

In the spreadsheet the individual payload components were represented by nodes from a finite element model (FEM) developed as part of a structural design study (5). The FEM model consists of 4489 nodes which are arranged in 13 zones representing the major payload components.

The acceleration due to any one node has been calculated assuming that the total mass of a component has been equally distributed amongst its nodes. The distance of the element from the center of payload 1 optics bench is given by the FEM model co-ordinates and an arithmetic translation to a co-ordinates system centered on the center of payload 1 optics bench, (Figure 1)

The spreadsheet was used to calculate accelerations and acceleration variation for two conditions:

Case 1: Isothermal - all nodes at uniform temperature of 295K

a) Calculate residual acceleration at centre of Payload 1 optics.

b) Plot variation in residual acceleration along the 3 orthogonal axes

Case 2: Thermo-mechancially distorted - The LISA thermal model discussed above was used to calculate the on orbit steady state temperature distribution. This temperature distribution was applied to the FEM model and the nodes allowed to adjust position according to local temperature and material propertiesas described in (5). Consequent new node locations from the FEM were then available for the gravitational effects spreadsheet so that a) and b) could be re-calculated.

The acceleration at any point due to any one node is available within the spreadsheet, however in the following summary only the influence of major payload items has been tabulated. Xo,Yo,Zo denote the location at which the overall influence is summed. Repeated use of the spreadsheet over several positions for Xo,Yo,Zo enables a map of the variation in acceleration around the notional axes origin to be constructed.

The resultant acceleration seen at the payload 1 optics bench, as given in Tables 4 and 5, is found to be dominated by the lumped masses, in a large extent due to the single node representation of the payload 1 telescope giving an acceleration of $\sim 1.0 \times 10^{-8}$ m/s^{-2}. Although significant contributions at ~ few $\times 10^{-9}$ m/s^{-2} are also present due to other electronics units modelled as lumped masses. The variation in residual acceleration within 1 cm of centre of payload 1 optics bench for the undistorted isothermal case is shown in figure 2.

The thermal induced structural distortions seen in FE model described in detail in (5) are typically a few micron and these give rise to an overall acceleration difference between case 1 and case 2 of 1.2805e-13 m/s^{-2}.

TABLE 4. Residual acceleration magnitudes (m/s^{-2}), and direction cosines due to the payload components listed.

Zone	Residual acceleration at *Xo, Yo, Zo*	Direction cosines		
		x	*y*	*z*
P1 baffle tube	3.85451121E-10	-5.14155131E-01	2.08424021E-03	-8.57694676E-01
P2 baffle tube	1.16941744E-10	8.80612708E-01	2.79478201E-03	-4.73828501E-01
Y tube	7.39239936E-10	8.19870087E-01	-7.90445304E-09	5.72549597E-01
Radiator plate	1.37891805E-09	4.94130450E-01	-8.68915071E-01	-2.86652763E-02
P1 Cylinder	2.48782914E-10	-4.99810242E-01	-2.47356013E-07	-8.66134933E-01
P2 cylinder	3.58660881E-10	9.99415867E-01	-1.22453972E-08	-3.41749001E-02
Optics bench 1	3.44491957E-12	-1.65991618E-01	0.00000000E+00	-9.86127164E-01
Optics bench 2	1.27842216E-09	9.99379102E-01	1.03411961E-18	3.52336513E-02
Therm shield 1	1.42092790E-10	-5.00029443E-01	-1.50018695E-08	-8.66008404E-01
Therm shield 2	1.35500615E-11	9.70612014E-01	-3.93393845E-09	-2.40649782E-01
Electronics	5.91924393E-10	5.88042865E-01	2.74745846E-07	8.08829765E-01
Flex pivots	2.77223683E-10	6.14668843E-01	-3.31746909E-18	7.88785277E-01
Telescope 1	6.13994785E-11	9.50795179E-01	-4.58821618E-05	-3.09820149E-01
Telescope 2	3.99085829E-10	-5.00110631E-01	-2.40071388E-06	-8.65961522E-01
Zone 11 +12	1.44699746E-10	8.16039192E-01	-1.80711322E-01	-5.49020450E-01
Lumped masses	7.22201026E-9	5.42987415E-01	-6.10876369E-01	-5.76189838E-01
OVERALL	9.97291066E-09	7.08150340E-01	-5.65024660E-01	-4.23403153E-01

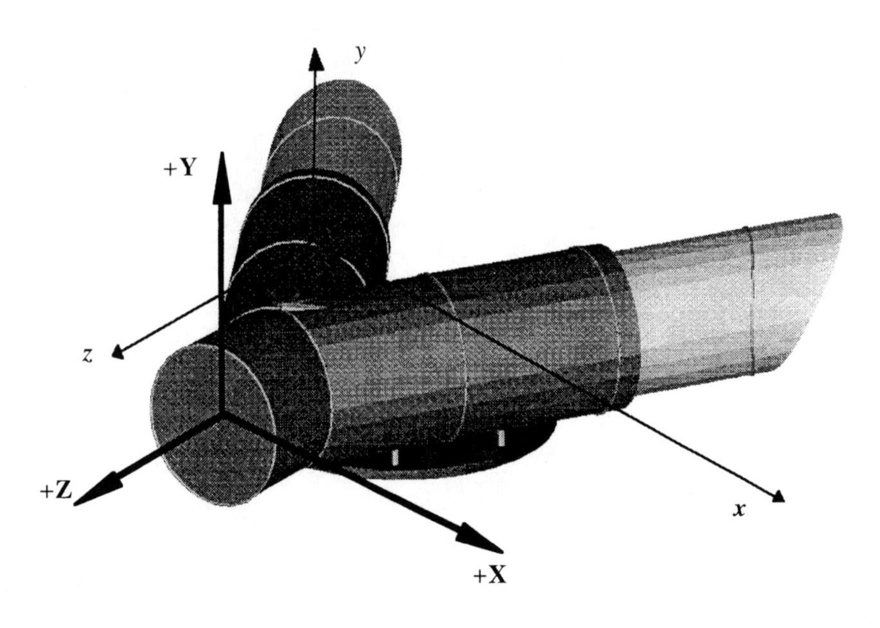

FIGURE 1. The LISA payload is in the form of tubes in the shape of a letter Y, as sketched above with co-ordinate definitions. XYZ is the co-ordinate system used in the FEM (5), *xyz* applies to gravitational study and hence Fig 2, and tables 4 and 5.

TABLE 5. Residual acceleration magnitudes (m/s^{-2}), and direction cosines due to the payload components distorted by thermo-mechanical effects.

Zone	Residual acceleration at *Xo, Yo, Zo*	Direction cosines x	y	z
P1 baffle tube	3.85434147E-10	-5.14155824E-01	2.08428241E-03	-8.57694260E-01
P2 baffle tube	1.16939125E-10	8.80609992E-01	2.79478194E-03	-4.73833548E-01
Y tube	7.39246789E-10	8.19866617E-01	-3.95897546E-07	5.72554565E-01
Radiator plate	1.37890787E-09	4.94130101E-01	-8.68915255E-01	-2.86657163E-02
P1 Cylinder	2.48780209E-10	-4.99799559E-01	4.68925889E-08	-8.66141098E-01
P2 cylinder	3.58658523E-10	9.99415798E-01	-1.61967762E-07	-3.41769337E-02
Optics bench 1	2.72342088E-12	-7.69386450E-02	1.31843536E-02	-9.96948654E-01
Optics bench 2	1.27841733E-09	9.99379182E-01	-1.37494687E-07	3.52313768E-02
Therm shield 1	1.42101278E-10	-5.00032115E-01	2.97649392E-09	-8.66006861E-01
Therm shield 2	1.35502265E-11	9.70613917E-01	-1.20763360E-07	-2.40642106E-01
Electronics	5.91933505E-10	5.88041283E-01	2.48689214E-07	8.08830915E-01
Flex pivots	2.77224123E-10	6.14668705E-01	-9.66951111E-08	7.88785384E-01
Telescope 1	6.13989145E-11	9.50794292E-01	-4.59368435E-05	-3.09822872E-01
Telescope 2	3.99077019E-10	-5.00112749E-01	-2.24196322E-06	-8.65960298E-01
Zone 11 +12	1.44700041E-10	8.16035408E-01	-1.80709593E-01	-5.49026645E-01
Lumped masses	7.22189566E-09	5.43006671E-01	-6.10885164E-01	-5.76162365E-01
OVERALL	9.97278261E-09	7.08203783E-01	-5.65026828E-01	-4.23310862E-01

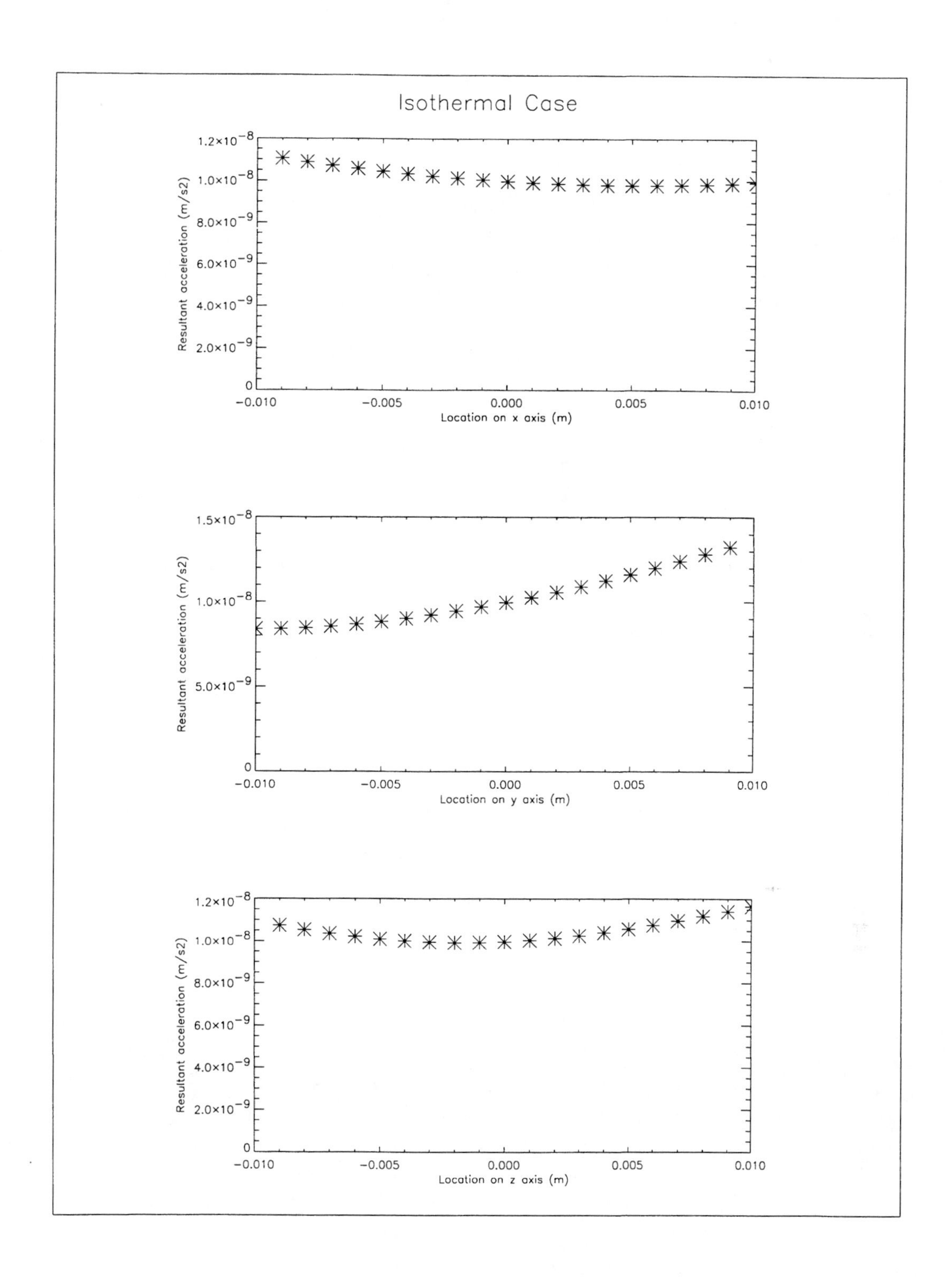

FIGURE 2. Variation in residual acceleration within 1 cm of centre of payload 1 optics bench for the undistorted isothermal case.

CONCLUSIONS

This study has shown that the steady state and transient thermal requirements as defined in (1) and (2) may be met but that margins are small, in the steady state for the electronics and in the transient state for the stability of the optical bench. Therefore improvements to the analysed thermal design are required and several possibilities have been considered for further investigation, including the use of active control at low frequencies.

The method used is simple to apply and could be easily extended to include the spacecraft structure, and used to optimise payload and spacecraft element locations to minimise the influence of self gravity effects.

REFERENCES

1. *LISA Pre-Phase A report.*, 2nd Edition, Publication MPQ-233, Max-Planck Institute for Quantum Optics, Garching, Germany, July 1998.
2. *LISA Payload Definition Document*, Version 3, Publication MPQ-233, Max-Planck Institute for Quantum Optics, Garching, Germany, April 1997.
3. Folkner W.M., et. al., Thermal stability analysis for a heliocentric gravitational radiation detection system
4. *LISA - thermal analysis*, ESTEC report PF/JWL/L5032, March 29, 1994.
5. Whalley M. S, Turner R .F., Sandford, M. C., Structural Design of the LISA Payload - these proceedings.

SOURCES OF GRAVITATIONAL WAVES

Supermassive Black Holes Then and Now

D. Richstone

Institute of Advanced Study, Princeton,
and Dept. of Astronomy, University of Michigan

Abstract. Recent surveys suggest that most or all normal galaxies host a massive black hole with 1/100 to 1/1000 of the visible mass of the spheroid of the galaxy. Various lines of argument suggest that these galaxies have merged at least once in our past lightcone, and that the black holes have also merged. This leads to a merger rate of massive black holes of about 1/yrs.

INTRODUCTION

Supermassive black holes have been a prime candidate for the probable energy sources of quasars, the most energetic objects in the universe, since the discovery of quasars. Over the last decade local surveys have suggested that quasars are present in most galaxies in the present universe [1,2]. The demographics of these objects are so fundamental to an estimate of their merger rate that we repeat the key points below.

Except where noted, All quantities in this paper are computed for a Friedman-Robertson-Walker Universe with $\Omega = 1$ and $H_0 = 80$ km s^{-1}Mpc^{-1}. Distances to nearby MBHs come from many sources, but are always rescaled to this Hubble constant.

STATISTICS OF MASSIVE BLACK HOLES IN GALAXY CENTERS

In about 15 cases, high resolution spectroscopy and imaging, coupled with detailed modelling has led to clear evidence for the presence of a massive dark object (MDO) in the centers of nearby galaxies (including our own). The common denominator in all these studies is the identification of test particles (stars or gas clouds), which orbit the object of mass (M) at a distance r at a speed v given by $v^2 = \alpha GM/r$. The estimate of α requires a detailed model, but often $\alpha \sim 1$. In the fortunate cases of a disk of stars or gas, the analysis is straightforward and fairly unambiguous. In the more complicated case of an anisotropic distribution of stellar orbits it is necessary to construct a detailed model. The favored technique is based on orbit superposition and is summarized in a number of articles [3,4]. These well-defined cases are listed in [1] and labelled in Figure 1.

In three cases (NGC 4258, the Galaxy and M32), it is possible to reject many alternatives to a black hole (such as clusters of neutron stars or black holes) for the observed MDO. The basic argument [5,6] against aggregate models is that the requirement that the evaporation time be less than the probable system age (a Hubble time) sets an upper limit on the mass of the constituent objects that in these cases is near $0.1 M_{\odot}$. Brown dwarfs or planets (or white dwarfs) of this mass or less would rapidly merge. There are no known stellar remnants of any sort of this mass. The MDO's might be clusters of low mass black holes or uninteracting elementary particles (of an unknown variety), but the formation of the former and the collapse of clusters of either to a dense state would both require major new theories. Based on these three objects, for the rest of this paper we assume that all MDO's are massive black holes (bh).

In addition to these very carefully studied cases, we [7] have combined HST images with ground–based spectra to analyze another 20 objects using two–integral distribution function based methods (this is inherently riskier than the orbit superposition methods used for the better data). Combining these analyses suggest that every normal galaxy has a massive black dark object at the present epoch, and that the black hole mass is proportional to the bulge mass of the galaxy (the visible mass of the entire galaxy if the galaxy is an elliptical). The relation between bh mass and the bulge luminosity is $M_{\bullet} = 2 \times 10^7 (L_{bulge}/5 \times 10^9 L_{\odot})^{1.2}$.

CP456, *Laser Interferometer Space Antenna*
edited by William M. Folkner
 1-56396-848-7/98/$15.00

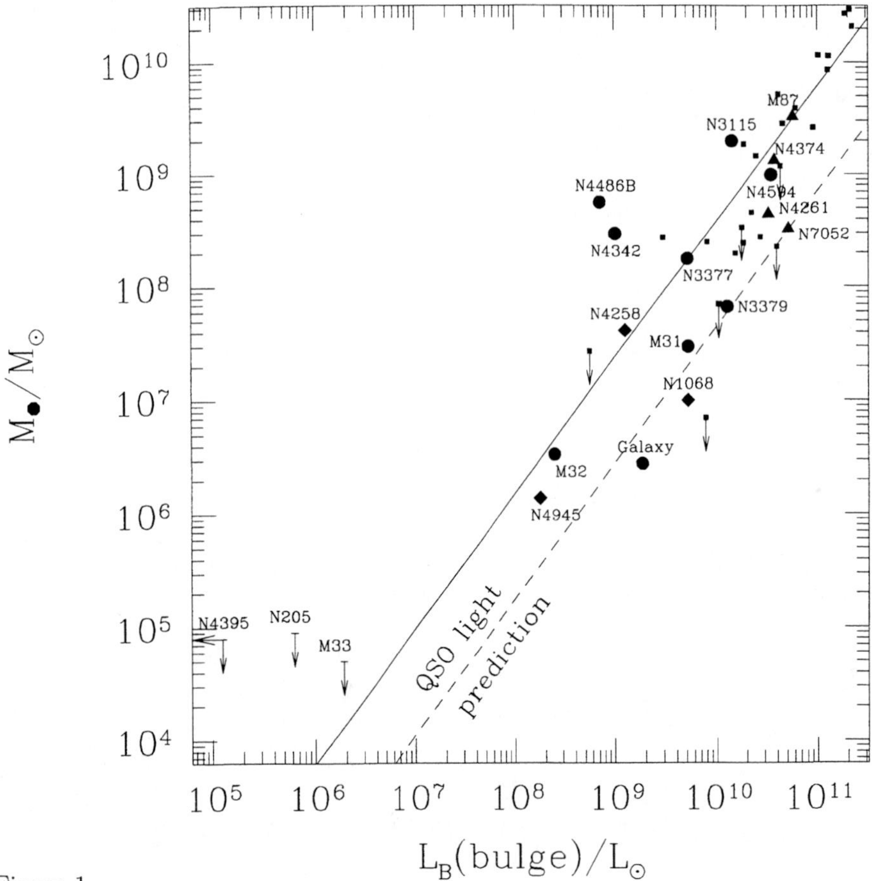

Figure 1. —
Mass estimates of the candidate MBHs in galaxies with dynamical information plotted against the bulge luminosity of their host galaxy. The labeled points are the results of painstaking observation and detailed modelling. The symbols indicate the how $M_\bullet$ was derived: kinematics of gas — triangles; dynamics of stars — filled circles; masers — diamonds; or two-integral modelling using ground-based stellar kinematics — small squares. Arrows indicate upper limits on $M_\bullet$. The solid line is a model with $M_\bullet = 0.005 M_{bulge}$ and $M_{bulge} = 5 \times 10^9 M_\odot (L_{bulge}/10^9 L_\odot)^{1.2}$. The distribution of $M_\bullet$ is roughly Gaussian in $\log(M_\bullet/M_{\rm bulge})$ with mean -2.27 ($M_\bullet/M_{bulge} = 0.005$) and standard deviation 0.5. The dashed line is the quasar light prediction of eqn 3 apportioned according to the bulge mass: $M_\bullet = 2 \times 10^7 (L_{bulge}/5 \times 10^9 L_\odot)^{1.2}$.
The small offset from the observed black-hole/bulge-mass relation indicates that the present integrated density in MBHs is broadly consistent with the integrated luminosity produced by AGNs over the life of the Universe. This offset may reflect a radiative efficiency of average quasar accretion less than 0.10. This figure is reproduced from reference 1.

Because quasars were populous in the youthful Universe, but have mostly died out, the Universe should be populated with relic black holes whose average mass density ρ_u matches or exceeds the mass–equivalent of the energy density u emitted by them [8]. The integrated comoving mass–equivalent density in quasar light (as emitted) is

$$\rho_u = 1/(\epsilon c^2) \int_0^\infty \int_0^\infty L\Phi_Q(L|z) dL \frac{dt}{dz} dz = 2 \times 10^5 \left(\frac{0.1}{\epsilon}\right) M_\odot \,\mathrm{Mpc}^{-3}. \tag{1}$$

where Φ_Q is the comoving density of quasars of luminosity L, and t is cosmic time and ϵ is the radiative efficiency. This density can be compared to the luminous density in galaxies. Using Loveday's estimates of the parameters of a Schechter luminosity function

$$\Phi_G(L)dL = \phi^* (\frac{L}{L^*})^{-1} e^{-L/L^*} d(\frac{L}{L^*}) \tag{2}$$

with $\phi^* = 1.4 \times 10^{-2} h^3 \mathrm{Mpc}^{-3}$ gives a luminous density of $j = 1.1 \times 10^8 L_\odot \,\mathrm{Mpc}^{-3}$ [9], we obtain the ratio of the mass in relic MBHs to the light of galaxies ($h = 0.8$):

$$\Upsilon = \frac{\rho_u}{j} = 1.8 \times 10^{-3} \left(\frac{0.1}{\epsilon}\right) \left(\frac{M_\odot}{L_\odot}\right). \tag{3}$$

We can compare the estimate of [7] to the prediction of the total luminosity in quasars (above) by apportioning the quasar–predicted mass according to the mass of each galaxy. The quasar light underpredicts the observed black hole masses by about a factor of 5, suggesting that a large fraction of black hole growth may occur at radiative efficiencies significantly less than 0.1.

AN ATTEMPT TO QUANTIFY THE MERGER RATE

The previous section suggests that every galaxy hosts a massive black hole. In the hierarchical model of galaxy formation elliptical galaxies form and grow as a result of generations of mergers of comparably massive progenitors. The exact nature of the progenitors and the epoch of the mergers are both uncertain, but several lines of argument suggest that there is a high merger rate of galaxies containing bh's in our past light cone.

The *number* density of galaxies above a luminosity of $0.01\,L^*$ is (from eqn 2)

$$n_{0.01} = \phi^* \int_{0.01}^\infty x e^{-x} \, dx = 5.6 \times 10^{-2} h^3 \mathrm{Mpc}^{-3} \tag{4}$$

Multiplying this by the Hubble volume $4\pi c^3/(3H_0^3)$ gives an estimate of 6×10^9 galaxies in our past lightcone. Dividing by $t_0 = 8 \times 10^9$yrs gives a merger rate since redshift $z = 1$ of $0.7h\mathrm{yrs}^{-1}$, if each galaxy undergoes one merger in that time. One might expect a comparable contribution to the merger rate from higher redshifts, at least up to $z \sim 3$ where the quasars are most numerous, suggesting that the massive bh population formed then or earlier [1]. There are several lines of argument that this merger rate is reasonable. The simplest approach to a galaxy merger rate is to use the Press–Schechter [10] formalism to estimate the change in the number of objects at a mass of about $10^{12} M_\odot$ since $z = 3$. For a fairly standard normalization of $\sigma_8 = 1$ at present and a bias near 1 [11], the number of collapsed objects has increased by about a factor of order unity in an $\Omega_0 = 1$ cosmology, and about 1/3 in a $\Omega_0 = .2$ Universe. The merger rate at higher redshifts is higher. A better calculation based on semi–analytic galaxy formation models and "conditional" Press-Schechter formalism suggests a growth factor of ~ 10 [12] since $z = 3$.

A second argument can be made from the observations of the "Lyman break objects", which suggests that the brightest objects seen at $z \sim 3$ are a factor of 10 less massive than bright objects today and considerably more numerous [13–15].

If the galaxies merge, do the massive black holes contained in them merge as well? Many of the calculations needed to answer this question have been carried out in a somewhat different context [16]. Dynamical friction will carry a massive black hole of $10^7 M_\odot$ or more into the center of a host galaxy in less than a Hubble time from far out in the galaxy. Smaller black holes, cloaked in sufficient numbers of bound stars from their parent pre–merger galaxy, will similarly be carried to the center. Two massive black holes in the center of such a galaxy will form a hard binary which decays increasingly slowly due to stellar scattering, until gravitational radiation becomes important. For galaxies with central densities like the Milky Way, black holes more massive than $10^6 M_\odot$ will reach a high enough binding energy to decay by gravitational radiation in less than a Hubble time. A similar look at this problem in a variety of galaxy types would be valuable.

Finally, there is some observational information that can be brought to bear on the question of mergers of black holes in our past light cone. Although the observed mass density of supermassive black holes is only 5 times greater than that predicted from the integral of the quasar light, the *number* of black holes of $10^8 M_\odot$, corresponding to Eddington luminosities of $\sim 10^{46}$ ergs/sec, is about $10^{-3}\mathrm{Mpc}^{-3}$ [1], while the number of quasars with luminosities of 10^{46} ergs/sec, at the peak of quasar numbers at $z \sim 3$ is only about $10^{-6}\mathrm{Mpc}^{-3}$ [17]. This discrepancy can be resolved in one of two ways. The obvious one is that quasars shine only for about 10^6yrs. This seems implausible as they could then only accrete (even at super Eddington rates) a few percent of their mass in this time, and must gain the rest in a manner invisible to us. Alternatively, they may have merged a few times since the quasar era. Even two generations of merging (producing a factor of 4 change in mass of a typical black hole since the quasar epoch) goes a long way to resolving the "numbers crisis" because we must then identify the quasars that powered the bright quasars at early epochs with much more massive black holes today. Since the galaxy luminosity function (and by hypothesis, the bh mass function) falls exponentially at high mass this modest growth factor serves to bring the numbers at high and low redshift into line (see [1]).

Thus it seems likely on several grounds that the supermassive black hole population has undergone a few mergers since the quasar epoch. If this is so, the bh merger rate in our lightcone could easily exceed 1/yrs. As noted in other talks at this meeting, these mergers should be observable for masses of at least $10^6 M_\odot$. Since our best current understanding is of yet higher mass black holes, it would be desirable to maintain or improve LISA's performance at the lowest frequency, where the heaviest objects will radiate. On the other hand, LISA may give us the best handle on the mergers of the low mass objects, and may provide the only information we will get on the mergers of protogalaxies before $z \sim 3$.

I'm grateful to the "Nukers" (E. A. Ajhar, R. Bender, G. Bower, A. Dressler, S. M. Faber, A. V. Filippenko, K. Gebhardt, R. Green, L. C. Ho, J. Kormendy, T. Lauer, J. Magorrian & S. Tremaine) and also to John Bahcall, Pawan Kumar, J. P., Ostriker, Martin Rees and David Spergel, for discussions of these topics. I thank the Ambrose Monell Foundation, the Guggenheim Foundation and NASA for financial support.

REFERENCES

1. Richstone, D., *et al.* Nature, in press, Oct 1998.
2. Kormendy, J. & Richstone, D. Inward bound — the search for massive black holes in galactic nuclei. Ann Rev Astron and Astroph **33,** 581 — 624 (1995).
3. Richstone, D. O. & Tremaine, S. ApJ **327,** 82 — 88 (1988).
4. van der Marel, R. P., Cretton, N., de Zeeuw, P. T. & Rix, H.-W. ApJ **493,** 613-631 (1998).
5. Goodman, J. & Lee, H.-M. ApJ **337**, 84 — 90 (1989).
6. Maoz, E. ApJ Letters **447**, L91 — L94 (1995). Updated in Maoz, E. ApJ Letters **491,** L181 — 184 (1998).
7. Magorrian J. *et al.* AJ **115,** in press (1998).
8. Chokshi A. and Turner, E. L. MNRAS **259,** 421 — 424 (1992).
9. Loveday, J., Peterson, B. A., Efstathiou, G. and Maddox, S. J. ApJ **390,** 338 — 344 (1992).
10. Press, W. H. & Schechter, P. L. ApJ **187,** 425 (1974).
11. Strauss, M. A., Willick, J. A. Phys Reports, **261,** 271 — 431 (1995).
12. Somerville, R., Primack, J. & Faber, S.M. ApJ in press (1998).
13. Steidel, C. C., Giavalisco, M., Pettini, M., Dickinson, M. & Adelberger, K. L. ApJ Letters **462,** L17 — L21 (1996).
14. Lowenthal, J.D. *et al.* ApJ **481,** 673 — 688 (1997).
15. Trager, S.C., Faber, S.M., Dressler, A. & Oemler, A. ApJ **485,** 92 — 99 (1997).
16. Xu, G. &Ostriker, J. P. ApJ **437,** 184 (1994).
17. Schmidt, M., Schneider, D. P. & Gunn, J. E. AJ **110,** 68 — 77 (1995).

Supermassive black holes as sources for LISA

Martin G. Haehnelt

Institute of Astronomy, Madingley Road, Cambridge CB3 0HA, UK.

Abstract. I briefly discuss some issues relevant for the formation of supermassive black holes and give estimates of the event rates for the emission of gravitational waves by coalescing supermassive black hole binaries. I thereby use models which take into account recent improvements in our knowledge of galaxy and star formation in the high-redshift universe. Estimated event rates range from a few to a hundred per year. Typical events will occur at redshift three or larger in galaxies lying at the (very) faint end of the luminosity function at these redshifts.

INTRODUCTION

Supermassive black holes (SMBH's) are amongst the prime targets for LISA and LISA will be primarily sensitive to events involving SMBH's in the mass range $10^{4-6}\mathrm{M}_\odot$ over a wide range range in redshift [1–3] (see also the contributions by Blandford and Sigurdsson these proceedings). The evidence for the existence of SMBH's more massive than that has been steadily increasing over the last years. The two most convincing cases are currently our own galactic centre and NGC4258 [4–6]. In both cases the inferred deep potential wells and high mass densities leave little room for alternative explanations other than the presence of a SMBH [7]. Our best estimate of the overall mass density in black holes still comes from the integrated flux emitted by optically bright QSO's which are generally believed to be predominantly powered by SMBH's. From this a present-day black hole mass density of $\sim 1.5 \times 10^5 \mathrm{M}_\odot\,\mathrm{Mpc}^{-3}$ was inferred [8,9] which corresponds to about $5 \times 10^7 \mathrm{M}_\odot$ per L_* galaxy. This estimate has recently been complemented by an investigation of a large sample of black hole masses for nearby galaxies which gives a factor three to ten higher value suggesting that the mass of the typical black hole of a galaxy could be as high as 0.6% of the stellar mass contained in its bulge [10,11] (see also Richstone these proceedings for a discussion).

THE FORMATION OF SUPERMASSIVE BLACK HOLES

A variety of more or less detailed scenarios has been suggested for the formation of SMBH's (see Rees 1984 [12] for a review). These scenarios involve one or several of the following basic processes leading to the concentration of mass,

- the dynamical evolution of a dense cluster of stellar objects,
- the build-up of a supermassive black hole by merging of supermassive black holes of smaller mass,
- and the viscous evolution and eventual collapse of a self-gravitating gaseous object (barred or non-barred disc, supermassive star).

While all these processes will occur the first two have serious difficulties when it comes to explain the existence of typical SMBH's observed in optically bright QSO's or nearby galaxies. I will briefly discuss the options one by one. The main problem with the "stellar route" to a SMBH is the rather long dynamical relaxation timescale $t_{\mathrm{rel}} \sim 3 \times 10^{10}\, v_{300}^{-3}\, N_8^2\, m_*\, \log{(N/2)}^{-1}\,\mathrm{yr}$, where v_{300} is the 1D-velocity dispersion of the stellar cluster and N_8 is the number and m_* the mass of stellar objects. Typical masses of black holes powering high-redshift QSO's are $\sim 10^{8-9}\mathrm{M}_\odot$. As the stellar cluster from which these could form has to be even more massive the relaxation

CP456, *Laser Interferometer Space Antenna*
edited by William M. Folkner
 1-56396-848-7/98/$15.00

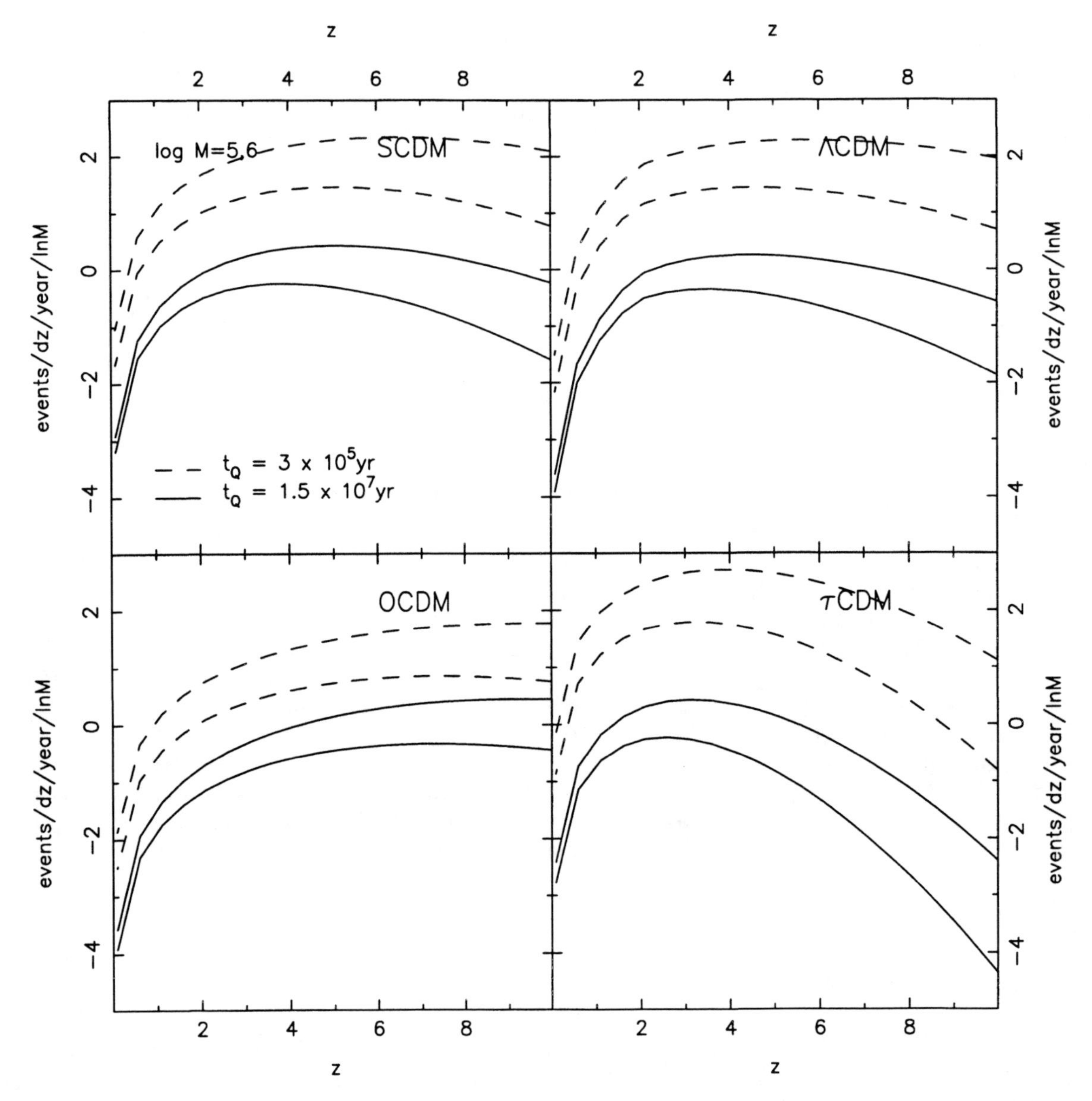

FIGURE 1. Event rates for the emission of gravitional waves involving supermassive black holes above a certain mass as indicated on the plot. The four panels are for different variants of the cold-dark-matter cosmogony (see table 1 for parameters). One event per newly-formed dark matter halo is assumed. Dashed and solid curves are for two different QSO lifetimes used to calibrate the models as indicated in the upper left panel. The upper and lower curves are for $10^5 M_\odot$ and $10^6 M_\odot$, respectively.

timescale is prohibitively long. It is generally very difficult to concentrate the mass in an efficient manner once the gas has fragmented into stars.

The "merging scenario" for the formation of SBH's has become increasingly attractive with the accumulating observational evidence for the hierarchical build-up of large galaxies predicted by CDM-like structure formation scenarios. In hierachical cosmogonies typical present-day galaxies have formed by merging of about ten smaller galaxies between redshift three and now and each of these "progenitors" will have formed from even smaller sub-units at higher redshift. When these galaxies merge the putative black holes at their centre will generally merge as well [13]. If merging were indeed the dominant process for the build-up of the mass of SMBH's the problem of the formation of present-day SMBH's could be deferred to the problem of the formation of much smaller mass SMBH's in galactic sub-units at very high redshift [16]. These would, however, have to form with the high efficiency inferred from the large black-hole mass to stellar-bulge mass ratio of present-day galaxies

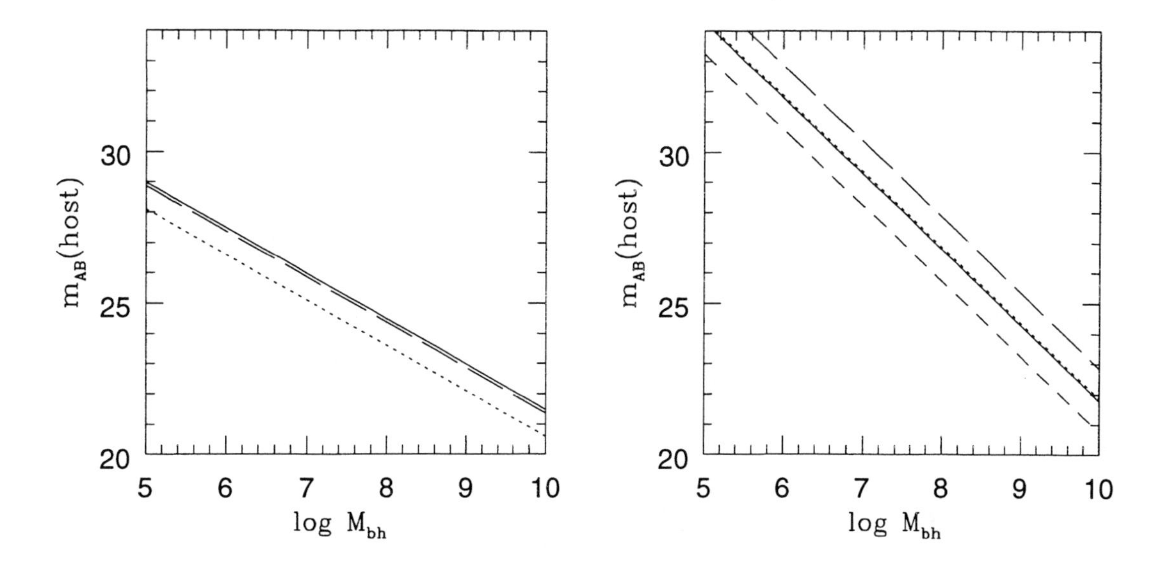

FIGURE 2. Typical apparent brightness of host galaxies at z=3. The right and left panel assume a QSO lifetime of 1.5×10^7 yr and 3×10^5 yr, respectively. Different lines are for different cosmological models.

in small protogalactic clumps with shallow potential wells. As argued by Haehnelt& Rees [14] and Haehnelt, Natarajan & Rees (HNR98) [15] the black-hole formation efficiency should be larger in the deeper potential wells of the larger galaxies forming around redshift three making this an unlikely but not impossible option.

This leaves the last option where most of the mass is accreted in gaseous form — the fastest and most efficient way to concentrate mass in SMBH's. The typical timescale for the concentration of the mass will be 10^{7-8} yr, the dynamical time scale of the galactic nucleus. The viscosity is mainly due to gravitational instabilities inevitably present in a self-gravitating disc. There is no fundamental limit for the efficiency of this process but the concentration of the gas has to compete with the consumption of the gas by star formation which will occur on a similar timescale. Unfortunately the expected gradual build-up of the mass is not likely to produce gravitational waves efficiently [3]. It will, however, be accompagnied by the frequent merging characteristic for hierarchical cosmogonies and probably also by the infall of SMBH's of intermediate mass which may form via the stellar route in the central star cluster at the galactic nucleus. In the next section we will give estimates of the expected event rates and briefly discuss some of the uncertainties involved.

EVENT RATES IN HIERARCHICAL STRUCTURE FORMATION SCENARIOS

Structure in the universe is generally believed to originate from small density fluctuations of some sort of collisionless dark matter (DM). These fluctuations are assumed to be Gaussian distributed and can be specified by their spatial power spectrum. The dynamical evolution of the dark matter, the mass function of collapsed DM halos and their merging rates can be reliable predicted once the cosmological model is specified. The main difficulty in a comparison with astronomical data is to predict the distribution of gas and stars relative to that of the dark matter. Here we would like to know the space density and merging rate of SMBH's to estimate the gravitational wave event rates due to coalescing SMBH binaries. Haehnelt & Rees and HNR98 have demonstrated that with simple relations between DM halo and black hole mass observed high-redshift galaxies and QSO's can be explained reaonably well. The models proposed in HNR98 predict the "formation rate" of newly-formed DM halos and thus allow us to get a crude estimate of the formation rate of SMBH's as function of their mass. However, these estimates depend on the uncertain lifetime of optically bright QSO's as they are calibrated by a comparison with their space density. The typical formation of a DM halo involves the merging of several DM halos and this formation should be accompanied by one or several merging events

TABLE 1. The model parameters of the CDM variants explored: σ_8 is the *rms* linear overdensity in spheres of radius $8\,h^{-1}$ Mpc and Γ is a shape parameter for CDM-like spectra. h is the Hubble constant in units of the $100\mathrm{km\,s^{-1}}$ and Ω_0 and Ω_Λ are the total energy density and that due to a cosmological constant, respectively.

MODEL	σ_8	h	Ω_0	Ω_Λ	Γ
SCDM	0.67	0.5	1.0	0.0	0.5
OCDM	0.85	0.7	0.3	0.0	0.21
ΛCDM	0.91	0.7	0.3	0.7	0.21
τCDM	0.67	0.5	1.0	0.0	0.21

of SMBH's. Figure 1 gives estimate of event rates and makes the rather conservative assumption that each newly-formed DM halo produces one gravitational wave event. The solid curves are for a lifetime of 1.5×10^7yr and assume a non-linear relation between black hole and DM halo mass as discussed by HNR98. Typical event rates are one per year about the same as those obtained by Haehnelt (1994) [2] for similar assumptions. The four panels are for different cosmogonies which span the range of currently viable models. The relevant parameters are given in table 1. Obviously the numbers have to be convolved with the sensitivity curve of LISA. LISA should detect the coalescence of equal mass binaries of $10^{5-6}\mathrm{M}_\odot$ out to maybe redshift ten. Even though at high redshift a sufficient signal-to-noise ratio will probably only be achieved for considerably less than a year. The coalescence of unequal mass SMBH binaries will only be detectable at significantly lower redshift. One should also keep in mind that little is known observationally about black holes of such small mass especially at high redshift and that the predictions rely on an extrapolation by the models from the typical black hole of mass $10^8\mathrm{M}_\odot$ or larger observed around redshift three. Furthermore at low redshift (below $z \sim 2$) the numbers in Figure 1 are likely to underestimate the merging rate of SMBH's as no attempt was made to model the late merging of galaxies in DM halos which formed at earlier times.

As demonstrated by the dashed curves event rates would be about a factor thirty higher if the lifetime of QSO's were as short as 3×10^5 yr. In this case a linear relation between DM halo mass and black hole mass is required [15]. The lifetime of the QSO's will also affect the predicted host-galaxy luminosity and the clustering properties of the QSO's. Constraints on both of these should be soon improved by the planned new large QSO surveys (2DF,SLOAN) and the uncertainty in the QSO lifetime reduced or even removed.

Figure 2 shows the predicted apparent brightness of typical host-galaxies at $z = 3$ as a function of black hole mass for the cosmological models in Fig. 1 . The left panel is for the long lifetime and the right panel for the short lifetime. Note that especially for short QSO lifetimes most of the predicted coalescences should occur in extremely faint galaxies.

DISCUSSION

The existence of supermassive black holes in a major fraction of all galaxies seems firmly established. Most of the mass in these SMBH's will have found its way beyond the event horizon in the gas-rich nuclei of the progenitors of present-day galaxies at high redshift. The emission of gravitational waves from coalescing supermassive binary black holes formed during the merger of such proto-galaxies should occur frequently enough to be detected during the lifetime of LISA. Typical events should occur in proto-galaxies at the (very) faint end of the luminosity function. The event rates are expected to increase with redshift with a rather broad peak at redshift three or larger and a slow decline at higher redshift The details of this decline are very uncertain and depend strongly on how efficiently small black holes form in shallow potential wells. The biggest uncertainty, however, is the number of detectable coalescences per newly-formed halo. Each new halo will be formed by the merging of a number of smaller halos each of which will contain one or more SMBH's. How many coalescences occur will depend on whether the black hole binary formed in one merger has already coalesced when the next black hole sinks to the centre. Otherwise sling-shot ejection is possible [13]. Furthermore the black holes will be embedded in a nuclear star cluster formed from gas on its way into the SMBH. These star cluster will not contribute much to the total mass in SMBH's. Nevertheless, they still could form with high velocity dispersion and black holes of 100 to $10^4 M_\odot$ might build up efficiently by the coalescence of stars [17]. At lower redshifts these would also be detectable by LISA when the coalesce with the central SMBH.

I finally conclude, that even a pessimist who assumes a rather long QSO lifetime and only one binary coalescence per newly-formed halo should expect a couple of SMBH binary coalescences during the lifetime of LISA while an optimist might expect to see up to several hundred of these exciting events.

ACKNOWLEDGMENTS

I would like to thank Martin Rees for many stimulating discussions on the issues discussed here.

REFERENCES

1. LISA Pre-phase A report, 1998
2. Haehnelt M.G., 1994, MNRAS, 269, 199
3. Rees M.J., 1997, Class. Quantum Grav., 14, 1441
4. Genzel R., Eckart A., Ott T., Eisenhauer F., 1997, MNRAS, 201, 219
5. Watson W.D., Wallin B.K., 1994, ApJ, 432, L35
6. Miyoshi M., Moran M., Hernstein J., Greenhill L., Nakai N., Diamond P., Inoue N., 1995, Nature, 373, 127
7. Maoz D., 1995, ApJ, 455, L115
8. Soltan A., 1982, MNRAS, 200, 115
9. Chokshi A., Turner E. L., 1992, MNRAS, 259, 421
10. Magorrian J., et al., 1998, AJ, 115, 2285
11. van der Marel R.P., 1998, ApJ, submitted
12. Rees M.J., 1984, ARAA, 22, 471
13. Begelman M.C., Blandford R.D., Rees M.J., 1980, Nature, 287, 307
14. Haehnelt M.G., Rees M.J., 1993, MNRAS, 263, 168
15. Haehnelt M.G., Natarajan P., Rees M.J., 1998, MNRAS, in press (HNR 1998)
16. Haiman Z., Loeb A., 1998, ApJ, 499, 520
17. Quinlan G.D., Shapiro S.L., 1990, ApJ, 356, 483

Supermassive Black Hole Quasar Remnants

Elihu Boldt* and Darryl Leiter†

*Laboratory for High Energy Astrophysics
NASA Goddard Space Flight Center
Greenbelt, MD 20771

†Foreign Science and Technology Center
Charlottesville, VA 22901

Abstract. Studies of the cosmic X-ray background indicate that it arises mainly from accretion powered AGNs and that the present-epoch mass density in the form of supermassive black holes must be at least an order of magnitude more than represented by currently active Seyfert galactic nuclei. The black hole mass spectrum of these Seyfert nuclei extends up to $\sim 2\times10^8$ solar masses. Evidence summarized here suggests that much of the local mass density of compact galactic nuclei is associated with inactive quasar remnants which are now black holes that are substantially more massive than Seyfert nuclei. It is emphasized that such quasar remnant black holes are sufficiently massive to preclude the tidal disruption of any infalling stars, solar sized as well as compact. Any Keplerian orbits about these supermassive black holes would have periods exceeding 7 hours.

COSMIC X-RAY BACKGROUND

By correlating surface brightness fluctuations of the Cosmic X-ray Background (CXB) with IRAS galaxies Barcons et al. (1) find that the present-epoch 2-10 keV luminosity density is dominated by Seyfert 1 galaxies ($L_x > 10^{42}$ ergs/s). Padovani, Burg and Edelson (2) have determined that the local mass density in the form of Seyfert 1 nuclei is $\sim 6\times10^2\ M_\odot/(\mathrm{Mpc})^3$, half of which arises from black holes of mass $M > 3\times10^7\ M_\odot$ (where $M_\odot \equiv$ solar mass) and essentially none from any possible active galactic nucleus (AGN) black holes of mass $M > 2\times10^8\ M_\odot$. For an accretion powered X-radiation efficiency $\leq 10\%$ {i.e., $L_x/[c^2(dM/dt)] \leq 0.1$} the observed CXB energy flux implies the build-up of a local mass density $> 14\times10^3(1+\langle z\rangle)\ M_\odot/(\mathrm{Mpc})^3$, where $\langle z\rangle \geq 1$ (3). This density is clearly much larger than that for Seyfert 1 nuclei. In order to account for the flux and spectrum of the CXB it is necessary to invoke a supplementary source population that somehow makes a substantial redshifted contribution to the observed CXB without making much of a contribution to the local 2-10 keV luminosity density (3, 4). This needed additional component (over and above that from Seyfert 1 nuclei) could arise from a population not at all represented locally [e.g., precursor AGNs (2) necessarily at large redshifts] and/or Seyfert 2 nuclei whose emission below ~10 keV is strongly attenuated by absorption (5, 6). The unified Seyfert AGN model (5) would imply that the present-epoch mass spectrum for Seyfert 2 nuclei would be the same as for Seyfert 1, renormalized by the ratio of their local number densities (n_2/n_1). In particular, for these scenarios, using $n_2/n_1 < 4$ (5, 6) implies a total local Seyfert mass density $< 3\times10^3\ M_\odot/(\mathrm{Mpc})^3$, much less than that implied by the CXB flux. In such models, the bulk of the local mass density related to the CXB would be accounted for by dormant Seyfert remnants having the same mean mass ($\sim 2\times10^7\ M_\odot$) as the AGNs.

QUASAR REMNANTS

Using observations of quasar populations at large redshifts and typical broad-band spectra, Chokshi and Turner (7) have estimated the bolometric background flux expected from all such sources. Assuming this emission is powered by accretion, with a 10% efficiency, they conclude that the present-epoch mass density in quasar remnants is $\geq 2\times10^5\ M_\odot/(\mathrm{Mpc})^3$ and that at least half this density arises from black holes of $M > 4\times10^8\ M_\odot$ (for $H_o \approx 50$ km s^{-1}/Mpc), substantially greater than a Seyfert nucleus (whose mass spectrum resides well below $4\times10^8\ M_\odot$). The expected local

CP456, *Laser Interferometer Space Antenna*
edited by William M. Folkner
 1-56396-848-7/98/$15.00

number density of quasar remnants having masses $>10^9$ $M_\odot$ is estimated to have a value that turns out to be comparable to that for Seyfert 1 AGNs. In particular, there should be at least a dozen such quasar remnants within 50 Mpc. Kormendy and Richstone (8) have obtained direct stellar-kinematic evidence for massive dark objects (MDOs) at the centers of several inactive galaxies and estimate that the local mass density in such objects is $\sim 14\text{x}10^4$ $M_\odot/(\text{Mpc})^3$. In this connection we note that the number of MDOs within 50 Mpc already identified by Magorrian et al. (9) as being more massive than 10^9 $M_\odot$ is 8. This is a lower limit to the total number of such supermassive MDOs within this volume since their sample of MDOs at the centers of nearby galaxy bulges is incomplete, albeit sufficiently large for the correlations sought by them (10). Using the luminosity function for field galaxies (11) to estimate the incompleteness of their sample suggests that the corrected number of supermassive MDOs could well be an order of magnitude greater than that already observed.

MASS CONSTRAINTS

To preclude the tidal disruption of a star (mass M^*, radius R^*) near a supermassive black hole ($M>>M^*$) the stellar object must reside at a sufficiently large distance ($R > R_t$) from the black hole origin, where R_t is the tidal radius, estimated by Rauch and Tremaine (12) as

$$R_t \approx 2R^*(M/M^*)^{1/3}. \tag{1}$$

For a compact neutron star (13, 14), we have

$$R^*/(R_g)^* < 6, \tag{2}$$

where $(R_g)^*$ is the stellar gravitational radius, given by

$$(R_g)^* \equiv GM^*/c^2. \tag{3}$$

From equations (1-3) we obtain the constraint

$$R_t/R_g < 12(M^*/M)^{2/3}, \tag{4}$$

where R_g is the black hole gravitational radius , viz:

$$(R_g) \equiv GM/c^2. \tag{5}$$

Since M is many orders of magnitude greater than M^* (by definition of supermassive), equation (4) then implies that the tidal radius of the black hole is much smaller than its gravitational radius. Hence, an infalling compact neutron star would clearly avoid tidal disruption all the way to the event horizon. We should note, however, that if the central mass were also a compact stellar object (i.e., a black hole or another neutron star for which $M \approx M^*$), tidal effects could be important (15, 16), especially for those neutron stars where $R^* \geq 5(R_g)^*$.

In general, the constraint that $R_t < R$, needed to preclude tidal disruption, places a *lower* limit on the black hole mass required, as follows:

$$M/M_\odot > 9\text{x}10^8\ [F(M^*, M)]\ (R/R_g)^{-3/2}, \tag{6}$$

where F is a dimensionless size factor, on the order of unity for usual non-compact stars, given in terms of R^* and M^* as follows

$$F = (R^*/R_\odot)^{3/2}(M^*/M_\odot)^{-1/2}, \tag{7}$$

where $R_\odot$ is the solar radius.

For most stars, Rauch and Tremaine (12) assume $R^* \propto (M^*)^{2/3}$ (reasonable for both high and low mass main-sequence stars). For such stars equation (7) yields

$$F = (M^*/M_\odot)^{1/2} \approx 1. \tag{8}$$

ORBITAL FREQUENCY LIMITS

For a Kerr hole, where $R \geq R_g$, we note from equation (6) that the critical mass is $9\text{x}10^8$ $M_\odot$: The Kepler orbital periods (τ) associated with this mass are

$$\tau \geq 2[(2\pi R_g)/c] = 6\text{x}10^4 \text{ seconds} \tag{9a}$$

$$\rightarrow \nu \leq 2\text{x}10^{-5} \text{ Hz} \qquad \text{for a Kerr hole.} \tag{9b}$$

For a Keplerian orbit around a Schwarzschild hole, where $R \geq 6R_g$, the critical mass is $6\text{x}10^7 M_\odot$. The orbital periods (τ) associated with this mass are

$$\tau \geq (6)^{3/2}[(2\pi R_g)/c] = 3\text{x}10^4 \text{ seconds} \tag{10a}$$

$$\rightarrow \nu \leq 4\text{x}10^{-5} \text{ Hz} \qquad \text{for a Schwarzschild hole.} \tag{10b}$$

These frequency *upper* limits (equations 9b & 10b) are somewhat lower than the limiting frequency (~10^{-4} Hz) currently envisaged as being the lowest to be accessible with LISA.

ACKNOWLEDGMENTS

E.B. acknowledges valuable discussions with Pranab Ghosh, Demos Kazanas, John Magorrian, and William Zhang.

REFERENCES

1. Barcons, X., Franceschini, A., De Zotti, G., Danese, L., and Miyaji, T., *ApJ* **455**, 480 (1995).
2. Padovani, P., Burg, R., and Edelson, R., *ApJ* **353**, 43 (1990).
3. Boldt, E., and Leiter, D., *Nuc. Phys. B (Proc. Suppl.)* **38**, 440 (1995).
4. Boyle, B. J., Georgantopoulos, I., Blair, A. J., Stewart, G. C., Griffiths, R. E. et al., *MNRAS* **296**, 1 (1998).
5. Madau, P., Ghuisellini, G., and Fabian, A. C., *MNRAS* **270**, L17 (1994).
6. Comastri, A., Setti, G., Zamorani, G., and Hasinger, G., *A&A* **296**, 1 (1995).
7. Chokshi, A., and Turner, E. L., *MNRAS* **259**, 421 (1992).
8. Kormendy, J., and Richstone, D., *ApJ* **393**, 559 (1992).
9. Magorrian, J., Tremaine, S., Richstone, D., Bender, R., Bower, G. et al., *AJ* **115**, 2528 (1998).
10. Magorrian, J., private communication (1998).
11. Efstathiou, G., Ellis, R., and Peterson, B., *MNRAS* **323**, 431 (1988).
12. Rauch, K. R., and Tremaine, S., *New Astronomy* **1**, 149 (1996).
13. Zhang, W., Smale, A., Strohmayer, T., and Swank, J., *ApJ* **500**, L171 (1998).
14. Miller, M. C., Lamb, F., and Psaltis, D., *ApJ*, in press (1998).
15. Lai, D., Rasio, F., and Shapiro, S., *ApJ* **406**, L63 (1993).
16. Uryu, K., and Eriguchi, Y., *MNRAS* **296,** L1 (1998).

High mass ratio sources of low frequency gravitational radiation

Steinn Sigurdsson

Institute of Astronomy, Madingley Road, Cambridge CB3 0HA, UK
&
Department of Astronomy & Astrophysics,
Pennsylvania State University, University Park, Pa 16802

Abstract. The gravitational radiation emitted during the final stages of coalescence of stellar mass compact objects with low mass massive black holes is a signal that may be detected by LISA with some confidence, while also offering the possibility of measuring strong field effects in relativity. I consider the uncertainties in estimating the rate of such events, and the possibility of enhanced coalescence rates of low mass black holes with low mass massive black holes due to ongoing star formation in centres of galaxies. I also conjecture that x–ray precursors to coalescence may conceivably be observable in some cases.

The proposed LISA mission has peak sensitivity to periodic strains due to gravitational radiation near radiation frequencies of 10^{-3}–$10^{-2.5}$ Hz, with hoped for strain sensitivities of $\sim 10^{-23}$ [1]. The characteristic frequency to which LISA is sensitive is comparable to the orbital frequency at the innermost stable orbit of Schwarzschild black holes of mass $M_{BH} \sim 10^6 M_\odot$. With high sensitivity to periodic sources, one of the more promising "guaranteed" sources for LISA is the gravitational radiation from the final stages of coalescence of low mass (1–100 $M_\odot$) compact objects with low mass massive black holes ($M_{BH} \sim 10^{6\pm1} M_\odot$).

Previous papers [2–4] considered the likely rate for detectable signals from degenerate compact objects in the cusps of normal galaxies, coalescing with central black holes such as the one inferred to be present in the Milky Way. Coalescence happens when a compact object moving in the density cusp around the central black hole scatters or diffuses in angular momentum until it gets close enough to the central black hole that gravitational radiation losses are strong enough to drive the system to coalescence before the compact object scatters back up to a higher angular momentum orbit away from the central black hole [5–7]. Estimates for detectable signal rates are of the order one per year, from these sources, but the estimates are sensitive to systematic uncertainties in the population contributing to the signal, with very large (many orders of magnitude) cumulative formal uncertainties in the expected signal rate. Since high rates of coalescence are self–limiting (through depletion of low mass compact objects at high coalescence rates; or the growth of the primary through coalescences, to the point where the frequency of the innermost stable orbit is too low for LISA to be sensitive to further coalescences), the formal uncertainty in the detectable rate is skewed to lower rates of coalescence than the "canonical" estimates in the literature. The true rate of detectable coalescences is sensitive to the assumptions about the mass function of the primary black hole, the number density of stars in regions where the appropriate mass black holes are found, and the mass and number of compact remnants available for coalescence in a Hubble time in such regions.

One major uncertainty is the mass function of massive black holes at the low mass end of the range. On the one hand, the two best determined central black hole masses, in the Milky Way and M32, are squarely in the range of masses of interest for signals detectable by LISA ($M_{BH} \approx 2 \times 10^6 M_\odot$) [9,10]; on the other hand, dynamical detection and measurements of masses of such low mass massive black holes are not possible beyond the local group, so the global mass function is poorly constrained. Three lines of evidence suggest that the volume density of low mass massive black holes may be high enough ($n \gtrsim 10^{-3}\, Mpc^{-1}$) for coalescences detectable by LISA to be common in the local universe: the apparent approximate scaling of central black hole masses with spheroid luminosity [8] suggests low mass massive black holes should be common in spirals with

CP456, *Laser Interferometer Space Antenna*
edited by William M. Folkner

weak bulges and in dwarf ellipticals; the ubiquity of low luminosity emission line systems in galactic nuclei suggests central black holes in the appropriate mass range are common [11]; and, the principle of mediocrity suggests that the local group should not be special and hence the local black hole mass function typical (unless there's anthropic selection against being near super–massive central black holes!). As a note of caution, it is worth remembering that the luminosity of AGNs depends as much on photon efficiency and the supply of mass to the central black hole at near Eddington rates, as on the actual black hole mass, and the weak AGNs seen in nearby spirals may be from black holes too massive for LISA; and, contrary to the spirit of the principle of mediocrity, M32 like dwarf elliptical appear to be rare in the local universe [12] and thus the local group is in fact special in having an one.

Assuming that low mass massive black holes exist in sufficient numbers, then the compact remnant of the stellar population near the black hole may coalesce with the central black hole. In particular, white dwarfs, neutron stars and low mass ($m_{bh} \sim$ 5–100$\mathrm{M}_\odot$) black holes present in the inner parsec around the central black hole will coalesce with the central black hole through emission of gravitational radiation. The typical starting orbit will be highly eccentric, and some fraction of the low mass compact objects will be lost through scattering straight into the central black hole, but calculations [2,3] suggest an appreciable rate for coalescences detectable by LISA. The primary uncertainties here are in the number density of remnants in the centers of galaxies. Previous work estimated 5-10% of the population would be white dwarfs, but a different choice of initial stellar mass function, with a smaller number fraction of low mass stars, and allowing white dwarfs to form from higher mass zero age stars [13] could imply a considerably higher number fraction of white dwarfs, possibly as high as 30% (Hogan this meeting). This would imply the actual rate of white dwarf coalescence with central black holes is higher by an order of magnitude than past estimates. Neutron stars coalescence with central black holes is unlikely to be a significant contributor to the LISA rate, as natal kicks efficiently remove neutron stars from the inner parsec of the galaxies of interest. Low mass black holes are, however, of major interest as sources for LISA.

It is plausible that low mass black holes form from main sequence stars, with the progenitor zero age masses being $\gtrsim$ 25–30$\,\mathrm{M}_\odot$. Binarity of the progenitor will affect the critical mass at which black holes, rather than neutron stars, form. The initial–final mass relation for such black holes is unknown, but it is likely that more massive stars form more massive black holes. Assuming a Salpeter initial mass function, roughly one star in ten thousand leaves a low mass black hole with masses, m_{bh} from as low as $2\,\mathrm{M}_\odot$ up to $\gtrsim 100\,\mathrm{M}_\odot$. The amplitude of gravitational radiation scales linearly with m_{bh}. LISA does better detecting coalescence of more massive low mass black holes at larger distances, if the gain in volume is greater than the reduced prevalence of the more massive low mass black holes. It seems likely that the initial population of low mass black holes in galactic nuclei is rapidly depleted through coalescence [4], in which case LISA may mostly detect coalescences from the first epoch of star formation at $z \sim 1$. At higher redshifts, the redshifting of the gravitational radiation to lower frequencies may reduce LISA's ability to detect coalescences, even for $m_{bh} > 100\,\mathrm{M}_\odot$, except for $M_{BH} < 10^6\,\mathrm{M}_\odot$. Of course we also expect central black holes to have been somewhat less massive at early times, which reduces the amplitude of gravitational radiation somewhat, but increases the gravitational radiation frequency at coalescence.

There are then two "best bet" scenarios for LISA detection of low mass black hole coalescence with low mass massive black holes. If there was significant **nuclear** star formation at $z \lesssim 1$, then LISA may see strong signals from the initial flurry of coalescence of the initial population of low mass black holes with the central black hole. Alternatively, if there is replenishment of low mass black holes in the nuclei of Milky Way like spirals, the rate of coalescences detectable by LISA will be correspondingly higher.

There are two ways the low mass black hole population in spiral nuclei may be replenished. Either by formation of massive stars *in situ*, or by the transfer of low mass black holes from outside the inner nucleus. Recent observations suggests that the Milky Way has in fact formed massive stars in its nucleus in the recent past. Cotera et al [14] find several O or Wolf–Rayet stars in the inner parsec of the galaxy, some of which may be massive enough to form black holes. Certainly the Pistol Star [15] in the inner 100 pc of the Milky Way must collapse to a black hole. The number of high mass stars suggests an approximate rate of black hole replenishment of $\sim 10^{-6}$ low mass black holes per year, with the mean rate of coalescence with the central black hole then necessarily $\sim 10^{-6}\,\mathrm{y}^{-1}$. This implies a LISA event rate of $1\,\mathrm{day}^{-1}$! More likely the current nuclear star formation rate is higher than the mean, but even if the duty cycle of such enhanced star formation is only 1%, the LISA event rate is enhanced proportionately.

Alternatively, a cluster of stars formed outside the inner parsec of a spiral nucleus, may, if compact enough, be brought to the centre by dynamical friction [16]. In this context, the Arches cluster [17] at an estimated distance of $\sim$ 50 pc from the Milky Way centre provides a fascinating instance. The cluster is very young, and quite massive $\gtrsim 10^{4.5}\,M_\odot$. It has $O(10)$ stars massive enough to form black holes, and if it remains bound, would be brought to the centre of the Milky Way in less than a billion years (formally in $\sim 6 \times 10^8$ y). If it does survive the tidal field of the nucleus, then this represents a black hole replenishment rate of $\sim 10^{-8}\,y^{-1}$. However, the Arches cluster is simply the youngest and therefore brightest of several clusters in the inner 50 pc. In a few hundred million years we would not detect it as a promising source of new black holes. Naively, given its age of a few million years, the Arches cluster should have ~ 100 older counterparts in the inner 50–100 pc of the Milky Way, implying an enhancement in the LISA rate of a factor of 100 or so, with the same caveats as before. In reality, the duty cycle for formation of such clusters may be low, for the same reasons the central star formation rate may cycle, but, again, if the duty cycle is 1% or higher, such processes may dominate the total coalescence rate.

On a more speculative note, it is worth considering whether precursor signals to low mass black hole coalescence may be seen. In a recent preprint, Fabian et al [18] found a variable periodicity in the Seyfert galaxy IRAS18325-5926. One possible explanation is perturbation of the accretion disk of the central black hole by a companion. Initial reports of a change in the period suggested a possible scenario where the companion might be a low mass $\sim 100\,M_\odot$ black hole in an eccentric orbit about a few$\times 10^5 M_\odot$ primary. More data is required, and this particular observation may be due to a quasi–periodic disk oscillation rather than external forcing, but it does raise the issue of whether a black hole coalescence might be preceded by related electro–magnetic emission. On a larger scale the ~ 12 year periodicity of blazar OJ287 has been conjectured to be due to a supermassive black hole binary pair. The mass–ratio for OJ287 is then 100:1, with the variation in luminosity more than a factor of hundred. The simplest model for luminosity changes postulates a low mass, thin accretion disk, dominated by the gravitational field of the central black hole, with the companion black hole perturbing the instantaneous mass flux through the inner few Schwarzschild radii. Either an enhancement or a depletion of the mass flux into the central black hole, for a period of comparable to or greater than the orbital period of the gas, modulated at the orbital period of the companion, could be detectable in x–rays. The simplest scenario implies the strength of the perturbation scales as $(m_{bh}/M_{BH})^2)$, so scaling to OJ287 a $\gtrsim 100\,M_\odot$ low mass black hole in orbit about a $\lesssim 10^{5.5}\,M_\odot$ primary might produce a detectable ($\gtrsim 10\%$) signal in x–rays or optical emission. If the coalescence rate is as high as $10^{-6}\,y^{-1}$ and the precursor signal can be seen $\sim 10^3$ years before coalescence, then only a few thousand low luminosity AGNs would have to be monitored for periodic modulation with secularly decreasing periods in order to have a reasonable hope of seeing a precursor for a LISA signal.

In conclusion, the prospects for a detection by LISA of low mass black holes coalescing with low mass massive black holes in the local universe are good and looking better.

ACKNOWLEDGMENTS

I would like to thank M. Rees for helpful discussions and C. Hogan for pointing out the possibility of higher white dwarf fractions.

REFERENCES

1. Schutz, B. F., Classical and Quantum Gravity, 13, A219-A238 (1996)
2. Hils, D., Bender, P.L., ApJL, 445, L7–10 (1995)
3. Sigurdsson, S., Rees, M.J., MNRAS, 284, 318–326 (astro–ph/9608093) (1997)
4. Sigurdsson, S., Classical and Quantum Gravity, 14, 1425–1429 (astro–ph/9701079) (1997)
5. Frank J., Rees M.J., MNRAS, 176, 633–647 (1976)
6. Rees M.J., Science, 247, 817–823 (1990)
7. Quinlan G.D., Hernquist L., Sigurdsson S., ApJ, 440, 554–564 (1995)
8. Kormendy J., Richstone D., ARAA, 33, 581–624 (1995)
9. Genzel, R., Eckart, A., Ott, T., Eisenhauer, F., MNRAS, 291, 219–234 (1997)
10. van der Marel, R.P., de Zeeuw, T., Rix, H.–W., Quinlan, G.D., Nature, 385, 610–612 (1997)
11. Ho, L.C., Filippenko, A.V., Sargent, W.L.W., ApJ, 487, 568–578 (1997)
12. Ziegler, B.L., Bender, R., A&A, 330, 819–822 (1998)
13. Elson, R.A.W., et al. ApJL, 499, L53–56 (1998)
14. Cotera, A.S., Figer, D.F., Blum, R.D., preprint (1998)
15. Figer, D.F., et al., ApJ, in press (1998)

16. Polnarev, A.G., Rees, M.J., A& A, 283, 301–312 (1994)
17. Serabyn, E., Shupe, D., Figer, D.F., Nature, in press (1998)
18. Fabian, A.C., et al., preprint (astro-ph/9803075) (1998)
19. Sillanpää, A., et al., A& A, 315, L13–16 (1996)
20. Sundelius, B., Wahde, M., Lehto, H.J., Valtonen, M.J., ApJ, 484, 180–185 (1997)

General Relativity as Seen in X-rays: What Can LISA Tell Us That We Don't Already Know?

Omer Blaes

Department of Physics, University of California at Santa Barbara, Santa Barbara, CA 93106

Abstract. If it performs as well as planned, LISA could provide tremendous information on physics and astronomy in the strong field regime of general relativity near black holes. X-ray astronomy has already provided us with tantalizing glimpses of phenomena occurring in the accretion flow onto compact objects in binary systems and active galactic nuclei. Iron K-alpha photons emitted by the accreting plasma provide direct evidence for relativistic motions and gravitational redshifts. In addition, there is evidence for the existence of the event horizon, as well as claims of general relativistic effects being responsible for certain classes of time variability. I review these exciting results, and compare and contrast them with the complimentary information which will be obtained by LISA.

INTRODUCTION

LISA has the potential of revolutionizing experimental gravitational physics and astrophysics. To place this potential in context, it is perhaps useful to briefly review what we are learning about general relativity in astronomical sources from current observational techniques. The exciting results come almost exclusively from the field of X-ray astronomy: many of the black hole and neutron star sources known in the sky emit copious X-ray radiation through the process of accretion of plasma. Three recent examples stand out as being particularly noteworthy: evidence for the existence of black hole event horizons, evidence of relativistic frame-dragging, and the mapping of the circular orbit structure of the Kerr metric. As exciting as these results are, none of them are yet on a solid foundation. This is partly because the interpretation of X-ray data relies on an understanding of complex magnetohydrodynamical and radiative transfer processes in accretion flows onto compact objects, and this understanding is at present incomplete, to say the least. This brings up one of the contrasting features between X-ray astronomy and gravitational wave astronomy: there should be many compact object systems where gravitational wave emission will be much simpler to interpret.

SOFT X-RAY TRANSIENTS: PROOF OF BLACK HOLE EVENT HORIZONS?

Soft X-ray transients arise in binary systems containing a black hole or neutron star which accretes material from a relatively ordinary companion star. In quiescence, i.e. most of the time, the system has a low X-ray luminosity L_{min}. Every few decades or so, however, the system undergoes a several month long outburst in which the X-ray luminosity becomes very large, L_{max}. The difference between these two states is clearly due to different rates of accretion of mass onto the central compact object. Whether that compact object is a black hole or a neutron star is partly inferred from the mass, when it can be measured. In some systems the mass of the compact object is consistent with that of a neutron star, whereas in others the inferred mass is found to be greater than 3 $M_\odot$, implying a black hole. In addition, neutron stars are inferred if the system also undergoes so-called type I X-ray bursts, which are thought to be due to thermonuclear flashes on the neutron star surface.

CP456, *Laser Interferometer Space Antenna*
edited by William M. Folkner
 1-56396-848-7/98/$15.00

The accretion flow in the outburst state is generally thought to take the form of a standard optically thick accretion disk [1]. The X-rays arise from thermal emission from the disk and, in the case of neutron stars, from material landing on the surface of the star. In both black holes and neutron stars the efficiency of conversion of rest mass energy to luminosity is probably comparable, of order 10 percent. The state of the flow in the low accretion rate quiescent state is not so clear, but an increasingly popular suggestion is that it is an "advection dominated accretion flow", or ADAF [2,3]. In such flows the gravitational energy released by accretion is stored in the ions of the plasma, with very little transferred to the electrons. Much of the heat is therefore advected inward toward the central object instead of being radiated away. ADAFs onto a black hole can therefore have much lower radiative efficiencies than standard accretion disks, as much of the heat flows down through the event horizon. If the flow is onto a neutron star, however, the efficiency of the system remains high because material will still radiate as it strikes the stellar surface.

Narayan, Garcia, & McClintock (1997) [4] have recently noted two additional striking differences between black hole and neutron star soft X-ray transients. The first is that black hole systems have larger outburst luminosities than neutron star systems. This is easily explained as being due to the fact that the black holes in these systems have larger mass than neutron stars, permitting them to have higher Eddington luminosities. The second is that out of nine systems considered, the ratio of quiescent to outburst luminosity L_{min}/L_{max} is much larger for neutron stars than black holes. It could be that the quiescent accretion rate in black hole systems is always smaller than that of neutron star systems, but a more likely explanation perhaps is that the radiative efficiency of the black hole systems is much lower due to the existence of their event horizons, the defining feature of black holes.

However, these results are based on only a small number of sources. In addition, the X-rays are observed in finite and different bandpasses, and accretion models must be used to justify the assumption that the observed luminosity dominates the bolometric luminosity. The theoretical consistency of ADAF models is also uncertain, as there seem to be ways of thermally coupling ions and electrons on the infall time scale or of channeling the accretion power directly into the electrons through magnetohydrodynamic turbulence [5–7]. Finally, ADAFs or other types of quiescent accretion flows may be accompanied by outflows, which would also reduce the radiative efficiency without necessarily requiring an event horizon.

QPO IN LOW MASS X-RAY BINARIES: LENSE-THIRRING PRECESSION?

X-ray binary systems are known to sometimes exhibit quasi-periodic oscillations (QPO) in their X-ray luminosity on a wide range of time scales. QPO on very short, millisecond time scales has recently been detected in a number of sources by the *Rossi X-Ray Timing Explorer*, which is interesting because this is comparable to the dynamical time scale of the accreting compact object. Pairs of kilohertz QPO have been detected in the quiescent emission from neutron star low mass X-ray binaries. The frequencies of these QPO, ν_b and ν_k, can vary with time, but the difference $\nu_s = \nu_k - \nu_b$ between them is remarkably stable. This difference frequency is itself exhibited by QPO during type I X-ray bursts from these systems (e.g [8]). It is very likely that ν_s is the spin frequency of the neutron star, with the QPO arising perhaps from inhomogeneous burning on the surface during the thermonuclear flash. The higher frequency member of the pair of QPO seen in quiescence, ν_k, could be the frequency of the innermost circular orbit of the accretion disk, with oscillations arising because of inhomogeneous blobs of X-ray emitting material orbiting at that radius. The lower frequency member, ν_b, is then a beat frequency between ν_k and ν_s, seen perhaps because of modulation of the accretion rate onto the central star due to the interaction between the stellar magnetic field and the inner region of the disk.

If this interpretation of the observations is correct, then one can use the measured spin frequency of the neutron star and orbital frequency of the inner disk to calculate the Lense-Thirring precession frequency ν_{LT} of a test particle in orbit at the same radius but inclined to the equator of the neutron star. The only unknowns are the mass and moment of inertia of the neutron star, but these are reasonably well-determined from models. The precession frequency ν_{LT} turns out to be $\sim$ 10 Hz, and amazingly, QPO are also observed at comparable frequencies in these same sources. Stella & Vietri (1998) [9] have therefore suggested that these QPO are due to Lense-Thirring precession of an accretion disk which is inclined to the equatorial plane of the star. Corrections must be made to ν_{LT} due to the classical retrograde precession produced by the neutron star quadrupole moment, but this frequency turns out to be smaller than ν_{LT}. Lense-Thirring precession has also been proposed as an explanation for QPO observed in black hole systems [10].

A problem with this interpretation of the observations is that viscosity in the accretion disk should align the plane of the inner disk with the equatorial plane of the compact object, even if the disk is inclined further out [11]. However, gravitomagnetic bending modes of the right frequency do in fact exist in the innermost regions of a viscous disk [12]. These modes are damped, and the question is whether or not they could be excited to sufficiently high amplitude to explain the observations.

Another problem with the Lense-Thirring interpretation is that there are other ways of producing QPO of comparable frequency, some of which are also due to general relativistic effects. For example, general relativity can cause the epicyclic frequency to have a maximum at some radius outside the inner edge of the disk, thereby trapping hydrodynamical gravity wave modes (e.g. [13]). Whatever is the correct explanation, it is likely that general relativity is playing some significant role.

MAPPING OF CIRCULAR ORBITS IN KERR METRIC USING THE FE Kα LINE

Accretion flows onto supermassive black holes in active galactic nuclei also produce copious X-rays. In Seyferts, the X-rays are generally thought to arise from Compton scattering by hot electrons of soft photons emitted by an accretion disk. Some of these X-rays shine back on the disk, where they ionize K shell electrons from iron atoms, causing the iron to fluoresce in Kα line photons as L shell electrons drop to the K shell. This line is observed to be a powerful line in Seyferts. The *Advanced Satellite for Cosmology and Astrophysics* (*ASCA*) has sufficient spectral resolution to measure the line profile, and it has turned out to be very broad [14–16]. Although the interpretation is not universally accepted, it appears likely that this is due to a combination of gravitational redshifts and Doppler broadening at relativistic speeds in the accretion flow very near the black hole. The profile can be well fit by simple power-law radial emissivities of the line in disks around Schwarzschild or Kerr black holes. In fact, one might hope to use the profile to measure the spin of the black hole, due to the fact that the innermost stable circular orbit (ISCO) can exist much closer to the horizon for large spins, producing much broader line profiles. Unfortunately, the spatial form of the line emissivity is not known, and if it declines weakly with radius then the line profiles for Kerr holes become similar to those of Schwarzschild holes. In addition, there may be sufficient iron line emission from the accretion flow inside the ISCO to permit Schwarzschild holes to also produce extremely broadened profiles [17].

ASCA's limited effective area requires long integration times to measure the iron line profile, much longer than the light crossing time of the inner parts of the accretion disk. X-rays are known to vary on short time scales in Seyferts, suggesting that some sort of transient flaring is producing regions of hot electrons above the disk. A large effective area X-ray satellite which is capable of detecting enough photons to measure the line profile on short time scales could measure the line profile changing with time as X-rays from a given flare reach different locations in the disk. Such reverberation mapping will just become feasible with the proposed mission *Constellation-X*, although upcoming missions such as the *X-ray Multi-Mirror Observatory* will provide some interesting data as well.

Reynolds et al. (1998) [18] have explored this idea and shown that such data would contain tremendous diagnostic information. The time between the start of the flare and the first detection of iron line photons depends on the distance between the flare and the disk and the disk inclination angle. The breadth of the instantaneous line profile depends on the disk inclination angle, being quite narrow for face-on disks and very broad for edge-on disks. As time goes on, the iron line becomes increasingly narrow as the X-rays reach the outer parts of the disk. A time-delayed redshifted response in the profile is caused by the Shapiro delay acting on photons passing near the hole. Transient high energy peaks can be produced by highly ionized (hydrogen and helium-like) iron where the Kα photons are at higher energy than for neutral iron.

In principle, such reverberation mapping can map the orbit structure of a Kerr spacetime, at least for circular equatorial orbits, thereby allowing a test of general relativity. However, the iron lines will also depend on the disk inclination angle, the illumination geometry of the flare, and the possible presence of simultaneous, multiple flaring regions. The instantaneous line profiles also depend on the ionization state of the gas in the disk, not only because the line energy depends on the ion species, but also because the line becomes resonant once electrons start being stripped from the L shell. Detailed photoionization and radiative transfer calculations may therefore be necessary to model the data. Finally, the results also depend on the geometry of the fluorescing, cool gas, which in the inner regions of accretion flows in Seyferts may not be smoothly distributed as in simple disk models (e.g. [19]).

CONCLUSIONS

As exciting as they are, each of the three cases I have discussed relies on understanding the complex physical processes occurring in accretion flows - flows where even the geometry is uncertain because direct imaging is impossible at the present time. Instead of QPO and line emission from orbiting gas which exists at all radii, LISA can detect gravity wave emission from the much simpler and cleaner system of a single star in orbit around a supermassive black hole, providing a much simpler test of general relativity. Accretion flows probe mainly equatorial circular orbits, although warping of the disk and photon geodesics probe other orbits off the equatorial plane. A single star need not be in a circular orbit, although in actual fact it may very well be if it is in orbit around an accreting black hole because of drag forces from the disk. However, unlike in X-ray astronomy, the accretion flow need not be there for LISA to be able to detect something. LISA views sources which are "lit up" in gravity waves, and could well see gravity wave emission from black holes which produce very little in the way of electromagnetic radiation. In addition to providing possibly cleaner tests of general relativity, it will also give us a fairer census (or at least biased in a different way from electromagnetic astronomy) of the presence of black holes in the universe.

Astronomical observations have revealed enormously energetic, transient phenomena involving dynamic spacetime curvature (supernovae and possibly gamma-ray bursts), but the electromagnetic data has little to say about dynamic general relativity. This is likely to always be the case because such sources are optically thick, so that photons lose memory of the processes going on at the heart of these sources. This is not a problem for gravitational waves.

The most important conclusion that one should draw from modern X-ray astronomy is that X-ray phenomena in the universe are incredibly rich and diverse, far more so than was predicted prior to the launch of X-ray satellites. The development of X-ray astronomy and astronomy in the other nonvisible bands of the electromagnetic spectrum is perhaps the most compelling argument for opening up a new observing window on the universe through gravitational waves.

REFERENCES

1. Shakura, N. I., & Sunyaev, R. A., *A&A* **24**, 337-355 (1973).
2. Abramowicz, M. A., Chen, X., Kato, S., Lasota, J.-P., & Regev, O., *ApJ* **438**, L37-L39 (1995).
3. Narayan, R., & Yi, I., *ApJ* **452**, 710-735 (1995).
4. Narayan, R., Garcia, M. R., & McClintock, J. E., *ApJ* **478**, L79-L82 (1997).
5. Begelman, M. C., & Chiueh, T., *ApJ* **332**, 872-890 (1988).
6. Quataert, E., *ApJ* **500**, 978-991 (1998).
7. Gruzinov, A. V., *ApJ* **501**, 787-791 (1998).
8. Strohmayer, T. E., Zhang, W., Swank, J. H., Smale, A., Titarchuk, L., & Day, C., *ApJ* **469**, L9-L12 (1996).
9. Stella, L., & Vietri, M., *ApJ* **492**, L59-L62 (1998).
10. Cui, W., Zhang, S. N., & Chen, W., *ApJ* **492**, L53-L57 (1998).
11. Bardeen, J. M., & Petterson, J. A., *ApJ* **195**, L65-L67 (1975).
12. Marković, D., & Lamb, F. K., *ApJ*, submitted (1998).
13. Nowak, M. A., Wagoner, R. V., Begelman, M. C., & Lehr, D. E., *ApJ* **477**, L91-L94 (1997).
14. Tanaka, Y., et al., *Nature* **375**, 659-661 (1995).
15. Mushotzky, R. F., Fabian, A. C., Iwasawa, K., Kunieda, H., Matsuoka, M., Nandra, K., & Tanaka, Y., *MNRAS* **272**, L9-L12 (1995).
16. Nandra, K., George, I. M., Mushotzky, R. F., Turner, T. J., & Yaqoob, T., *ApJ* **477**, 602-622 (1997).
17. Reynolds, C. S., & Begelman, M. C., *ApJ* **488**, 109-118 (1997).
18. Reynolds, C. S., Young, A. J., Begelman, M. C., & Fabian, A. C., *ApJ*, submitted (1998).
19. Krolik, J. H., *ApJ* **498**, L13-L16 (1998).

Gravitational Radiation from Close Double White Dwarfs

Ronald F. Webbink* and Zhanwen Han†

*Department of Astronomy, University of Illinois, Urbana, Illinois 61801, U.S.A.
†Centre for Astrophysics, University of Science and Technology of China, Hefei, 230026, China
and Yunnan Observatory, Academia Sinica, Kunming, 650011, China

Abstract. We use recent population synthesis models for the production of close double white dwarfs [1] to calculate their gravitational wave spectrum. The properties of that spectrum are nearly independent of the assumptions involved in that model, depending in normalization only on the frequency of primordial binaries of intermediate mass, orbital periods of a few years, and mass ratios near unity. The gravitational wave energy density at lower frequencies in the LISA window will provide a robust measure of the galactic type Ia supernova rate.

Roughly 3600 individual close white dwarf binaries should be resolvable at frequencies above the confusion limit at 3.6 mHz, with 90 percent of these systems showing detectable orbital evolution due to gravitational radiation within a 5-year mission lifetime. For these binaries, distances and chirp masses can be obtained independently, making it possible to study their intrinsic and spatial distributions separately. In an exceptional one or two cases, it should be possible to detect $\ddot{\nu}$ within a 5-year mission, providing an important probe into tidal heating processes preceding a merger. The probability of observing an actual merger within a 5-year mission is small (about 15%), but not negligible. Only a small fraction of mergers can produce supernovae. However, it will be possible to identify supernova progenitors from the general population of close double white dwarfs, and to predict their outburst dates with uncertainties as small as a century, or even a few years in the most favorable of cases.

INTRODUCTION

Although the existence of a class of interacting binary white dwarf stars has been recognized since the identification of HZ 29 = AM CVn as its first member [2,3], the realization that short-period double white dwarfs should be created in great profusion [4] came only when the role of common-envelope evolution in the formation of cataclysmic variables came to light [5,6]. Theoretical estimates of their formation rate have repeatedly indicated that they should dominate the galactic gravitational wave background in the frequency range from 0.1 mHz to 100 mHz [7,8].

The suggestion that mergers of such close double white dwarfs are the principal production channels for type Ia supernovae (SN Ia), R CrB stars, and other objects of astronomical interest [9,10] has spurred a number of surveys of known white dwarfs to verify the existence of these objects.

TABLE 1. Spectroscopic Surveys for Close Double White Dwarfs

Survey	P_{search} (hr)	N_{search}	N_{CDWD}
Robinson and Shafter [11]	0.5-3	44	0
Tytler and Rubenstein [12]	>0.7	120	0
Bragaglia et al. [13]	0.5-10^2	54	1
Foss, Wade and Green [14]	3-10	25	0
Saffer, Livio and Yungelson [15]	2-20	107	18

CP456, *Laser Interferometer Space Antenna*
edited by William M. Folkner
 1-56396-848-7/98/$15.00

TABLE 2. Confirmed Close Double White Dwarfs

WD	Name	P_{orb} (d)	M_1	M_2	Ref
0135–052	L 870-2	1.56	0.47	0.52	[16]
0957–666	L 101-26	0.061	0.37	0.32	[17,18]
1101+364	Ton 1323	0.1446	0.31	0.27	[19]
1202+608	Feige 55	1.49	0.40	$\geq$0.22	[20]
1241–010	PG 1241-010	3.35	0.31	$\geq$0.37	[21]
1317+453	G 177-31	4.8	0.33	$\geq$0.42	[21]
1713+332	GD 360	1.12	0.35	$\geq$0.18	[21]
2032+188	GD 231	2-10	0.36		[21]
2331+290	GD 251	0.14-0.20	0.39	$\geq$0.32	[21]

General spectroscopic surveys for close double white dwarfs have returned generally disappointing results (Table 1). Only the most recent of these surveys [15] has turned up large numbers of candidate binaries, and these await confirmation: at least 12 of their 18 candidate double white dwarfs are common to previous surveys, none of which found any of these 12 to be velocity variable (though Marsh, Dhillon and Duck [21] subsequently found two of them binaries—WD 1317+453 and WD 1713+332). Much more successful have been searches focused on white dwarfs with anomalously large photometric radii or anomalously low surface gravities from spectral line-fitting (Table 2). This last criterion identifies white dwarfs of such small mass that they cannot have arisen from single stars by normal evolution within a Hubble time.

If a clear measure of the frequency of close double white dwarfs has yet to emerge observationally, population synthesis models of binary evolution by most authors (*e.g.*, [1,4,10,22–25] point consistently to frequencies of a few (1-10) percent among observable white dwarfs). We examine below the implications of a recent study [1] for the galactic gravitational wave background from close double white dwarfs, using it to identify the principal uncertainties in their expected contribution in the LISA bandpass. This work incorporates new insights into the role of stellar winds in binary star evolution [26], and into the energetics of common envelope ejection [27], avoiding the need for super-efficient ejection invoked by many of these studies (see the discussion in [23]). However, as we shall see, the predicted gravitational wave signal in the LISA bandpass from close double white dwarfs is remarkably insensitive to the details of evolutionary models.

FORMATION CHANNELS

Close binary white dwarfs may be formed through a number of evolutionary channels, but those of short enough orbital period to be efficient gravitational wave emitters come predominantly through one of two evolutionary channels.

The first of these channels involves a phase of stable Roche lobe overflow, followed by a reverse phase of mass transfer in which the binary passes through common envelope evolution. This channel typically requires initial mass ratios near unity (equal masses for the two binary components) in order to stabilize mass transfer. The first phase of mass transfer strips the donor (primary) star to its hydrogen-exhausted core, transferring much, perhaps most, of the initial mass of the donor to its (secondary) companion, with the binary separation expanding as a result. When the second (reverse) phase of mass transfer is encountered, the secondary is typically more massive than the initial primary, and the orbital period longer, so it reaches interaction possessing a core more massive than the remnant primary core, which now becomes the accretor. The mass ratio differs far from unity, dynamical instability ensues, and the system enters common envelope evolution. The orbit of the binary contracts by a factor typically of 20 to 100—more if the common envelope ejection efficiency (see below) is much less than unity. In the resulting close double white dwarf system, the second white dwarf formed (the hotter, observable one) is the more massive component.

The second channel proceeds through two successive common envelope phases. The initial binary in this case tends to have a more disparate initial mass ratio and longer initial orbital period, both of these circumstances favoring the onset of dynamical time scale mass transfer and common envelope evolution. The first phase of mass transfer strips the primary star to its core, ejecting its envelope from the binary by dissipating the orbital energy of the remaining binary, and leaving the binary significantly more compact. When the secondary star evolves to fill its Roche lobe, mass transfer is again unstable (unless the initial mass ratio was very much less than unity), and a second common envelope phase ensues, producing a second episode of dramatic orbital contraction. In this case, the second white dwarf formed is typically the less massive component of the close double white dwarf system.

STANDARD MODEL

For purposes of exploring the gravitational wave signal of close double white dwarfs, we adopt model 4 from the population synthesis study by Han [1] as our standard model. This model was judged to satisfy various observational constraints best. Its salient properties are as follows:

The primordial binary distribution assumes a constant star formation rate, with one star (or binary system) formed per year in the Galaxy in the mass range $0.8\ \mathrm{M}_\odot \leq M \leq 8\ \mathrm{M}_\odot$, M being the mass of the primary component, M_1, in the case of a binary system. Primary component masses are assumed to follow a Miller and Scalo [28] initial mass function, with secondary masses following a flat distribution of mass ratios,

$$n(q) = 1 \ ,$$

where $q \equiv M_2/M_1$, and $0 < q \leq 1$. The distribution of binary separations is logarithmic, with a short-period cutoff:

$$n(A) = \begin{cases} 0.070/A & (A > 10\, R_\odot) \\ 0.070/A\,(A/10\, R_\odot)^{1.2} & (A \leq 10\, R_\odot) \end{cases}$$

The galactic space distribution of sources is assumed to follow an exponential disk:

$$n(R, z) = \frac{1}{2\pi H_R^2 \beta} \exp(-R/H_R) \exp(-|z|/\beta) \ ,$$

with vertical scale height $\beta = 90$ pc, radial scale length $H_R = 4$ kpc, and solar galactocentric distance $R_0 = 8$ kpc.

The treatment of binary star evolution involves certain parameters which can, at present, be constrained only loosely (if at all) by theory, and which must therefore be constrained empirically. These include the common envelope ejection efficiency,

$$\alpha_{CE} \equiv E_{env}/\Delta E_{orb} = 1 \ ,$$

where the initial binding energy of the envelope, E_{env}, includes its recombination energy. The fraction of mass lost in Roche lobe overflow accreted by the companion is taken to be,

$$\alpha_{RLOF} \equiv -\dot{M}_2/\dot{M}_1 = 0.5 \ .$$

Matter lost from the system carries the specific angular momentum of the donor star. Finally, a tidally-driven enhancement in the stellar wind of the donor star is assumed:

$$\dot{M}_1 = -4 \times 10^{13} \mathrm{g\ s}^{-1} (\mathrm{R}_1/\mathrm{R}_\odot)(\mathrm{L}_1/\mathrm{L}_\odot)(\mathrm{M}_1/\mathrm{M}_\odot)^{-1} \{1 + \mathrm{B} \min[(\mathrm{R}_1/\mathrm{R}_{\mathrm{L}_1})^6,\ 1/2^6]\} \ ,$$

with $B = 1000$. Further details can be found in Han [1].

GRAVITATIONAL WAVE SPECTRUM

Figures 1 and 2, respectively, show standard model distributions with gravitational wave frequency of the numbers of galactic close double white dwarf sources, and of their total gravitational wave strain amplitude at the Sun. Monte Carlo results from the population synthesis models have been smoothed successively by gaussians in $\log \nu$, in merger time scale, and in orbital frequency at merger, in such a way as to preserve statistically significant features of the distributions. Also shown are model distributions varying each of the principal assumptions detailed above. These assumptions significantly influence the distribution of low-frequency (long orbital period) sources, but above $\nu_{GR} = 10^{-4}$ Hz (*i.e.*, in the LISA observing window) the spectra converge to essentially the same functional form—those of a steady-state ensemble of sources spiraling upward in frequency until mergers begin to deplete their population above $\nu_{GR} = 10^{-2}$ Hz.

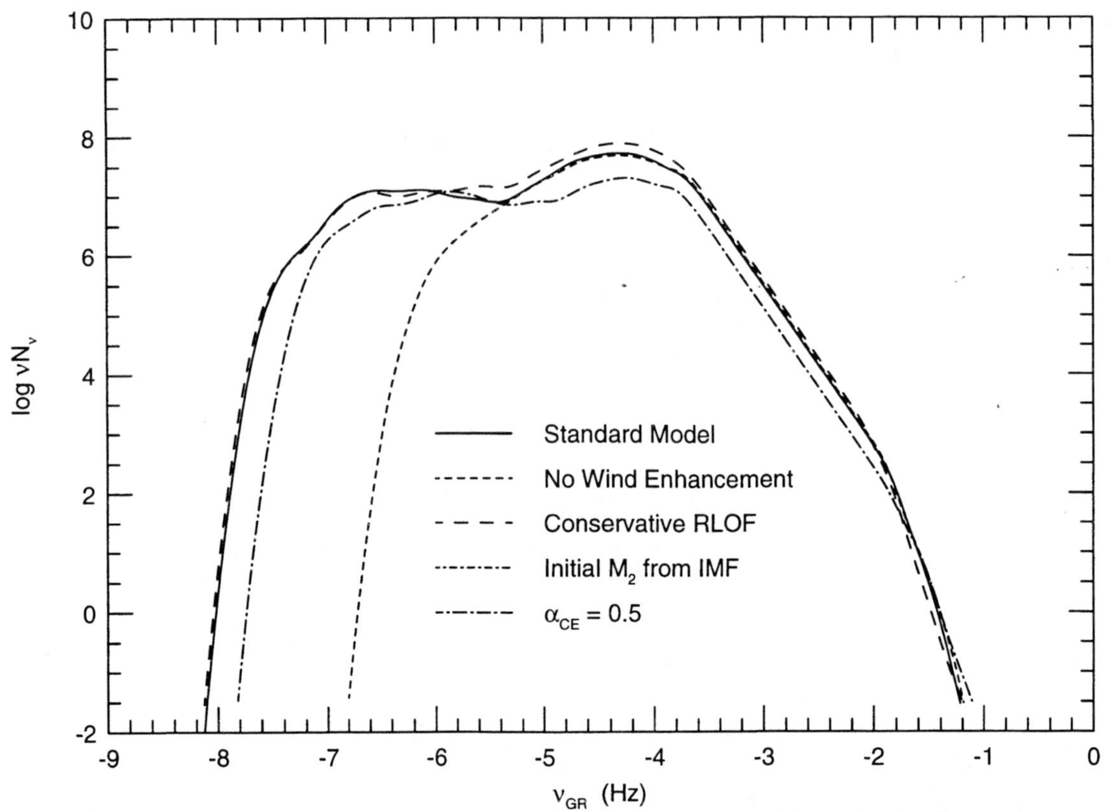

FIGURE 1. Total number distribution of galactic close double white dwarfs with gravitational wave frequency. Our standard model is indicated by the solid line; this model, and the variants upon it (in order from top to bottom in the legend) correspond respectively to models 4, 2, 10, 11, and 12 in Han [1].

The remarkable thing about the distributions in Figures 1 and 2 is that, except for a clear dependence on the primordial mass ratio distribution, the model spectra differ insignificantly from each other in amplitude, as well as in functional form. This result is a consequence of the fact that the gravitational wave spectrum above 0.1 mHz samples only the short-period tail of the distribution of close double white dwarfs as they emerge from their final episode of common envelope evolution. These systems are, in turn, the progeny of the short-period extreme of a broader distribution of primordial systems contributing to the total close double white dwarf birthrate. In that short-period extreme, the transformation of primordial binaries through the rigors of mass exchange and common envelope evolution is virtually homologous, at least insofar as it is currently understood. Thus, varying parameters in the treatment of binary evolution may move the window in initial parameter space (the relative span of component masses and separation) that give rise to a specific final state, but they do not materially alter the size of the window. It is only the density of primordial systems in the window of initial parameter space that matters, and the largest uncertainty in this regard is (by far) the distribution of initial component mass ratios. Primordial binaries with mass ratios very different from unity do not survive to the close double white dwarf state, so the surviving population is depressed if the initial mass ratio distribution favors small companion masses at the expense of comparable masses.

WHAT WE CAN LEARN FROM LISA

Because of spacecraft motion, LISA can be expected to separate individual sources throughout the Galaxy only where their spectral number density falls to $\nu N_\nu \lesssim c/v_\oplus = 10^4$. In our standard model, this occurs for $\nu > 3.6$ mHz. At lower frequencies source signals overlap and become increasingly confused, although very nearby systems may still stand out above the general background. Although a nuisance to the observation of other astrophysically interesting sources, this quasi-continuum also provides a key measure of the galactic white dwarf merger rate, and with it the galactic SN Ia rate. (Surveys of conceivable binary evolutionary

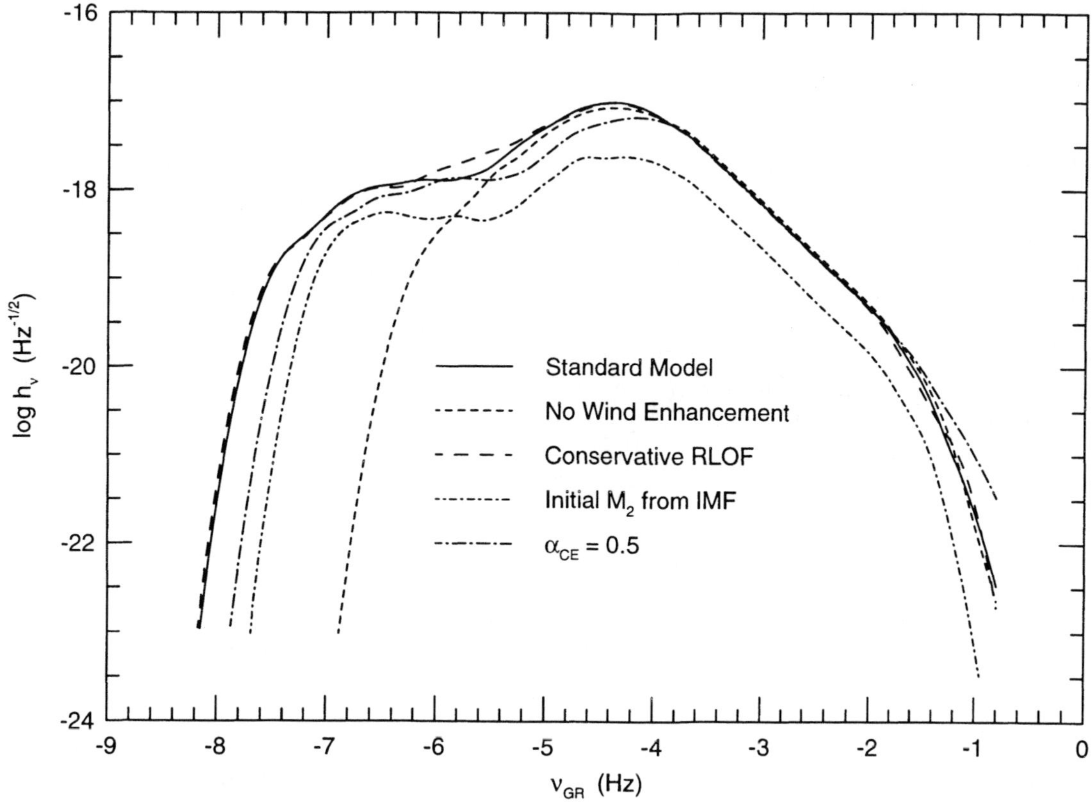

FIGURE 2. Spectral density of the gravitational wave amplitude due to close double white dwarfs. The standard model and its variants are labeled as in Figure 1.

channels leading to type Ia supernovae continue to identify mergers of massive white dwarfs as the most likely progenitors—see [29].) The standard model adopted here predicts a SN Ia rate of 0.0036 yr^{-1} [1]; but, remarkably, what dispersion is evident in Figure 2 among the predicted spectra of different models virtually disappears if the local energy density in gravitational radiation above 0.1 mHz (where close double white dwarfs are expected to dominate [8]) is normalized to the model SN Ia rate. The strain or energy density distributions are subject to large stochastic variations in their high-frequency limits, as they are dominated by relatively few, luminous sources evolving rapidly toward coalescence. The quasi-continuum from, say $\nu_{GR} = 10^{-4}$ Hz to $\nu_{GR} = 10^{-2.5}$ Hz, thus provides a more robust measure of the merger rate; *in this frequency range*, we find a ratio of gravitational wave energy density to SN Ia rate $u_{GR}/\dot{N}_{SN\ Ia} = 5.5 \times 10^{-12}$ erg cm^{-3}/yr^{-1}.

Above 3.6 mHz, some 3.6×10^3 close white dwarf binaries are expected to be individually resolvable, including even the intrinsically weakest sources on the far side of the galactic disk. For resolvable sources, measurement of the strain amplitude and polarization yields directly the source orbital inclination i and the quantity $m^{5/3}/d$, where $m \equiv M_1^{3/5} M_2^{3/5} (M_1 + M_2)^{-1/5}$ is the "chirp" mass and d the distance to the source. While very few sources can be expected to be optically identifiable, the dynamical range of chirp masses predicted from the population synthesis models in this part of the frequency spectrum is relatively small, $\langle m \rangle = 0.42\ M_\odot$ with a standard deviation $\sigma_m = 0.14\ M_\odot$, adequate for the estimation of rough distances to individual sources. With a more sophisticated treatment, incorporating model distributions of chirp masses, it will be possible to construct large-scale models of the stellar distribution in the galactic disk, taking advantage of its transparency to gravitational radiation. This is possible optically (and even in the near-infrared, where stellar emission still dominates) only along a relatively few lines of sight through the galactic disk.

In fact, the great majority of resolvable sources evolve so rapidly under gravitational radiation losses that their orbital evolution should be detectable within a 5-year lifetime for the LISA mission. If we suppose, conservatively, that frequency evolution becomes detectable when the phase shift with respect to a best-fit linear (constant-frequency) ephemeris exceeds $\pi/2$ radians, then in a time span τ, $\dot{\nu}$ becomes detectable when $\dot{\nu} > 3/\tau^2$. Under pure gravitational radiation losses, $\dot{\nu}_{GR} \sim m^{5/3} \nu_{GR}^{11/3}$; by the above criterion, frequency evolution becomes detectable within a 5-year mission span for sources with $\nu_{GR} \geq 2.3\,(m/M_\odot)^{-5/11}$ mHz, a condition satisfied by 90 percent of sources in the resolvable part of the frequency spectrum, $\nu_{GR} > 3.6$ mHz. For these systems, m and d can be derived individually from $\dot{\nu}_{GR}$ and the strain amplitude and polarization, and it becomes possible to study their intrinsic properties and galactic distribution in much greater detail.

For a very small number of systems nearing merger, it will be possible to detect not only $\dot{\nu}_{GR}$, but $\ddot{\nu}_{GR}$ as well. Again assuming that $\ddot{\nu}$ becomes detectable when the accumulated phase shift with respect now to a best-fit quadratic ephemeris exceeds $\pi/2$ radians, the detection threshold becomes $\ddot{\nu} > 30/\tau^3$. Again assuming pure gravitational radiation losses, $\ddot{\nu}_{GR} \sim m^{10/3}\nu_{GR}^{19/3}$ should be detectable in a 5-year mission for systems with $\nu_{GR} \gtrsim 26$ mHz. We have made allowance here for the fact that, at such high frequency, systems in the low-mass tail of the chirp mass distribution will already have merged, leading to a higher mean value, $\langle m \rangle = 0.57$ $M_\odot$, among survivors at this frequency; roughly a third of these survivors are sufficiently massive to produce SN Ia on merging. The expected number of systems above this limiting frequency in our standard model is very small ($\langle N \rangle = 1.34$), but measurement of $\ddot{\nu}$ will be important because its departure from the value expected from pure gravitational radiation decay,

$$\ddot{\nu}_{GR} = \frac{5\,\dot{\nu}_{GR}^2}{3\,\nu_{GR}} \,.$$

provides a probe of non-gravitational orbital energy losses from systems on the verge of merging. Tidal heating during spinup of the incipient donor star as it approaches merger can in principle be quite severe [30,31], accelerating the orbital energy and angular momentum losses driving its evolution, and possibly driving it to high enough luminosity to make optical identification possible. The extent of such tidal heating is also clearly of importance in understanding ignition conditions during the dynamical merger believed to give rise to SN Ia.

There is a small, but non-negligible chance that an actual white dwarf merger will occur within the Galaxy during the LISA mission lifetime. The total double white dwarf merger rate in our standard model is 0.0296 yr^{-1} [1]. In the majority of cases, this merger accelerates to a truly dynamical time scale within a year or so of the less massive white dwarf first filling its Roche lobe [30,32].

The likelihood is fairly remote (of order 1.8% in our standard model) that a white dwarf pair sufficiently massive to give rise to a type Ia supernova will merge during a 5-year mission. Only slightly less exciting is the prospect that the extraction of chirp masses and distances from among the ~3200 resolvable sources with detectable period derivatives should yield of order 500 SN Ia progenitors (identifiable by their large chirp masses: $m > 0.60$ $M_\odot$). It will thus become possible not only to identify individual supernova progenitors (at least to within the angular resolution of LISA), but also to estimate their individual lifetimes until merger. There is an irreducible uncertainty to the lifetimes to merger, because the onset of merger depends not only on chirp mass m, but on binary mass ratio (which is indeterminate from the gravitational wave signature); this uncertainty varies from ~120 yr at $m = 0.6$ $M_\odot$ to <10 yr for $m > 1.0$ $M_\odot$. Nevertheless, the prospect of predicting the location and outburst date of an incipient supernova is unprecedented in astronomy.

ACKNOWLEDGMENTS

The authors thank P.P. Eggleton for facilitating their collaboration. This work was supported in part by US National Science Foundation grant AST96-18462, and by the Pandeng Scheme and the Research Fund of Academia Sinica and the Postdoctoral Research Fund of the Chinese Educational Commission.

REFERENCES

1. Han, Z., *Monthly Notices R. Astr. Soc.* **296**, 1019 (1998).
2. Smak, J., *Acta Astr.* **17**, 255 (1967).
3. Warner, B., and Robinson, E. L., *Monthly Notices R. Astr. Soc.* **159**, 101 (1972).
4. Webbink, R. F., in *White Dwarfs and Variable Degenerate Stars, I.A.U. Colloq. No. 53*, ed. H. M. Van Horn and V. Weidemann, Rochester: Univ. Rochester, p. 426 (1979).
5. Paczyński, B., in *The Evolution of Close Binary Systems, I.A.U. Symp. No. 73*, ed. P. Eggleton, S. Mitton, and J. Whelan, Dordrecht: D. Reidel, p. 85 (1976).
6. Ostriker, J. P., unpublished paper presented at I.A.U. Symp. No. 73 (1975).
7. Evans, C. R., Iben, I., Jr., and Smarr, L., *Astrophys. J.* **323**, 129 (1987).

8. Hils, D., Bender, P. L., and Webbink, R. F., *Astrophys. J.* **360**, 75; **369**, 271 (1990).
9. Webbink, R. F., *Astrophys. J.* **277**, 355 (1984).
10. Iben, I., Jr., and Tutukov, A. V., *Astrophys. J. Suppl.* **54**, 335 (1984).
11. Robinson, E. L., and Shafter, A. W., *Astrophys. J.* **322**, 296 (1987).
12. Tytler, D., and Rubenstein, E., in *White Dwarfs, I.A.U. Colloq. No. 114*, ed. G. Wegner, Berlin: Springer, p. 524 (1989).
13. Bragaglia, A., Greggio, L., Renzini, A., and D'Odorico, S., *Astrophys. J. Lett.* **365**, L13 (1990).
14. Foss, D., Wade, R. A., and Green, R. F., *Astrophys. J.* **374**, 281 (1991).
15. Saffer, R. A., Livio, M., and Yungelson, L. R., *Astrophys. J.* **502**, 394 (1998).
16. Saffer, R. A., Liebert, J., and Olszewski, E. W., *Astrophys. J.* **334**, 947 (1988).
17. Bragaglia, A., Renzini, A., and Bergeron, P., *Astrophys. J.* **443**, 735 (1995).
18. Moran, C. K. J., Marsh, T. R., and Bragaglia, A., *Monthly Notices R. Astr. Soc.* **288**, 538 (1997).
19. Marsh, T. R., *Monthly Notices R. Astr. Soc.* **275**, L1 (1995).
20. Holberg, J. B., Saffer, R. A., Tweedy, R. W., and Barstow, M. A., *Astrophys. J. Lett.* **452**, L133 (1995).
21. Marsh, T. R., Dhillon, V. S., and Duck, S. R., *Monthly Notices R. Astr. Soc.* **275**, 828 (1995).
22. Yungelson, L. R., Livio, M., Tutukov, A. V., and Saffer, R. A., *Astrophys. J.* **420**, 336 (1994).
23. Han, Z., Podsiadlowski, P., and Eggleton, P. P., *Monthly Notices R. Astr. Soc.* **272**, 800 (1995).
24. Lipunov, V. M., Postnov, K. A., and Prokhorov, M. E., *Astr. Astrophys.* **310**, 489 (1996).
25. Iben, I., Jr., Tutukov, A. V., and Yungelson, L. R., *Astrophys. J.* **475**, 291 (1997).
26. Han, Z., Eggleton, P. P., Podsiadlowski, P., and Tout, C. A., *Monthly Notices R. Astr. Soc.* **277**, 1443 (1995).
27. Han, Z., Podsiadlowski, P., and Eggleton, P. P., *Monthly Notices R. Astr. Soc.* **270**, 121 (1994).
28. Miller, G. E., and Scalo, J. M., *Astrophys. J. Suppl.* **41**, 53 (1979).
29. Tutukov, A. V., Yungelson, L. R., and Iben, I., Jr., *Astrophys. J.* **386**, 197 (1992).
30. Webbink, R. F., and Iben, I., Jr., in *The Second Conference on Faint Blue Stars, I.A.U. Colloq. No. 95*, ed. A. G. D. Philip, D. S. Hayes, and J. W. Liebert, Schenectady: L. Davis Press, p. 445 (1989).
31. Iben, I., Jr., Tutukov, A. V., and Fedorova, A. V., *Astrophys. J.* **503**, 344 (1998).
32. Benz, W., Bowers, R. L., Cameron, A. G. W., and Press, W. H., *Astrophys. J.* **348**, 647 (1990).

Confusion Noise Estimate for Gravitational Wave Measurements in Space

Dieter Hils

JILA, National Institute of Standards and Technology and University of Colorado
Boulder, Colorado 80309
hils@jila.colorado.edu

Abstract. LISA's sensitivity for studying gravitational waves is expected to be confusion noise limited from roughly 0.1 - 3 mHz due to unresolvable signals from very many close binaries throughout the Galaxy. The most important contributors to the confusion noise appear to be close white dwarf binaries. However, there are additional binary systems that we have not considered before. These include systems of helium-rich secondaries in semidetached binaries with CO white dwarf primaries. Observational astronomers refer to these systems as AM CVn binaries. The theory of close binary evolution suggests at least two different scenarios that lead to the AM CVn stage. We will discuss both scenarios and will show that their inclusion probably will lead to only a small increase in the confusion noise.

INTRODUCTION

The large number of binary systems in our Galaxy produce a gravitational wave (GW) background that is called the binary confusion noise (Hils, Bender and Webbink 1990; Bender and Hils 1997; hereafter HBW90 and BH97). The binary confusion noise together with LISA's instrumental sensitivity will ultimately determine how well one is able to obtain information from important extragalactic sources involving MBHs. Some of the galactic binary sources that are responsible for the binary confusion noise are shown with their spectral amplitudes in Figure 1. Most of these sources are well understood and have been observed by astronomers in large numbers. For a detailed discussion on these sources see HBW90. The values for the three varieties of close white dwarf binaries (CWDBs) were recalculated in BH97 and are shown for 10% of the theoretical space densities found from the results of Webbink (1984).

CWDBs were first predicted theoretically some 15 years ago (Webbink 1984; Iben and Tutukov 1984; Tutukov and Yungelson 1986) but it is only recently that a few of these systems with times to coalescence less than a Hubble time have been observed (Marsh et al. 1995; Marsh 1995; Moran et al. 1997). Because of the small number of observed systems and because of selection effects that are difficult to correct for, it is not yet possible to deduce a reliable CWDB space density. In the near future, a better estimate may become available as more CWDBs are discovered and possible deviations from the theoretical distribution over period and component masses are observed.

However, it can be misleading just to look at h. We also need to know how many sources there are in a frequency resolution bin. For frequencies below roughly 1 mHz, there are so many binaries even in a 1 cycle/yr frequency bin that it will be hard to resolve individual sources. Because of the very steep decline of those numbers with frequency, at a somewhat higher frequency there will be bins that contain no galactic binaries. This opens the possibility to study very weak extragalactic sources. An estimate of the confusion noise level based on the sources discussed in HBW90 was shown in BH97.

A more recent estimate of the confusion noise level is shown in Figure 2, along with the currently adopted LISA sensitivity curve and some expected galactic binary signal levels (LISA Study Team 1998). The break in the confusion noise that occurs near 1.6 mHz marks the point where some frequency bins begin to be free of galactic sources. At even higher frequencies, there is a rapid transition to a lower amplitude, determined by all the extragalactic binaries.

CP456, *Laser Interferometer Space Antenna*
edited by William M. Folkner
 1-56396-848-7/98/$15.00

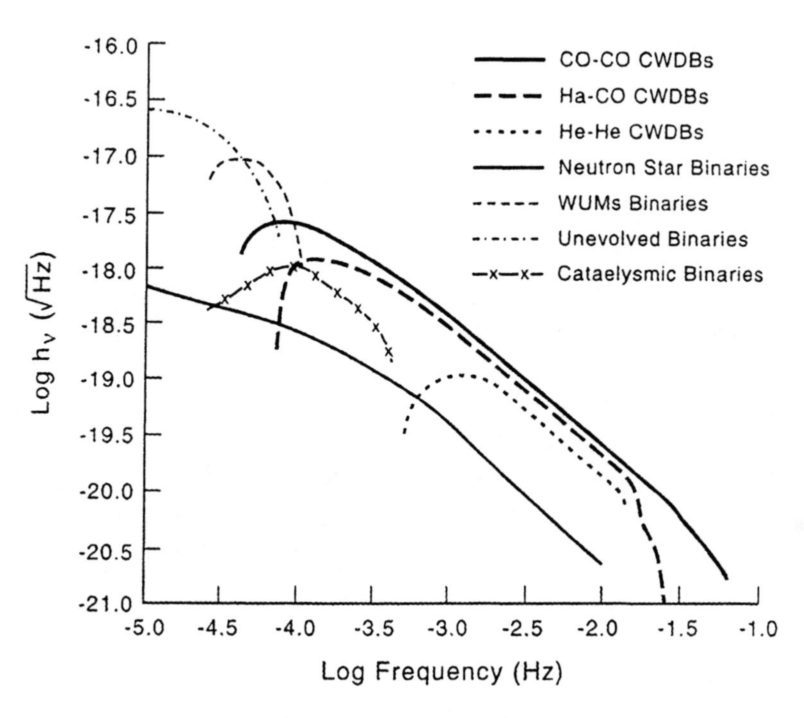

FIGURE 1. Root-mean-square estimates of the signal levels due to different types of galactic binaries. The three CWDB curves are for 10% of Webbink (1984).

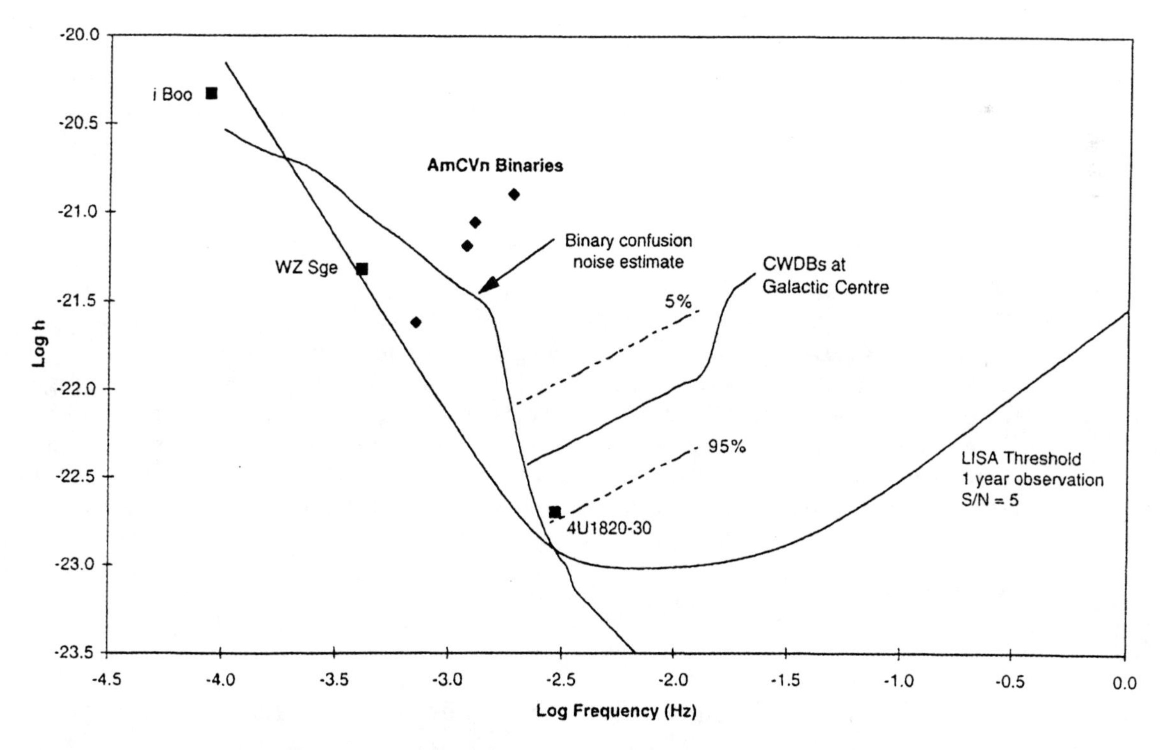

FIGURE 2. The signal levels and frequencies are given for a few known galactic sources, along with the expected LISA threshold sensitivity and an estimate of the binary confusion noise level. In addition, the range of levels for 90% of resolvable close white dwarf binary signals from our galaxy is shown.

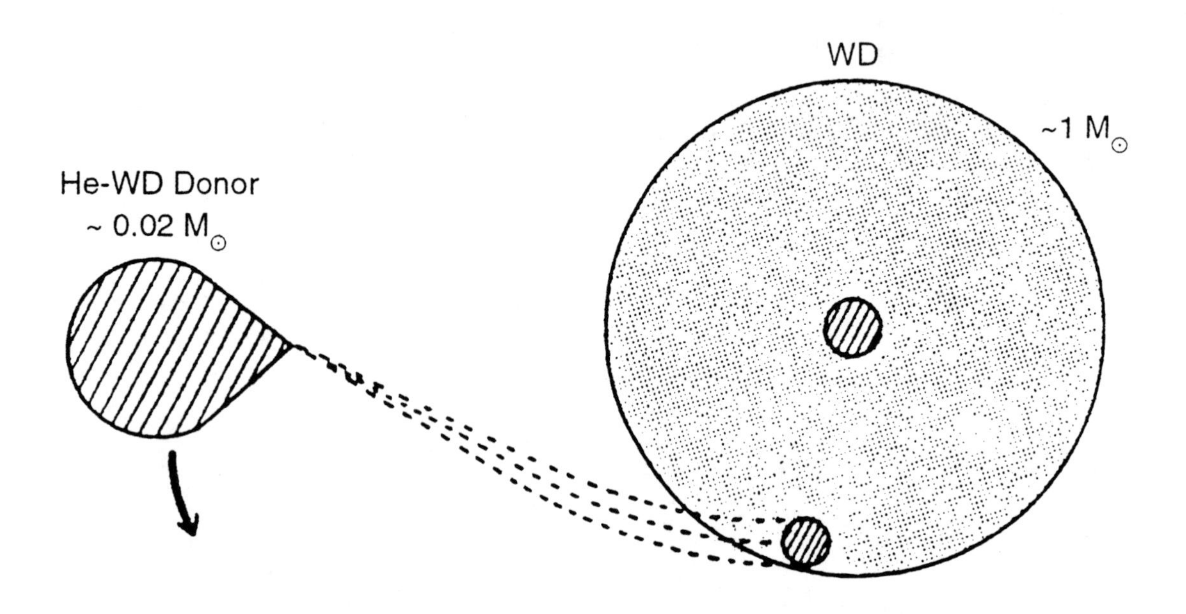

FIGURE 3. A schematic of an AM CVn system consisting of two white dwarfs. The lower mass helium white dwarf, is transferring mass to the more massive white dwarf through the inner Lagrangian point to an accretion disk, where a bright spot is formed.

Included in the confusion noise estimate shown in Figure 2 are preliminary results for an additional and potentially important class of binaries not shown in Figure 1. These are the helium cataclysmic variables (HeCVs). Our previous HeCV results were obtained based on one particular model proposed by Iben and Tutukov (1991; hereafter IT91). In the meantime, we have become aware of an additional HeCV model put forward by Tutukov and Yungelson (1996; hereafter TY96). In the course of this investigation it became apparent that our earlier analysis of the IT91 scenario was incomplete. In this paper I would like to present our findings for the IT91 and TY96 models and discuss the impact these results may have on the binary confusion noise.

WHAT ARE HECVS?

HeCVs are more commonly known under the name interacting white dwarf binaries and are sometimes also refered to as AmCVn binaries (Warner 1996). They have been mentioned recently by Warner (1996) and by Hellings (1996) as potential gravitational wave sources. Their characteristic signiture is that they show in their spectra only helium lines. The amplitude of some individual nearby AmCVn binaries are shown in Figure 2. The generally accepted model that can explain all the observations is shown in Figure 3. It shows a very low mass (0.02 $M_\odot$) helium WD which fills its Roche lobe and transfers helium rich material to a much more massive CO WD. The other details of the model need not concern us here. The more interesting question is how do you form such an exotic system in the first place. It turns out that if one wants to calculate the GW emission from these binaries, one needs to know their progenitors also. We will discuss two rival scenarios that also address the origins question.

As already indicated earlier, the first is by IT91, the second is by TY96. Both scenarios have enough in common that I will describe only the IT91 case. But I will indicate where the two models differ.

THE HECV SCENARIOS OF IT91 AND TY96

Figure 4a shows at time t=0 two main sequence (MS) stars of similar mass but with $M > 2.5\ M_\odot$. The initial binary period in the chosen example is $P=250^d$. At this stage, both stars are well inside their Roche lobes. After the primary has completed its MS phase and exhausted central helium, it begins to fill its Roche

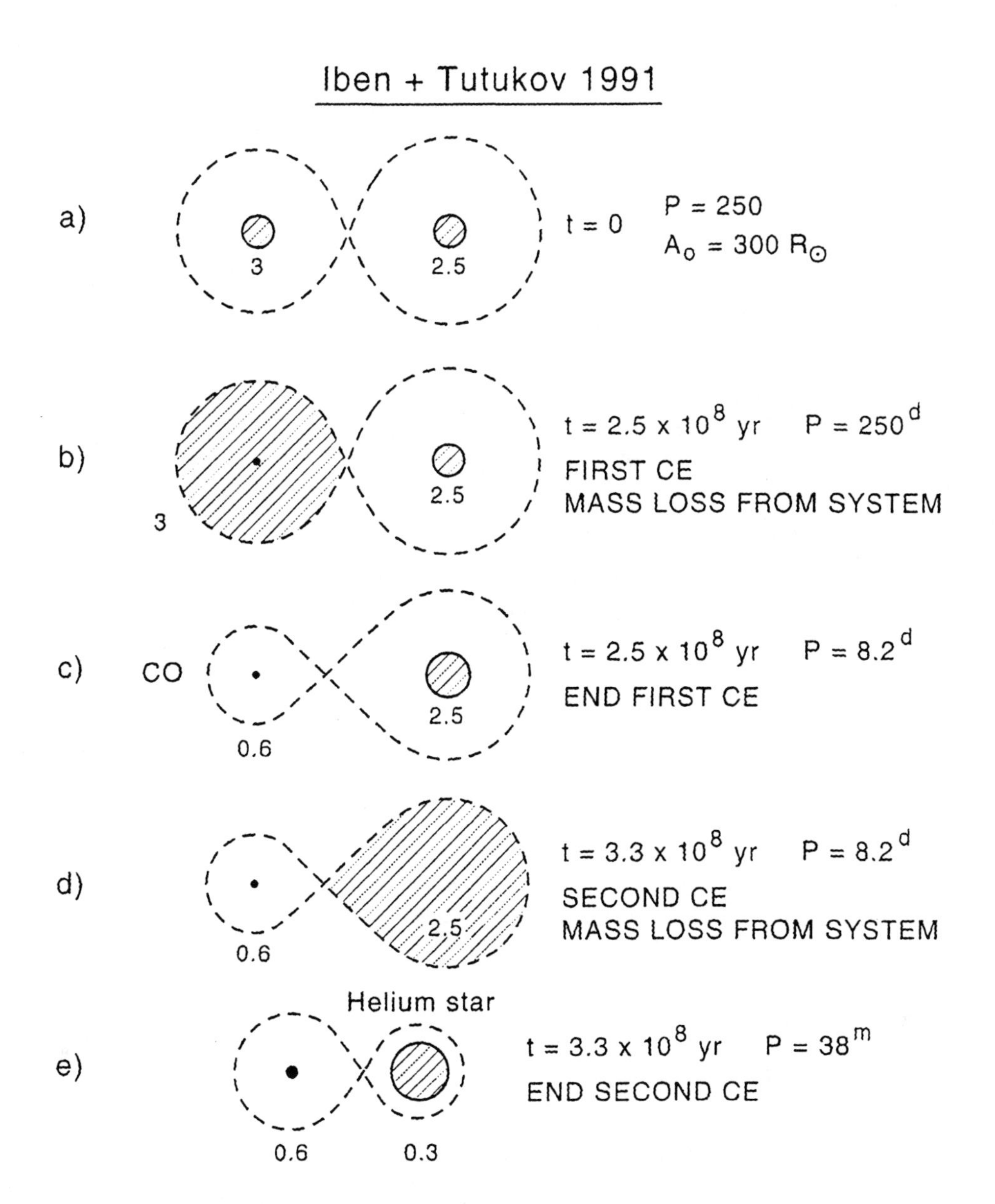

FIGURE 4. The evolution of a close binary system from a typical progenitor orbit to the HECV stage prior to the beginning of stable tidal mass transfer (IT91)

lobe (Figure 4b). At this point it has developed an electron degenerate CO core. The remaining hydrogen-rich outer layers of the star eventually form a common envelope (CE) that surrounds both stars. Drag forces with the orbiting pair will eventually drive off the CE from the binary system. What is left then is a much more tightly bound system that consists of a CO WD orbiting the original secondary (Figure 4c). The period now is 8.2 d. When the original secondary completes its MS phase it will already fill its Roche lobe before central helium ignition (Figure 4d). This is the beginning of a second CE episode which results in the formation of a nondegenerate helium star (the core of the original secondary) in orbit around the CO WD. The final period now is only 38 m (Figure 4e). During both CE events the matter inside the CE is assumed to be lost from the system.

We now have a binary system that is so tightly bound that emission of GWs will determine its further evolution. At the beginning, the helium star is inside its Roche lobe. However, the orbit keeps on contracting. Eventually, the helium star will fill its Roche lobe. This is the beginning of a long mass exchange phase (ME) during which the helium star transfers helium rich matter to the CO WD. At this stage it is commonly assumed that the further evolution of the binary follows from the condition that the secondary star fills its Roche lobe. In addition one assumes the conservation of total binary mass.

IT91 gave such a helium transfering system the name HeCV, in analogy to the well-known hydrogen transfering CVs, where a normal low mass MS star transfers hydrogen rich matter to a CO WD.

I now briefly indicate where the TY96 model differs. The main difference is in the mass of the secondary star. If the mass of the secondary is below 2.3 $M_\odot$, it will have developed an electron degenerate helium core by the time it ascends the giant branch after central hydrogen exhaustion. Depending on when it begins to fill its Roche lobe during this phase, a helium WD of mass greater than 0.13 $M_\odot$ will form. Therefore the two scenarios both have a CO WD accompanied by either a nondegenerate helium star in the one case or a helium WD in the other case.

GRAVITATIONAL WAVES FROM HECVS

We have calculated the gravitational wave flux of HeCVs based on the two scenarios by IT91 and TY96. Depending on initial conditions at birth, the first phase in the evolution of these binary systems, before Roche lobe contact is established, is driven by gravitational radiation. The duration of this dormant phase cannot be neglected in a calculation of the system's GW emission. Typically it takes less than 0.1 Gyr in the IT91 scenario before ME begins. This led us to believe, erronously as we later discovered, to neglect this phase all together in our first calculation.

Figure 5 will help to clarify the processes that have to be considered in a binary system where ME takes place driven by GWs. Here we show in a period–time diagram how the binary progresses from its birth to the ME phase.

We first discuss the IT91 case in Figure 5a. Beginning at t=0, the system is born with initial period P_{ff}. At period P_c, the nondegenerate helium star begins to fill its Roche lobe and ME begins. Except for a small change in slope, the binary orbit continues to contract as the secondary helium star looses mass (thereby shrinking in radius) to the more massive primary CO WD. The now following events have been discussed by Savonije et al. (1986). As the mass of the helium star drops below the point where helium is able to burn in the core, the rate at which the orbital period decreases begins to slow down as the star's radius begins to

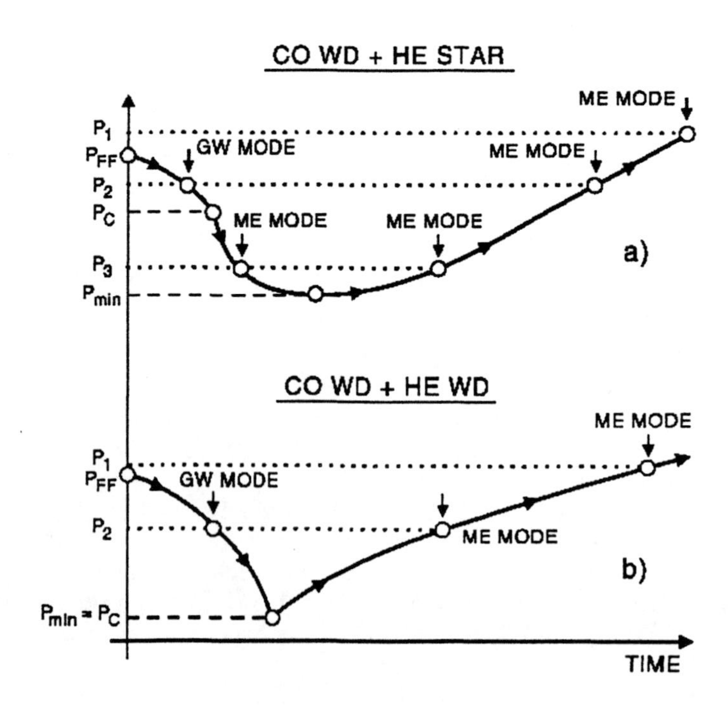

FIGURE 5. The two HECV scenarios shown in a period-time diagram. The two different modes for producing gravitational waves at frequency $2/P_i$ are shown. The HECV orbit after the second common envelope is P_{ff} and the orbital period at the beginning of mass transfer is P_c.

decrease more slowly. This is in response to the increasing star's central density, while at the same time its central temperature is falling. By the time the minimum period P_{min} is reached, the interior of the secondary has turned semi-degenerate. The period from then on begins to increase, as the radius of the secondary now has to increase with decreasing mass.

Suppose now that we want to observe at some GW frequency f_2 corresponding to a binary period $P_2 = 2/f_2$. As long as $P_2 > P_c$, P_2 can be reached by GW evolution alone. We call this the GW mode. The other possibility to reach binary period P_2 is by following the binary past the minimum period P_{min} until the period of the system has increased to P_2 again due to the ME process. Therefore we shall call this the ME mode. It goes without saying that optical astronomers are only interested in this phase of the HeCV system.

Suppose we observe at period $P_1 > P_{ff}$ instead. The only path for the binary to reach P_1 from P_{ff}, is via the ME mode.

Finally, consider a binary period $P_3 < P_c$. Here again there are two ME modes which have to be considered. The first ME mode operates during the quasi MS phase of the helium star whereas the second ME mode occurs later in time during the secondary's semi-degenerate phase.

The situation is only slightly different for the TY96 scenario. Because the secondary is a He WD at birth, its radius can only increase in response to mass loss. As a result, its period increases as soon as Roche lobe contact is established. This situation is shown in Figure 5b. Here again we have a pure GW mode for the case $P_2 < P_{ff}$ as well as a ME mode. If $P_1 > P_{ff}$, there is only the ME mode to consider.

Following IT91 we can calculate the birthrate of HeCVs as a function of progenitor parameters. We then evolve the system to the present epoch to find its orbital period and the mass of its two components. This is sufficient to estimate the GW emission. For the distribution of HeCV sources we assume an exponential galactic disk model which was described before (HBW90).

RESULTS

I first present the results for the IT91 case. Figure 6 gives number of HeCV sources per unit bandwidth as function of GW frequency for the GW and ME modes. It is evident that over most frequencies the ME mode dominates by more than an order of magnitude. The rapid drop off at the low frequency end of the ME curve is caused by the finite Hubble time, which puts a limit to the minimum secondary mass (corresponding to a maximum binary period) a system can have. Nevertheless, most HeCVs are at present at these very low

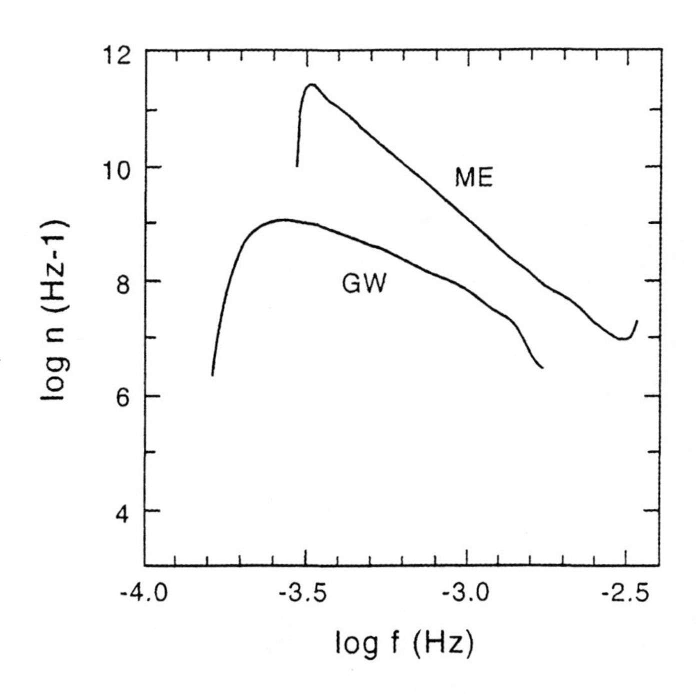

FIGURE 6. The number of systems in the IT91 scenario for the two different modes.

frequencies with a maximum near a frequency of $\approx$0.3 mHz. The reason for this accumulation is the very rapid decline of the ME rate with period ($dm_2/dt \propto P^{-8/3}$).

HeCV systems at the highest frequency (in our calculation at $\approx$3mHz) correspond to binary systems located at the minimum orbital period. Although the GW mode has far fewer systems, it extends towards somewhat lower frequencies than the ME mode. Those are binaries born with relatively long periods which then evolve by GW radiation slowly to shorter periods. On the other hand, the GW mode vanishes for frequencies log f > -2.75 Hz when the secondary helium star begins to fill its Roche lobe. This corresponds to a binary period P_c > P_{min}, which is why the ME mode extends to somewhat higher frequencies than the GW mode.

Figure 7 shows a similar picture emerge for the TY96 case except that for a helium WD secondary the highest frequency $f_{max} = 2 / P_c$ allowed is the same for both modes.

Figure 8 shows the results for the GW amplitude for the IT91 scenario. Here the situation is reversed. The 'dormant' GW mode dominates over the 'active' ME modes, again emphasizing the importance of this phase for the GW luminosity of HeCVs. The reason for this at first surprising result is that the observed flux (at frequency f) is proportional not only to the number of sources per unit bandwidth, n, but also to the average GW luminosity of a source. The GW luminosity itself is proportional to the chirp mass of the binary which is considerably bigger for the GW mode. Most of the ME mode systems have a very small chirp mass as the secondary's mass is eroded away down to a very small value. Overall, the luminosity factor wins out.

Figure 9 presents the corresponding results for the TY96 case, which again shows the GW mode exceeding the ME modes.

I now come to a comparison of the two scenarios. Figure 10 shows the total number of sources per unit bandwidth, where I now sum over the different modes. There are more systems with a He WD donor than there are with a semi–degenerate He star secondary. In addition, the HE WD systems occupy a some- what larger frequency range.

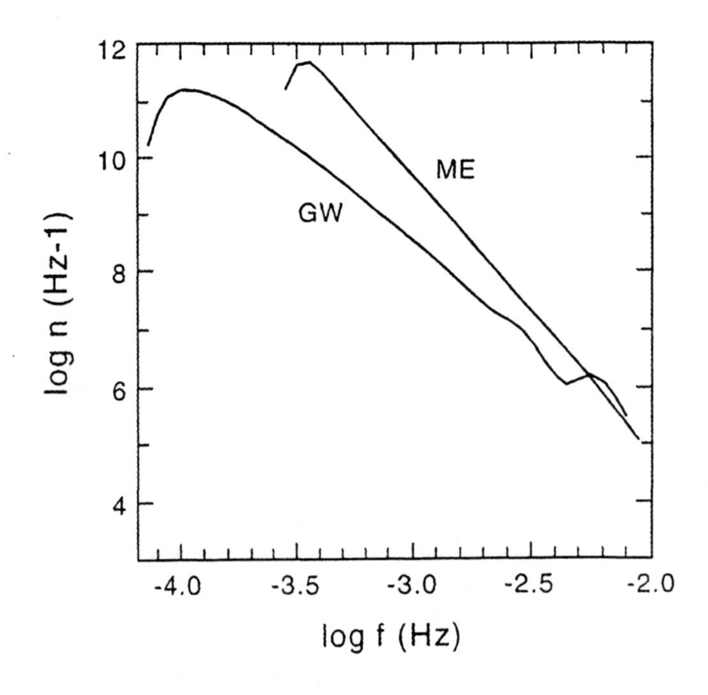

FIGURE 7. The number of systems in the TY96 scenario for the two different modes.

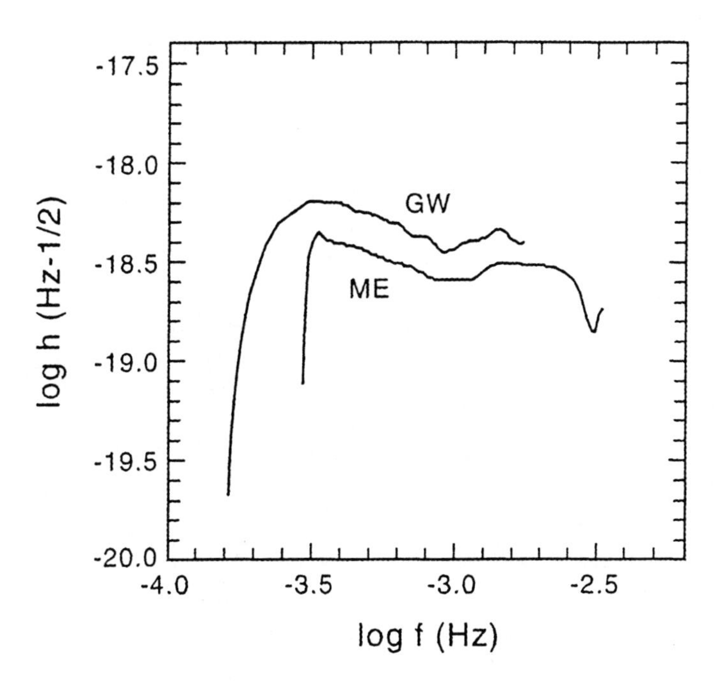

FIGURE 8. Root-mean-square amplitude in the IT91 scenario for the two modes.

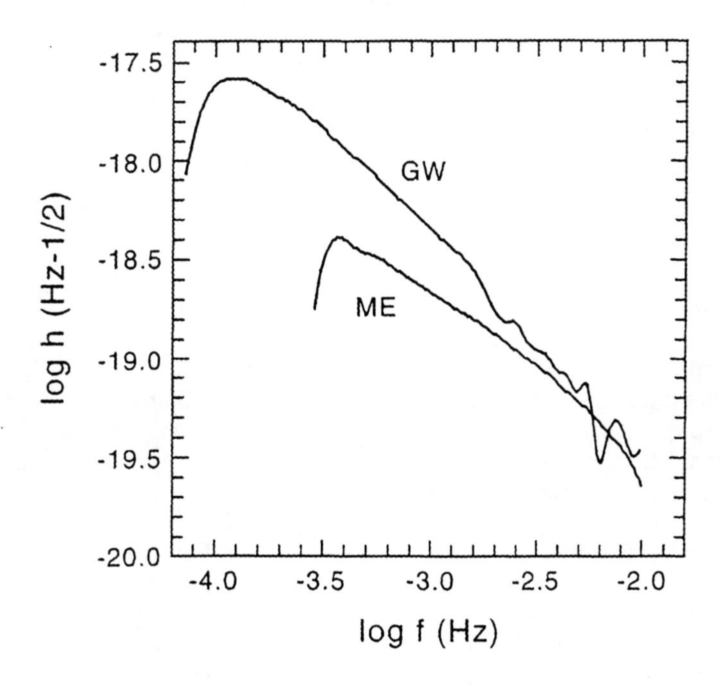

FIGURE 9. Root-mean-square amplitude in the TY96 scenario for the two modes.

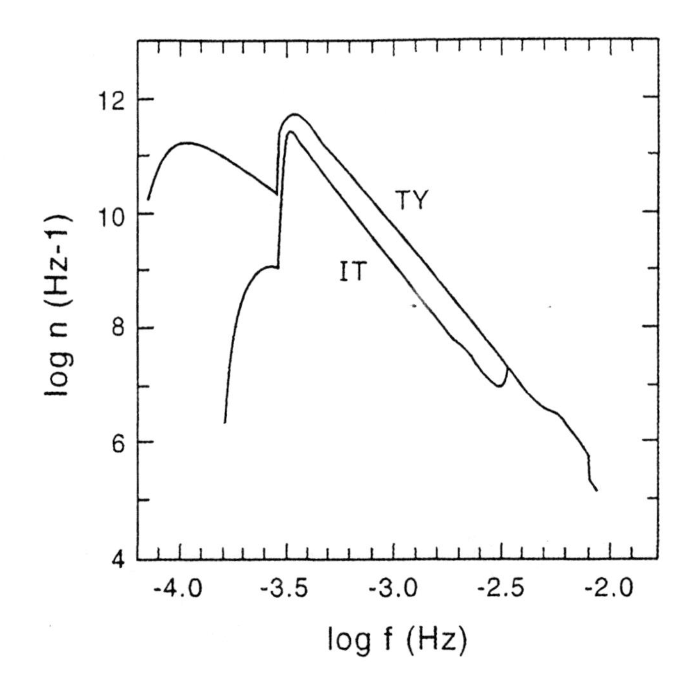

FIGURE 10. The number of sources for the two HECV scenarios.

DISCUSSION AND SUMMARY

How do the results for the HeCVs fit in with the rest of the important binary systems in our Galaxy? In particular, how will the inclusion of HeCVs modify our earlier estimate for the binary confusion noise shown in Figure 2? As I already indicated at the beginning, the most important binary component for any space based GW detector such as LISA, are the CWDBs. It therefore is prudent to compare the results for HeCVs with those for CWDBs. Figure 11 shows the amplitudes for the two HeCV cases together with the unnormalized results for CWDBs (HBW90). The TY96 case gives a larger amplitude than the IT91 case up to frequencies f $\approx$1 mHz. Thereafter, for frequencies f $>$ 3 mHz, the IT91 systems produce a somewhat larger amplitude.The HeCV amplitudes of TY96 are lower than the CWDB amplitudes by roughly half a decade. Therefore adding the HeCV amplitude in quadrature will have a negligible effect on the total amplitude.

The second important parameter influencing the confusion noise [see equation (4) in BH97] is the number of binary systems per unit bandwidth. In Figure 12 we make the comparison with the CWDBs. At 1 mHz there are almost ten times as many HeCVs as there are CWDBs. At 2 mHz, the two are nearly equal. Therefore adding the HeCVs to the total number of sources will have some effect.

The break in the confusion noise curve shown earlier in Figure 2 near 1.6 mHz occurs roughly at a frequency where there is one binary source per resolution bin (see BH97). We can make use of this observation to obtain a rough estimate of the changes to be expected based on the new total number of sources. To be consistent with our previous results shown in Figure 2, we also reduce the HeCV space density by the same factor ten applied to the CWDB data. We then find a shift in the break point in the confusion curve by approximately Δlog f $\approx$ 0.05 Hz towards higher frequencies. The lower crossing point between the confusion noise and LISA's sensitivity curve (near log f $\approx$ -2.5 Hz) hardly shifts at all. Therefore the inclusion of the new result will lead to only a minor change in the confusion noise.

Another issue is which of the two HeCV scenarios one is going to chose. Without much guidance from observations, this question probably cannot be resolved completely. One can however make several observations. All our results are for a CE parameter α = 1 (in the definition of IT91). The results for the IT91 scenario are sensitive to this choice. A CE parameter of α = 0.3 (preferred by others, who use a somewhat different definition of α, i.e., Han et al. 1995), would make the IT91 scenario impossible (TY96). The TY96 scenario in contrast is much less sensitive to α (TY96). It is for this reason that in a future determination of the binary confusion noise we will include HeCVs based on the TY96 scenario.

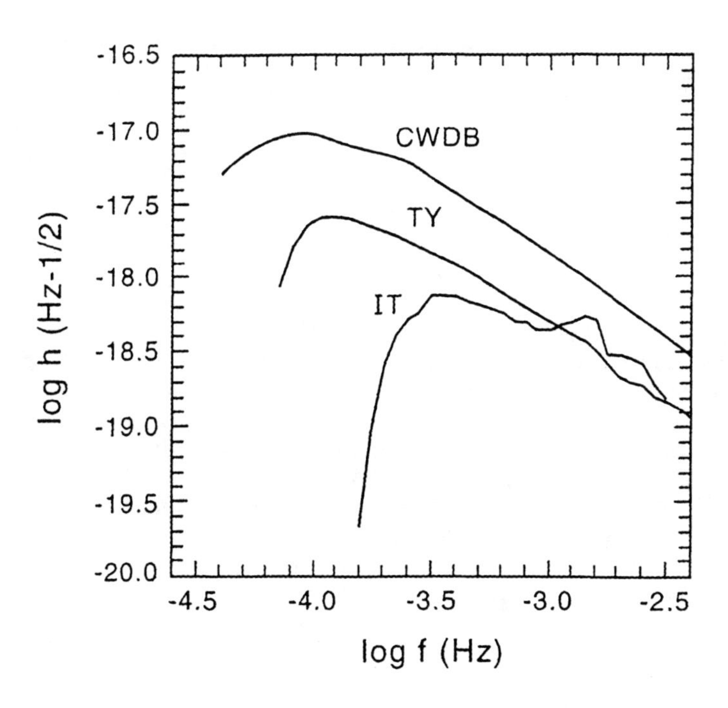

FIGURE 11. Root-mean-square amplitude for the two HECV scenarios in comparison with the amplitude of CWDB's.

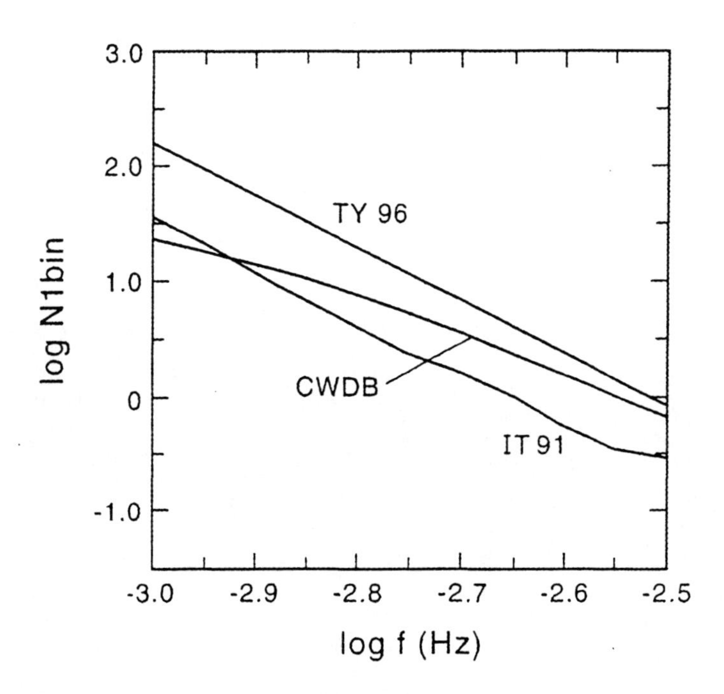

FIGURE 12. Number of sources per frequency resolution bin for the two HECV scenarios in comparison with the numbers of CWDB's.

ACKNOWLEDGMENTS

I am indebted to Peter Bender for his generous advice and many insightful discussions. Tuck Stebbins was kind enough to prepare Figure 2 and I thank Ron Webbink for a discussion on the common envelope parameter α.

REFERENCES

1. Bender, P. L., and Hils, D., *Class. Quantum Grav.* **14**, 1-6 (1997).
2. Han, Z., Podsiadlowski, P., and Eggleton, P. P., *Mon. Not. R. Astron. Soc.* **272**, 800-820 (1995).
3. Hellings, R. W. talk presented at Int. Conf. on Gravitational Waves: Sources and Detectors (Cascina, Italy, 19-23 March, 1996.).
4. Hils, D., Bender, P. L., and Webbink, R. F., *Astrophys. J.* **360**, 75-94 (1990).
5. Iben, I., Jr., and Tutukov, A. V., *Astrophys. J. Suppl.* **54**, 335-372 (1984).
6. LISA Study Team, LISA Pre-Phase A Report, 2nd Edition (Publication MPQ-233, Max-Planck Institute for Quantum Optics, Garching, Germany, July, 1998).
7. Marsh, T. R., Dhillon, V. S., Duck, S. R., *Mon. Not. R. Astron. Soc.* **275**, 828-840 (1995).
8. Marsh, T. R., *Mon. Not. R. Astron. Soc.* **275**, L1-L5 (1995).
9. Moran, C., Marsh, T. R., and Bragaglia, A., *Mon. Not. R. Astron. Soc.* **288**, 538-544 (1997).
10. Savonije, G. J., de Kool, M., and Van Den Heuvel, E.P.J., *Astron. Astrophys.* **155**, 51-57 (1986).
11. Tutukov, A. V., and Yungelson, L., *Soviet Astr.* **30**, 598-600 (1986).
12. Tutukov, A. V., and Yungelson, L., *Mon. Not. R. Astron. Soc.* **280**, 1035-1045 (1996).
13. Webbink, E. D., *Astrophys. J.* **277**, 355-360 (1984).
14. Warner, B., *Astrophys. Space Sci.* **225**, 249-270 (1995).

Cosmological Gravitational Wave Backgrounds

Craig J. Hogan

Astronomy and Physics Departments, University of Washington [1], Seattle, WA 98195-1580

Abstract. An overview is presented of possible cosmologically distant sources of gravitational wave backgrounds, especially those which might produce detectable backgrounds in the LISA band between 0.1 and 100 mHz. Examples considered here include inflation-amplified vacuum fluctuations in inflaton and graviton fields, bubble collisions in first-order phase transitions, Goldstone modes of classical self-ordering scalars, and cosmic strings and other gauge defects. Characteristic scales and basic mechanisms are reviewed and spectra are estimated for each of these sources. The unique impact of a LISA detection on fundamental physics and cosmology is discussed.

I INTRODUCTION

In relativistic Big-Bang cosmology, the Universe is optically thin to gravitational waves all the way back to the Planck epoch; once a wave is produced, absorption is negligible, with energy losses due to redshift alone. The cosmological background of gravitational waves thus contains directly observable information from the entire history of the macroscopic universe— an unobstructed view of our past light cone. Here I present a broad survey of ideas for processes in the early universe which may produce gravitational waves, including guesses at their amplitudes and frequency spectra, and what a detection by LISA might teach us about the early universe. I direct the reader to other recent documents and reviews for more detailed and comprehensive descriptions of many of these ideas and for more thorough surveys of the literature. [1–4]

As in so many areas of astrophysics, gravitational waves offer insights into cosmological processes quite distinct from those accessible by electromagnetic observations. Direct light paths come to us from a temperature $T \approx 3000K$ and a cosmic age $t \approx 0.5My$; the thermal spectrum of the primordial radiation was fixed when the last efficient photon-production processes froze out at $T \approx 500eV$ and a cosmic age of a few weeks. By contrast, the LISA gravitational wave backgrounds, if they exist, illuminate much earlier events, and probe directly the physics of very high energies: the electroweak, GUT, and Planck scales. LISA is sensitive to gravitational waves produced by cosmic strings in the temperature interval 10 keV— 10MeV, and cosmic ages from 10^{-2} to 10^4 seconds; our current knowledge of these epochs comes only from fossil low-bandwidth information such as the light element abundances. Large-scale relativistic flows of energy during phase transitions above 100 GeV or so (earlier than about 10^{-10} seconds) produce waves in the LISA band but have few other observable effects; it is possible that production of these waves may be associated with our only other relics of these epochs, the cosmic baryon number and perhaps the dark matter. The physics which shaped the metric on the largest scales of space and time, currently described by inflation models, probably also generated gravitational waves in ways akin to the generation of the perturbations which led to cosmic structure, perhaps at the GUT epoch. There are also ways in which many spacetime dimensions and other internal degrees of freedom now being explored for fundamental physics near the Planck scale could lead to intense gravitational wave backgrounds in the LISA band, produced at temperatures of 1 to 1000 TeV. Detection of a cosmological background by LISA would provide our first view of early mesoscopic gravitational phenomena about which no other trace survives, with a likely connection to the frontier problems of cosmological theory— the production of baryons, dark matter, and fluctuations in binding energy which led to cosmic structure— as well as the structure of fundamental fields at very high energies.

---

[1] This work was supported at the University of Washington by NASA.

CP456, *Laser Interferometer Space Antenna*
edited by William M. Folkner
 1-56396-848-7/98/$15.00

II REDSHIFTED HUBBLE FREQUENCY

As a reference point for the following discussion, it is useful to define a characteristic frequency associated with most classical production mechanisms for cosmological gravitational waves, the Hubble rate divided by the cosmological redshift:

$$\omega_0(z) \equiv \frac{H(z)}{(1+z)} \approx 2 \times 10^{-5}\mathrm{Hz}\frac{T(z)}{100\mathrm{GeV}}, \tag{1}$$

where T is the temperature of the universe. (There is a weak dependence $\omega_0 \propto g_*^{1/4}$ on the number of effective degrees of freedom g_* which varies only slowly in the Standard Model, from about $g_* \approx 60$ at 1 GeV to about 100 at 1 TeV.) This is the frequency of gravitational waves observed today which were produced on the horizon scale at temperature T. Gravitational waves in the LISA band can be produced by horizon-scale processes at temperatures between a few hundred GeV and a few hundred TeV, by relativistic processes well within the horizon at lower temperature, or by inflationary processes at much higher temperature.

III BROAD-BAND ENERGY

We specify the gravitational wave spectrum $\Omega_{GW}(f)$ in terms of energy density per unit of log frequency, in units of the critical density. After gravitational waves are produced they redshift in the same way as electromagnetic waves and other relativistic forms of energy. Therefore, the integrated energy density of the gravitational wave background(s) scale in proportion to the energy density of the sum of the other relativistic components. Currently this sum is (assuming massless neutrinos)

$$\Omega_{rel}(z=0) = \Omega_{\gamma+\nu+\bar{\nu}} = 0.7 \times 10^{-4} T_{2.726K}^4 h_{75}^{-2}. \tag{2}$$

We define a quantity

$$F(f) = \Omega_{GW}((1+z)f)/\Omega_{rel}(z), \tag{3}$$

the ratio of gravitational wave to other relativistic energy, which is approximately conserved. Notice that at high redshift Ω_{rel} is shared among many more relativistic degrees of freedom than it is today, but F is conserved so long as the coupled system of particles remains highly relativistic.

The quantity F roughly corresponds to the overall gravitational wave production efficiency— the fraction of mass-energy converted into gravitational waves. Since it seems unlikely that gravitational waves would be produced more efficiently than other forms of energy, a plausible upper limit on the cosmological background is $F < 1$ at all frequencies.

Compare this with the LISA sensitivity as estimated in the Yellow Book [4]. The rms amplitude of a fluctuating gravitational wave in a bandwidth f about a frequency f is

$$h_{rms}(f, \Delta f = f) = 10^{-15}[\Omega_{GW}]^{1/2} f_{mHz}^{-1} h_{75}; \tag{4}$$

however the background will be distinguishable from instrumental noise only over much narrower bands; the strain produced in one frequency resolution element after a year of observation is

$$h_{rms}(f, \Delta f = 3 \times 10^{-8}\mathrm{Hz}) = 5.5 \times 10^{-22}[\Omega_{GW}/10^{-8}]^{1/2} f_{mHz}^{-3/2} h \tag{5}$$

$$= 4.6 \times 10^{-20} F(f)^{1/2} f_{mHz}^{-3/2} h\ . \tag{6}$$

At a $f \approx$ few mHz, LISA's instrumental noise drops to as low as $h_{rms} \approx 10^{-24}$ in this band so one can contemplate detecting effects as small as

$$F \approx 10^{-8} \quad \text{or} \quad \Omega_{GW} \approx 10^{-12}; \tag{7}$$

indeed for backgrounds larger than this the cosmological background starts to dominate the noise over some frequency intervals, becoming a nuisance for other observations.

IV CURRENT OBSERVATIONAL CONSTRAINTS ON Ω_{GW}

Although within its band LISA achieves a sensitivity in F and Ω_{GW} far better than any other technique, we already have some meaningful constraints on $\Omega_{GW}(f)$ in other frequency intervals which impact the candidate sources for LISA.

The most sensitive in terms of F is cosmic background radiation anisotropy. [5,6] Tensor mode perturbations generate temperature fluctuations in the microwave background; on scales larger than about a degree these preserve roughly the amplitude they had entering the horizon $\delta T/T \approx h_{hor}$. Now $\Omega_{GW} \approx (f/H_0)^2 h_{now}^2$, so observed limits (and measurements, which may not be of tensor modes) of about $\delta T/T \approx 10^{-5}$ yield a constraint $\Omega_{GW}(f = H_0) \lesssim 10^{-10}$ on the current horizon scale. The limit becomes smaller at higher frequencies, the details depending on the cosmological model. Detailed constraints are placed on inflation models from the predicted tensor modes and their effect on anisotropy. In principle, though not yet in practice, polarization allows tensor and scalar sources to be distinguished observationally. [7,9]

A scale-free process such as inflation with h_{hor} = constant creates a flat background spectrum above about $f = (\Omega_{rel}(z=0)/\Omega)^{1/2}$, with $F(f)$= constant $\approx h_{hor}^2 \approx 10^{-10}$ with amplitude constrained by CBR anisotropy. In the scale-free case, CBR data provide the most sensitive limits on F— better than LISA. However, it is still interesting to consider direct limits on waves at higher frequencies since sources are never precisely scale-free and sometimes not even approximately so.

Pulsar timing measurements directly limit the background at frequencies determined by the observation timescale of a few years. The principle resembles that of interferometric detectors. The pulsar acts as a very steady clock with timing residuals δt of about a microsecond, and over few years the lack of deviations from steady ticking (aside from those expected from Newtonian accelerations of both us and the pulsar) constrains the strain amplitude to $h \lesssim \mu sec/10^8 sec \approx 10^{-14}$. After allowing for the fact that unknowns such as the Newtonian accelerations and the precise pulsar direction are "fitted out", the current limit [10] is $\Omega_{GW}(f = 10^{-8}\text{Hz})h^2 \leq 6 \times 10^{-8}$. Note that if many pulsars are added with accurate timing and good coverage over the sky, it is possible to extract a signature indicating a positive detection of gravitational waves.

For backgrounds that were already present at the epoch of cosmic nucleosynthesis $1sec < t < 100sec$, abundances of light elements provide another constraint. The presence of gravitational waves adds to the total energy density of in the same way as adding additional relativistic degrees of freedom; for example, an extra neutrino species adds the equivalent of $F \approx 1/6$. Although the precise limits are a matter of opinion riding upon continually changing debate over observational errors [11–15], it is clear that standard nucleosynthesis fails (primarily due to overproduction of helium-4) unless $F \lesssim 0.1$. This limit applies to most of the sources we will consider for the LISA band. Notice that the limit becomes stronger for scale-free backgrounds for which many octaves of f contribute to the density.

Chaotic universes with $F \approx 1$ are prone to forming many black holes. This is a disaster since the energy locked up in black holes does not redshift away and quickly comes to dominate the energy density. [16] At most $10^{-8}(T/GeV)^{-1}$ of the mass can convert can convert to holes at temperature T without exceeding the mass per photon in the universe today. So there is constraint on the production mechanism for backgrounds which approach the $F \approx 0.1$ nucleosynthesis bound— they must produce their waves efficiently in a well-regulated process that does not allow a wide dispersion of gravitational potentials since even a tiny fraction of matter in very deep potentials ($v \approx 1$) causes problems. The Goldstone modes discussed below offer an example of such a mechanism but the phase transition bubble collisions do not.

There are also limits from the spectrum of the background radiation. Gravitational waves create observable quadrupoles in the radiation field at each point, so the average radiation field is no longer thermal but a mixture of temperatures with a spread of the order of h_{rms} times the Hubble velocity $H(z)/f$ for f comparable to the scattering rate of photons at redshift z. The COBE/FIRAS limits on spectral distortions limit energy inputs after this epoch to the order of 10^{-4} [6]; but the corresponding limit on h_{hor} is competitive with the CBR anisotropy limits only over a narrow range of frequencies.

V GRAVITATIONAL WAVES FROM INFLATION

Inflation generates gravitational wave backgrounds by the parametric amplification of quantum fluctuations, the same process thought to create the scalar modes that lead to large-scale cosmic structure. [17,18] Backgrounds are created both by the fluctuations of the inflaton field— which make the familiar scalar modes— and by quantum fluctuations of the gravitational field itself.

The amplitude of scalar perturbations depends on details of the inflaton potential. These are tuned to yield h_{scalar} in agreement with some suite of observations including the CBR fluctuations and large scale structure. The best fit to the data for scalar modes is close to scale-free with amplitude $h_{hor,scalar} \approx 10^{-5}$, but may have a small "tilt" or slow variation of h_{hor} with scale. As they enter the horizon, there is some mixing between scalar and tensor modes, since the scalar perturbations lead to quadrupolar mass flows that act as sources for gravitational waves. Very roughly, the mixing will lead to $h_{hor,tensor} \approx h^2_{hor,scalar}$. In spite of the suppresion, the LISA band is so much higher in frequency than the direct observational constraints that even a small tilt can lead to essentially any amplitude in the LISA band.

In the case of gravitational field fluctuations, the amplitude is not dependent on the inflaton potential directly but essentially on just the expansion rate or density during inflation when waves of observed frequency f match the inflationary expansion rate:

$$F(f) \approx V_{inflation}(f)/m^4_{Planck} \tag{8}$$

which is close to scale-free and hence limited to $F \lesssim 10^{-10}$. Inflation very close to the Planck scale generally runs into difficulties with CBR anisotropy from these tensor sources.

The conclusion is that LISA may detect gravitational waves from inflationary fluctuations (from the inflaton fluctuations) if the tilt of the spectrum is favorable, but is unlikely to detect modes directly generated by quantum fluctuations of the graviton.

VI FIRST-ORDER PHASE TRANSITIONS: RELATIVISTIC FLUID FLOWS AND BUBBLE COLLISIONS

The universe may have undergone catastrophic phase transitions at various stages, associated with a sudden change in the ground state configuration of the vacuum fields. [19–25] The macroscopic description is similar whether the fields are associated with QCD, electroweak breaking, or supersymmetry breaking.

We imagine an order parameter ϕ with a free energy density or effective potential $V_T(\phi)$; this potential has two distinct minima corresponding to two distinct phases; the free energy difference between them vanishes at the critical temperature T_c, one phase being favored at higher and the other at lower temperature. The transition from one phase to the other cannot happen smoothly because of the activation barrier between them corresponding to the energy cost of creating a surface interface between phases. The system supercools by a small amount until the free energy difference between phases is sufficient to nucleate bubbles; thermal or quantum fluctuations must create a bubble of low-temperature phase large enough that the volume energy difference exceeds the cost of creating the surface between them; above this size bubbles grow by detonation or deflagration with the phase boundary propagating close to the speed of light. The release of latent heat heats the inter-bubble medium back to almost T_c and increases the pressure between bubbles, so (in the deflagration case) after the shocks from the bubbles meet, fresh nucleation slows and the transition finishes by the slow growth of already-nucleated bubbles. The universe can expand for a significant quiescent period near T_c with both phases coexisting; as it expands it fills more with the lower-density, low-temperature phase. The remnants of the high phase are eventually isolated as islands by the percolation of the low phase, and finally when there is no more high phase left normal cooling resumes.

The production of gravitational waves occurs because of relativistic flows of matter of different densities in the two phases; a substantial fraction of the matter is accelerated to close to the speed of light, and asymmetric shocks are formed as bubbles collide. Not all of the processes involved are computed accurately and many depend on the detailed input physics, but the main parameter is generic: the maximum fractional supercooling δ. Since this is the amount by which the universe expands during the nucleation of the typical bubbles, it determines the bubble size δ/H. Because the nucleation rate depends exponentially on δ, even a very strongly first order transition generically obeys [20]

$$\delta \lesssim \log[T/m_{Planck}] \approx 10^{-2} \tag{9}$$

From scaling we estimate the background from flows of scale δ/H, maximal density contrast, and $v \approx c$; it will be a broad-band background of with a peak at frequency $f_{peak} \approx \omega_0(T)/\delta$ and peak amplitude $F(f_{peak}) \approx \delta^2$.

For example, if there is a very strongly first-order phase transition at 100GeV to 1TeV, a background could be generated with a characteristic frequency of around 2 to 20 mHz and an amplitude as large as $\Omega_{GW} \approx 10^{-8}$, which is detectable. These parameters might be associated with electroweak symmetry breaking and/or supersymmetry breaking. Although a first order transition is not required by the Standard Model, many workers believe that it is first order because that or some other significant disequilibrium is required to create the baryon asymmetry. There is at least the possibility that LISA might make an important connection here, the first direct window on the process that created cosmic matter from radiation.

On the other hand, a very strong disequilbrium may not be required and the fractional supercooling could easily be orders of magnitude less than $\delta \approx 10^{-2}$, which would make the background undetectable. This seems likely to be the case for the QCD transition (which is probably not even first order, but had it been strong might have been detected at lower frequencies from pulsar timing).

VII SELF-ORDERING SCALARS: GRAVITATIONAL WAVES FROM GOLDSTONE MODES

Gravitational waves which may be generated by global excitations of new classical scalar degrees of freedom. Such fields often appear in effective theories derived from unified models such as supersymmetric theories and string theores.

We describe the behavior of active classical scalar fields with the simple Lagrangian density

$$L = \partial_\mu \phi \partial^\mu \phi / 2 - V(\phi) = \dot{\phi}^2/2 - V(\phi), \tag{10}$$

leading to the evolution equation

$$\ddot{\phi} + 3H\dot{\phi} - \nabla^2 \phi + \partial V / \partial \phi = 0, \tag{11}$$

We now suppose that there is more than one scalar component and that the effective potential $V(\vec{\phi})$ has some set of degenerate minima, no longer at just one ϕ but over some set of points far from the origin which all have $|\vec{\phi}| = \phi_0$. For each direction within this surface with $\partial V / \partial \phi = 0$ this wave equation describes massless "Goldstone modes", coherent classical massless modes which propagate at c and dissipate only by redshifting (via $3H\dot{\phi}$). [26,27]

Typically different states $\vec{\phi}$ are reached at different points in space by cooling down from some higher symmetry (*a la* Kibble) which generates spatial gradients in $\vec{\phi}$. These variations excite the Goldstone modes of the field, in general with a large initial amplitude, $\delta\phi \approx \phi$, and with random phase on all scales larger than the horizon. Modes larger than the horizon are essentially frozen in amplitude. The dominant energy density comes from modes just entering the horizon scale (when they have propagated about one wavelength), at which time they contribute a density of the order of $(\phi_0/m_{Planck})^2$ times the total density; after this time the amplitude and frequency of the waves redshift like other relativistic waves, and eventually dissipate. Since the waves induce coherent quadrupolar flows of energy on the horizon scale and close to the speed of light, they couple to gravitational radiation and create a gravitational wave background. Roughly a fraction $(\phi_0/m_{Planck})^2$ of the scalars' energy is radiated per oscillation time in gravitational waves on the horizon scale. If the other couplings of the field are not very weak the main energy loss may not be gravitational radiation and the gravitational wave background may be as small as $F(f = \omega_0(z)) \approx (\phi_0/m_{Planck})^4$. (Uncertainty arises here not only from other couplings in the Lagrangian but also from the gravitational coupling to the other cosmic matter fluids, which may be dissipative and reduce the final energy in the gravitational wave channel.) A scale ϕ_0 near the Planck scale is needed to produce a detectable background.

Unless some other coupling is added to damp the Goldstone oscillations at some point, this is a scale- free background and is subject to the constraints discussed above from lower frequencies, which already imply a fairly small background. This constraint can be avoided however if the theory contains fields which strongly damp the Goldstone modes after a certain epoch t which then reduces the low-frequency gravitational waves (i.e., those below $\omega_0(t)$). For example, a second phase transition could occur removing the degeneracy in the minima of V; the fields would everywhere relax to the single minimum, removing the source of subsequent Goldstone excitation.

The excitation of these modes happens on a timescale determined by the motion in $\vec{\phi}$ space normal to the surface of degenerate minima. If the Higgs masses corresponding to these directions of V are very small, the Kibble excitation may not occur until temperatures much lower than ϕ_0, which cuts off the spectrum at high frequencies.

Note that although the waves are generated classically at the 1— 1000 TeV temperatures characteristic for $\omega_0(T)$ to lie in the LISA band, the physics probed is on the scale of ϕ_0 which can be close to the Planck scale and reflects new fundamental fields close to the scale of quantum gravity. Multitudinous internal spaces and dimensions are now being contemplated for fundamental theory near the Planck scale ("M-theory" or string theory) [28]. The ground state is far from being understood but is often described using the kind of effective theory we have just sketched with a large set of degenerate minima and many internal degrees of freedom. Since the compactifications and symmetry breakings occur close to the Planck scale, we might expect the effective theory to contain scales ϕ_0 close to m_{Planck}. In this case the Goldstone modes are a plausible mechanism which may come close to saturating the $F \approx 0.1$ (nucleosynthesis) bound in the LISA band– a spectacularly strong background

$$\Omega_{GW} \approx 10^{-5} \tag{12}$$

which could have a signal-to-noise of 10^7! Such a strong detection would clearly enable many details to be studied and provide spectacular direct probe into degrees of freedom not seen in any other way. Although a classical macroscopic process, it would reveal fields linked to the unification of gravity and other forces. Even though it is quite possible that nothing of the sort occurs near enough to the Planck scale to produce a detectable background, we should bear this possibility in mind.

VIII COSMIC STRINGS

Strings are topologically stable defects in gauge fields, analogous to vortex lines in superfluids, within which the vacuum is trapped in the excited "false vacuum" state. They are formed again by the Kibble mechanism: the rapid quenching that occurs from the cosmic expansion prevents a global alignment of fields and guarantees plentiful defects. After forming they stretch, move, interconnect, form kinks and wave excitations, and break into loops in a complicated network teeming at close to the speed of light. Their main energy loss is by gravitational radiation. [29–34]

The gravitational interactions of strings are determined by a parameter μ, the mass per unit length, or equivalently a "deficit angle" $\delta = 4\pi G\mu$ for the conical space created by a straight string. For a symmetry breaking at scale m, μ is of the order m^2, leading to $\delta \approx (m/m_{Planck})^2$. Formation of strings is quite generic and may occur even during electroweak symmetry breaking if the topology of the Higgs sector is suitable, but the gravitational effects are usually only considered for GUT scale strings with $\delta \approx 10^{-5}$ which are heavy enough to produce large scale structure and CBR anisotropy. Current calculations show that strings predict a poor fit to these two datasets (too little structure for a given anisotropy [35]) but one should bear in mind that strings may still exist at smaller δ and produce gravitational waves. Strings have many very distinct observable effects; for example, a string in the plane of the sky creates a duplicated strip of images of width δ; galaxy images on the boundary of the strip have sharp, straight edges.

Although the early calculations [30,31] of the spectrum of the background were based on a rather simplified picture of the network, they agree remarkably well with recent predictions based on sophisticated simulations of the behavior of the network. In the LISA band the spectrum is flat with $F \approx \delta^{1/2}$; this is so strong that it is easily observable even if δ is too small to affect any other astrophysical observable. These may be the one type of object for which gravitational radiation is the most easily observable gravitational effect! Indeed even now the pulsar timing bound on gravitational waves is of comparable significance to CBR fluctuations in constraining μ.

The waves are produced by decay of string loops and kinks which occurs after about δ^{-1} oscillations, dominated at temperature T by frequencies about $\delta^{-1}\omega_0(T)$; the waves in the LISA band therefore were emitted at temperatures from 10 MeV down to about 10 keV.

Cosmic strings and Goldstone modes both rely on macroscopically excited new scalars excited by the Kibble mechanism. However, there are important differences. Strings derive from a gauge field (a local rather than a global symmetry breaking) and a nontrivial topology in the manifold of degenerate vacua (leading to the topological stability). A global field with nontrivial topology is also possible (global strings, monopoles, textures) with qualitatively similar results to the Goldstone estimates. In terms of gravitational wave production, strings are more efficient for a given ϕ_0 and make cleaner predictions for a wide variety of phenomena; on the other hand the Goldstone modes can occur with ϕ_0 close to the Planck scale and therefore can produce the most intense backgrounds. Gravitational waves can also be efficiently produced by "hybrid" defects. [36]

IX IMPACT OF A DETECTION ON COSMOLOGY

All of the sources considered above are well motivated from some physical point of view. However, the amplitudes for many of them are almost unconstrained. It is possible that LISA will never detect a cosmological background; it is also possible that an intense cosmological background dominates the LISA noise budget by a large factor, limiting its utility for studying many sources at low redshift. In the latter case, there will at least be a big payoff in completely new knowledge of the early universe. I have not discussed here the problems of distinguishing backgrounds from noise or separating cosmological backgrounds from others such as Galactic binaries. But assuming a cosmological background is detected, how are we to interpret it?

The sources discussed here all produce highly confused isotropic stochastic backgrounds of broad-band Gaussian noise. The first clue to interpretation will be the shape of the spectrum. The spectra of the sources we have considered fall into two broad categories: (1) Scale-free sources with Ω_{GW} approximately independent of f, or $h_{rms} \propto f^{-3/2}\Delta f^{1/2}$, over the LISA band. These include inflation, generic Goldstone modes, and cosmic strings. Even a tilted spectrum from inflation— the only inflationary contribution likely to be detectable— will be approximately flat over the LISA band. (2) Other sources with the imprint of some characteristic scale. These include some Goldstone models (those where features imprinted by damping or Higgs modes happen to lie in the LISA band), and waves from bubbles or other relativistic flows which will bear the imprint of the nucleation scale where the spectrum peaks. Distinguishing between these broad categories is possible with even moderate signal-to-noise detection because of the fairly broad band available, about a factor of a thousand in frequency. The intensity of the spectrum may give another clue: for example, a very intense scale-free background points to some kind of macroscopically active scalar.

LISA surveys a domain of cosmological history which has left few other direct observables. It is worth commenting that these other observables concerning the very early universe are either very large-scale (e.g., fluctuations on galaxy clustering scales and above from inflation) or very small-scale (e.g., abundances of nuclei determined by microscopic reaction rates and thermodynamics); whereas gravitational waves probe a possibly richly varied primordial "mesoscopic" phenomenology about which all other traces have been erased. A detection of a gravitational wave background would depart from the quiescent behavior we have been led to expect from the early universe by the observed small fluctuations, tiny spectral distortions, and abundances in agreement with homogeneous nucleosynthesis; it would give us insight into a nonlinear, chaotic or turbulent stage in the early history of the universe about which we currently have no clue.

REFERENCES

1. Allen, B., in *Les Houches School on Astrophysical Sources of Gravitational Waves*, eds. Jean-Alain Marck and Jean-Pierre Lasota, (Cambridge University Press, 1996); gr-qc/9604033
2. Battye, R. A. and Shellard, E. P. S. (1996); astro-ph/9604059
3. Flanagan, E. E., to appear in GR 15 (1999); gr-qc/9804024
4. LISA Study Team, *LISA Pre-Phase A Report* (1998)
5. Bond, J. R. and Jaffe, A. H., Proc. Roy Soc. A (1998); astro-ph/9809043
6. Smoot, G. and Scott, D., Eur. Phys. J. 3, 127 (1998); astro-ph/9711069
7. Seljak U., Pen U.-L., Turok N., Phys.Rev.Lett. 79, 1615 (1997); astro-ph/9704231
8. Kamionkowski, M., Kosowsky, A., and Stebbins, A., Phys. Rev. D 55, 7368 (1997); astro-ph/9611125
9. Kamionkowski, M., and Kosowsky, A., Phys. Rev. D57, 685 (1998); astro-ph/9705219
10. Kaspi, V., Taylor, J., and Ryba, M., ApJ 428, 713 (1994)
11. Copi, C. J., Schramm, D. N., and Turner, M. S., Science 267, 192 (1995)
12. Hogan, C. J. in *Critical Dialogues in Cosmology*, ed. N. Turok, (Princeton University, 1996); astro-ph 9609138
13. Sarkar, S., Rep. Prog. Phys. 59, 1493 (1996); hep-ph/9602260
14. Hata, N., Steigman, G., Bludman, S., Langacker, P., Phys. Rev. Lett. 55, 540 (1997)
15. Fiorentini, G., Lisi, E., Sarkar, S. and Villante, F. L., to appear in Phys. Rev. D (1998); astro-ph/9803177
16. Carr, B. J. ARAA 32, 531 (1994)
17. Kolb, E. W., in *Current topic in Astrofundamental Physics*, eds. N. Sanchez and A. Zichichi (World Scientific, 1997), p.162; astro-ph/9612138
18. Lidsey, J. E., Liddle, A. R., Kolb, E. W., Copeland E. J., Barreiro, T., and Abney, M., Rev. Mod. Phys. 69, 373 (1997)
19. Peierls, R. E., Singwi, K. S. and Wroe, D. PR 87,46, (1952)
20. Hogan, C. J. Physics Letters 133B, 172,1983

21. Hogan, C. J., M.N.R.A.S. 218,629 (1986)
22. Turner, M. S. and Wilcek, F., Phys. Rev. Lett. 65, 3080 (1990)
23. Ignatius J., Kajantie K., Kurki-Suonio H. and Laine M. Phys.Rev. D49, 3854 (1994); astro-ph/9309059
24. Kamionkowski, M., Kosowsky, A., and Turner, M. S., Phys.Rev. D 49, 2837 (1994)
25. Kurki-Suonio, H., and Laine, M., Phys.Rev.Lett. 77, 3951 (1996); hep-ph/9607382
26. Vilenkin, A., Phys. Rev. Lett. 48, 59 (1982)
27. Hogan, C. J., Phys. Rev. Lett. 74, 3105, (1995); astro-ph/9412054
28. Schwarz, J. H., Phys. Rep. (1998); hep-th/9807135
29. Vilenkin, A., Phys. Lett. 107B, 47 (1981)
30. Hogan, C. J., and Rees, M. J., Nature 311, 109 , 1984
31. Vilenkin, A., Phys. Rep. 121, 263 (1985)
32. Vilenkin, A. and Shellard, S., *Cosmic Strings and Other Topological Defects*, Cambridge University Press (1994)
33. Battye, R. A., Caldwell, R. R., Shellard, E. P. S., to appear in *Topological Defects in Cosmology*, Ed. F.Melchiorri and M.Signore; astro-ph/9706013
34. Avelino P.P., Shellard E.P.S., Wu, J.H.P., Allen B. to appear in Phys. Rev. Lett. (Sept. 1998); astro-ph/9712008
35. Turok N., Pen U.-L., Seljak U., Phys.Rev. D58 (1998); astro-ph/9706250
36. Martin, X. and Vilenkin, A., Phys.Rev.Lett. 77, 2879 (1996); astro-ph/9606022

Some Specific Sources in our Galaxy for the Laser Interferometer Space Antenna (LISA)

Odylio D. Aguiar*, José C. N. de Araújo*, Mara T. Meliani*, Francisco J. Jablonski* and Marcelo E. Araújo†

* *Instituto Nacional de Pesquisas Espaciais - INPE, Divisão de Astrofísica*
Av. Astronautas 1758, São José dos Campos, SP 12227-010, Brazil

† *Universidade de Brasília - UnB, Departamento de Matemática*
Brasília, DF 70919-900, Brazil

Abstract. We calculate the strain amplitudes and frequencies as should be observed by LISA, for some specific gravitational waves sources in our galaxy, such as cataclysmic and low-mass X-ray binaries and a possible supermassive black hole in the center of our galaxy.

WAVES FROM A SUPERMASSIVE BLACK HOLE

Introduction

There are two ways to produce gravitational waves from a supermassive black hole: exciting its quasinormal modes or by the orbiting of matter around the hole.

A supermassive black hole in the center of our Milk Way could produce bursts of waves interesting to the LISA cluster of spacecrafts, by excitation of its quasi-normal modes, if it has a mass in the range of 10^4-10^8 solar masses (1). These quasinormal modes can be excite if matter falls into the hole or pass around it.

On the other hand, if there is matter orbiting a supermassive hole, this could be potentially detected by LISA if the orbit has period in the range of 2s-5.5h ($M_{BH} < 4.2\times10^7\ M_\odot$). The signal, in this case, is continuous.

There are strong evidences for a supermassive black hole at the center of our galaxy (2,3). Here we analyze the consequences of this for LISA, supposing it has a mass of $2.6\times10^6\ M_\odot$.

Quasinormal Modes

The quasinormal-mode frequencies of a Schwarzschild black hole can be determined by a Bohr-Sommerfeld-type formula derived using the phase-integral method with three transition points (4).

Matter falling or passing around a hole can excite its quasinormal modes. When this happens, most of the energy couples to the lowest multipole modes, mainly the quadrupolar one (5), because the energy coupled is proportional to e^{-bl} (where l=2 is the lowest multipolar mode and b is function of the angular momentum).

The quadrupolar quasinormal modes for a supermassive black hole of $2.6\times10^6\ M_\odot$ are given by Table 1 (4).

CP456, *Laser Interferometer Space Antenna*
edited by William M. Folkner

 1-56396-848-7/98/$15.00

TABLE 1. Quadrupolar Quasinormal Modes for a Supermassive Black Hole of 2.6 x 10^6 $M_\odot$

n	ord	Re(f)	Im (f)	
0	3	4.7	0.6	x 10^{-3} Hz
1	3	4.4	1.7	x 10^{-3} Hz
2	1	3.8	3.0	x 10^{-3} Hz
3	1	3.1	4.4	x 10^{-3} Hz
4	3	2.6	5.9	x 10^{-3} Hz
5	3	2.1	7.4	x 10^{-3} Hz
6	3	1.6	9.1	x 10^{-3} Hz
7	3	1.2	11	x 10^{-3} Hz
8	3	0.78	14	x 10^{-3} Hz

The mode Figure of Merit is Q_{mode} = Re(f) / Im(f) = $\tau.\pi$Re(f), where τ is the amplitude decay time.

The characteristic dimensionless amplitude h_c is given by (1):

$$h_c \approx 2 \times 10^{-21} (M_F / M_\odot).(10 \text{ Mpc} / d)$$

where M_F is the mass of the body that falls and d is the distance to Earth.

This means that for a 2.6x10^6 solar-mass hole, the sudden fallen of 1.5 M_{Earth} ($\approx$ 4x10^{-6} $M_\odot$) is enough to produce an event detectable by LISA, which has a sensitivity of h $\approx$ 8x10^{-24} around (3-5)x10^{-3} Hz. Unfortunately, there is no reason to believe that the occurrence of this event is frequent.

Orbiting Material

The scenario for the emission of radiation by matter orbiting a hypothetical supermassive hole in the center of our galaxy is perhaps more optimistic. There are indications of the presence of sources and material close to the hypothetical supermassive black hole in Sgr A* (6,7). If enough material is in close orbit (period < 5.5 h) around the hole, LISA would be able to detect the waves from it. For example, according to Thorne (1), the orbit of a body of ~2x10^{-2} $M_\odot$ with a period of 5.5h around a 2.6x10^6 $M_\odot$ hole in Sgr A* would produce waves detectable by LISA. Closer orbits would need even less material. Only ~10^{-5} $M_\odot$ (or ~3.5 M_{Earth}) would be necessary if the orbital period was ~20.2 min. (last stable orbit for a 2.6x10^6 $M_\odot$ hole), mainly because LISA is more sensitive around 1.65x10^{-3} Hz (and its harmonics) than it is around 10^{-4} Hz.

GRAVITATIONAL RADIATION FROM BINARIES

Summary

General Relativity predicts that binary systems of stars produce gravitational waves (GW) of significant intensity. We here are particularly interested in the cataclysmic variable binaries. These systems emit low frequency gravitational waves, f_{gw} < 10^{-2} Hz (8).

We present, based on the 6th edition of the Catalogue of cataclysmic binaries, low-mass X-ray binaries and related objects (9), a list of binaries that LISA could detect.

Cataclysmic variables as sources of GWs

The binary systems are the most understood of all sources of gravitational waves (GWs) (see, e.g., (1)). Knowing the masses of the stars, the orbital parameters and their estimated distances, one can calculate the details of the GW produced.

Binary systems, as is well known, are extremely abundant in galaxies, thereby they are a very relevant class of sources of GWs. In the present work we study, in particular, the binary systems denominated cataclysmic variables, which are formed by a white dwarf and a low mass secondary star. The total number of such a kind of binary is estimated to amount 10^6 in the Galaxy. These systems produce low frequency GWs, namely, $f_{gw}<10^{-2}$ Hz, that could be detect by LISA.

In the 6th edition of the Catalogue of cataclysmic binaries (CBs), low-mass X-ray binaries and related objects (9), it is present a list of 318 cataclysmic variables. From this catalogue we are preparing a list of sources that could be detected by LISA. Unfortunately in that catalogue it does not appear all the necessary information to calculate and then plot a graph of the amplitude of the GW, h, as a function of f_{gw} ($= 2\ f_{orbital}$).

In Table 2 we present 47 cataclysmic binaries taken from that catalogue, for which the masses and the periods are known, with the distances being obtained from (10) and (11).

In Table 3 we present 35 cataclysmic binaries, where only the period are known, again the distances are taken from (10) and (11), and the masses are calculated using mass versus period relations found in (12) (see also (13)).

In Fig. 1 we show the results we obtained for h as a function of f_{gw} for the cataclysmic binaries presented in Tables 1 (open circles) and 2 (stars). We also plot the one-year sensitivity curves with signal-to-noise ratios of 1 and 5, as well as a curve for the stochastic background produced by the Galactic cataclysmic binaries (see, e.g., (13)).

Although the stochastic background produced by the binaries (cataclysmic or other kinds of binaries) could present an obstacle to any source of GWs in the frequency band 10^{-5} Hz $\leq f_{gw} \leq 0.01$ Hz. A particular cataclysmic binary, for example, could be detected after an integration time, t, such that

$$h_{c\ periodic} > (f_{gw}\ .\ t)^{-1/2}\ h_{c\ background}.$$

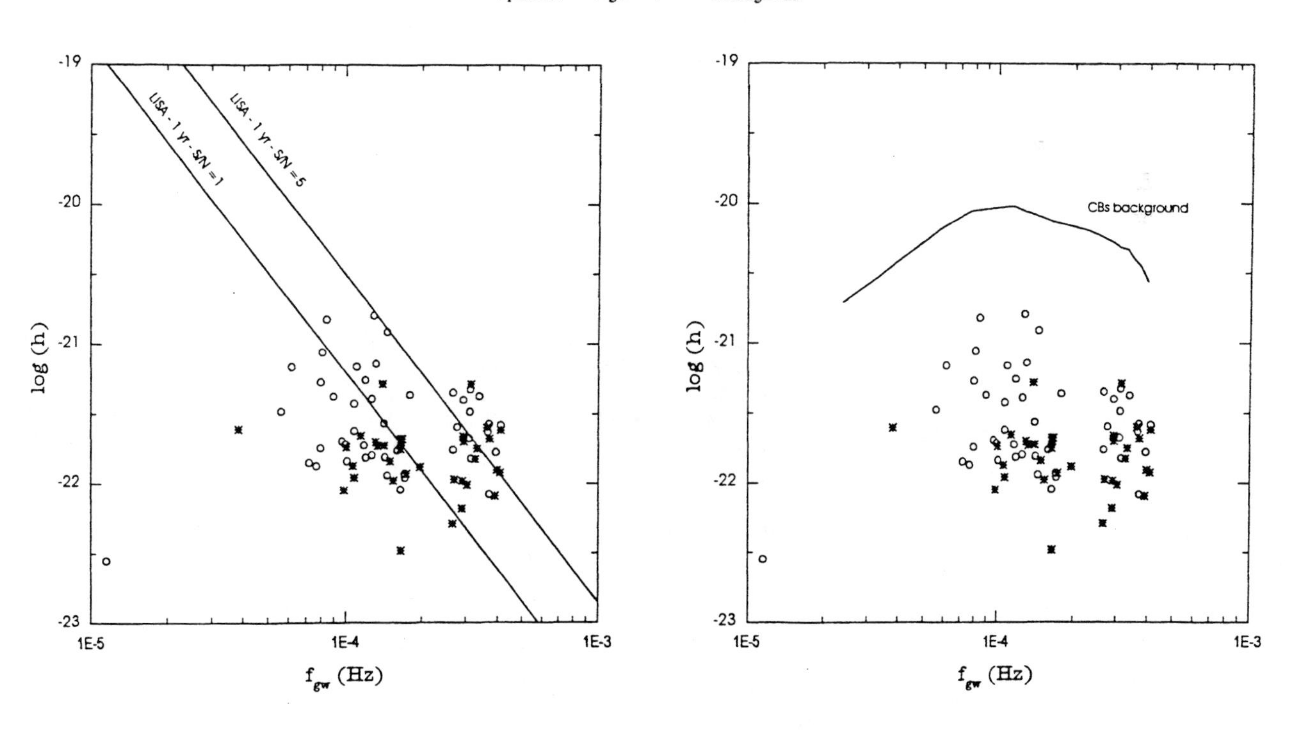

FIGURE 1. Graphic h versus f_{gw} for CBs present in Table 2 (open circle) and in Table 3 (stars). Superimposed are:
(a) the one-year sensitivity curves with signal-to-noise ratios 1 and 5;
(b) the curve of stochastic background by the Galactic cataclysmic binaries.

TABLE 2. (in this table all parameters are known)

Name	Dist. (pc)	Period (days)	M_1 ($M_\odot$)	M_2 ($M_\odot$)	f_{GW} (Hz)	h ($\times 10^{-22}$)
RXAnd	135	0.2099	1.14	0.48	1.10e-4	6.9
AEAqr	140	0.4117	0.79	0.50	5.63e-5	3.3
SSAur	200	0.1828	1.08	0.39	1.27e-4	4.1
ZCam	175	0.2898	0.99	0.70	7.99e-5	5.4
OYCar	100	0.0631	0.685	0.07	3.67e-4	2.3
HTCas	165	0.0736	0.61	0.09	3.14e-4	1.5
WWCet	100	0.1758	0.85	0.41	1.32e-4	7.3
ZCha	130	0.0745	0.84	0.13	3.11e-4	3.3
YZCnc	290	0.0868	0.82	0.17	2.67e-4	1.7
TVCol	500	0.2286	0.75	0.56	1.01e-4	1.5
TXCol	550	0.2383	1.30	0.57	9.71e-5	2.0
EMCyg	350	0.2909	0.57	0.76	7.96e-5	1.8
SSCyg	75	0.2751	1.19	0.70	8.41e-5	15
UGem	81	0.1796	1.26	0.57	1.29e-4	16
HHHer	250	0.2581	0.95	0.76	8.97e-5	4.2
AMHer	75	0.1289	0.39	0.26	1.80e-4	4.3
EXHya	105	0.0682	0.78	0.13	3.39e-4	4.2
VWHyi	65	0.0743	0.63	0.11	3.12e-4	4.7
WXHyi	265	0.0748	0.90	0.16	3.09e-4	2.1
DPLeo	450	0.0624	0.71	0.11	3.71e-4	0.83
TLeo	76	0.0588	0.16	0.11	3.94e-4	1.7
V426Oph	100	0.2853	0.90	0.70	8.11e-5	8.8
CNOri	295	0.1632	0.74	0.49	1.42e-4	2.7
RUPeg	174	0.3746	1.21	0.94	6.18e-5	6.9
GKPer	340	1.9968	0.90	0.25	1.16e-5	0.28
V1223Sgr	600	0.1402	0.50	0.40	1.65e-4	0.90
SWUMa	140	0.0562	0.71	0.10	4.12e-4	2.6
IXVel	150	0.1939	0.82	0.53	1.19e-4	5.5
TWVir	455	0.1827	0.91	0.40	1.27e-4	1.6
V1315Aql	300	0.1397	0.73	0.30	1.66e-4	1.9
V363Aur	600	0.3212	0.86	0.77	7.21e-5	1.4
V436Cen	210	0.0625	0.70	0.17	3.70e-4	2.7
HLCMa	210	0.2145	1.00	0.45	1.08e-4	3.7
BGCMi	700	0.1347	0.80	0.38	1.72e-4	1.1
ACCnc	800	0.3005	0.82	1.02	7.70e-5	1.3
CMDel	280	0.1620	0.48	0.36	1.43e-4	1.6
HRDel	285	0.2142	0.67	0.55	1.08e-4	2.4
DQHer	330	0.1936	0.60	0.40	1.20e-4	1.5
STLMi	128	0.0791	0.76	0.17	2.93e-4	4.0
IPPeg	124	0.1582	1.15	0.67	1.46e-4	12
VZScl	530	0.1446	1.00	0.40	1.60e-4	1.7
LXSer	340	0.1584	0.41	0.36	1.46e-4	1.1
SWSex	450	0.1349	0.58	0.33	1.72e-4	1.2
RWTri	270	0.2319	0.45	0.63	9.98e-5	1.9
UXUMa	250	0.1967	0.47	0.47	1.18e-4	1.9
EFPeg	172	0.0837	0.65	0.17	2.77e-4	2.5
HUAqr	111	0.0868	0.95	0.15	2.67e-4	4.5

TABLE 3. (the masses in this table are calculated according to (13))

Name	Dist. (pc)	Period (days)	M_1 ($M_\odot$)	M_2 ($M_\odot$)	f_{GW} (Hz)	h ($\times 10^{-22}$)
UUaql	225	0.1405	0.53	0.27	1.65e-4	1.8
VYAql	97	0.0635	0.52	0.09	3.65e-4	2.5
TTAri	185	0.1376	0.52	0.26	1.68e-4	2.1

BYCam	190	0.1398	0.53	0.26	1.66e-4	2.1
WXCet	185	0.0583	0.52	0.08	3.97e-4	1.3
V1500Cyg	1200	0.1396	0.53	0.26	1.66e-4	0.33
CQDraBc	100	0.1656	0.67	0.33	1.40e-4	5.2
DMDras	580	0.0870	0.52	0.14	2.66e-4	0.51
EFEri	94	0.0563	0.52	0.08	4.11e-4	2.4
BLHyi	128	0.0789	0.52	0.12	2.93e-4	2.2
AYLyr	52	0.0737	0.52	0.11	3.14e-4	5.1
MVLyr	322	0.1329	0.52	0.25	1.74e-4	1.2
CWMon	290	0.1762	0.73	0.36	1.31e-4	2.0
CqMus	290	0.0594	0.52	0.08	3.90e-4	0.81
V841oph	255	0.6042	0.89	0.86	3.83e-5	2.5
UVPer	115	0.0622	0.52	0.09	3.72e-4	2.1
AOPsc	420	0.1496	0.58	0.29	1.55e-4	1.1
VVpup	145	0.0698	0.52	0.10	3.32e-4	1.8
MRser	139	0.0788	0.52	0.12	2.94e-4	2.0
ANUMa	270	0.0798	0.52	0.12	2.90e-4	0.98
SUUMa	280	0.0764	0.52	0.12	3.03e-4	2.0
QQVul	320	0.1545	0.61	0.30	1.50e-4	1.5
ARAnd	269	0.1630	0.65	0.33	1.42e-4	1.9
FOAqr	325	0.2021	0.88	0.44	1.15e-4	2.2
WXAri	198	0.1393	0.53	0.26	1.66e-4	2.0
AFCam	425	0.2300	0.89	0.52	1.00e-4	1.8
ARCnc	681	0.2146	0.89	0.48	1.08e-4	1.1
ALCom	190	0.0567	0.52	0.08	4.08e-4	1.2
WWHer	430	0.0802	0.52	0.12	2.89e-4	0.66
DOLeo	878	0.2345	0.89	0.54	9.87e-5	0.90
RZLeo	174	0.0708	0.52	0.10	3.27e-4	1.5
TUMen	270	0.1172	0.52	0.21	1.98e-4	1.3
AYPsc	565	0.2173	0.89	0.48	1.07e-4	1.3
UZSer	300	0.1730	0.71	0.35	1.34e-4	1.9
DVUMa	277	0.0860	0.52	0.12	2.69e-4	1.1

ACKNOWLEDGMENTS

We would like to thank Deonísio Cieslinski for supplying us with useful information about Cataclysmic Binaries. J.C.N. de Araújo would like to thank FAPESP for support.

REFERENCES

1. Thorne, K. S., *300 years of Gravitation*, eds. Hawking, S. and Israel, W., Cambridge U. Press, 1987, ch. 9, pp. 330-458.
2. Ghez, A. M., "High Proper Motion Stars in the Vicinity of Sgr A*: Evidence for a Supermassive Black Hole at the Center of our Galaxy", Pierce Prize Lecture at the American Astronomical Society Meeting, San Diego, CA, June 8-11, 1998.
3. Hollywood, J. M. and Melia, F., *ApJ.* **443**:L17-L20 (1995).
4. Anderson, N.; Araújo, M. E. and Schutz, B. F., *Class. Quamtum Grav.* **10**, 757-765 (1993).
5. Oohara, K., *Dynamical Spacetimes and Numerical Relativity*, ed. Centrella, J. M., Cambridge U. Press, 1986, pp. 365-378.
6. Hollywood, J. M. et al., *ApJ.* **448**:L21-L24 (1995).
7. Melia, F., private communication (1998).
8. Douglass, D. H. and Braginsky, V. B. ,*General Relativity* , eds. Hawking, S.W. and Israel, W., Cambridge U. Press, 1979, ch. 3, pp. 90-137.
9. Ritter, H. and Kolb, U.,*A&A* **129**, 83 (1997).
10. Verbunt, F. and al., *A&A* **327**, 602 (1997).
11. Warner, B., *Cataclysmic Variables Stars*, Cambridge U. Press, 1995.
12. Patterson, J.,*ApJS.* **279**, 785 (1984).
13. Hils, D.; Bender, P. L. and Webbink, R.F., *ApJ.* **360**, 75 (1990).

ANALYSIS TECHNIQUES FOR GRAVITATIONAL-WAVE DETECTORS

LISA's Angular Resolution for Monochromatic Sources

Curt Cutler and Alberto Vecchio

The Max Planck Institute for Gravitational Physics, Potsdam, Germany

Abstract. We present results concerning the angular resolution of the LISA detector for monochromatic sources, such as white-dwarf binaries. We compare with the angular resolution achievable by OMEGA.

I INTRODUCTION

This contribution presents results on LISA's angular resolution $\Delta\Omega_S$ for sources that are essentially monochromatic, such as white-dwarf binaries. (See the contribution by Vecchio in this volume for a discussion of the LISA's parameter estimation accuracy for coalescing massive black hole binaries [1].) For any given detection, the position error box within which LISA can localize the source will depend on the gravitational wave frequency, f_{gw}, the source's angular position, its polarization, and its intrinsic amplitude. (For a binary, the polarization is determined by the direction $\hat{L}$ of its orbital momentum vector; possible polarizations are in 1-1 correspondence with possible directions $\hat{L}$.) Because the high dimensionality of parameter space makes it difficult to plot or visualize $\Delta\Omega_S$ as a function of all these parameters, we summarize our results for LISA in three histograms– Figs. 1, 2, and 3–showing the distribution of $\Delta\Omega_S$ for detectable sources having frequencies $f_{gw} = 10^{-4}, 10^{-3}$, and 10^{-2} Hz, respectively. Fig. 4 conveys the same information, but for the proposed OMEGA detector. Figs. 1-4 are the only new results presented here, and represent the essential content of this contribution.

LISA's angular resolution for monochromatic sources was first analyzed in detail by Danzmann et al. [2], Peterseim et al. (1996) [3], and Peterseim et al. (1997) [4]; in particular, [4] contains very useful contour plots showing how $\Delta\Omega_S$ varies with source position. The treatment in [2-4] had two shortcoming, however. First, in their calculation of the variance-covariance matrix, the source position angles θ_S, ϕ_S are taken to be the only unknown parameters. In reality, the (complex) amplitude, frequency, and polarization of the signal are also unknown, and must be extracted from the datastream. Correlations between the various parameters then increase the position uncertainty $\Delta\Omega_S$. Second, for simplicity, [2-4] considered only the single output obtainable from two arms of LISA. Since LISA has three arms, it produces two independent outputs, measuring both polarizations of the gravitational wave at any instant.

In a previous paper [5], Cutler explained the ideas and formalism for calculating $\Delta\Omega_S$, including information from all 3 arms and the correlations with other physical variables. In [5], he also presented sample results for a handful of source parameters chosen 'out of the air'. This handful of cases gave a fair indication of the rough magnitude of LISA's angular resolution, but gave little feeling for the full distribution of $\Delta\Omega_S$ over parameter space. We provide that here, plotting the distributions in Figs. 1-3. These figures also show how $\Delta\Omega_S$ varies over LISA's frequency range, which had not been done in [2-4].

II RESULTS

The histograms in Figs. 1-3 were compiled as follows. We fixed a frequency f_{gw} and intrinsic amplitude (the latter is equivalent to fixing $\mathcal{M}_c$, the binary's chirp mass), and assumed that space was uniformly filled with binaries having that frequency and amplitude, and having arbitrary orientations $\hat{L}$. We then randomly picked sources from this distribution, in Monte Carlo fashion, and tested whether the source was detectable, in the

CP456, *Laser Interferometer Space Antenna*
edited by William M. Folkner
 1-56396-848-7/98/$15.00

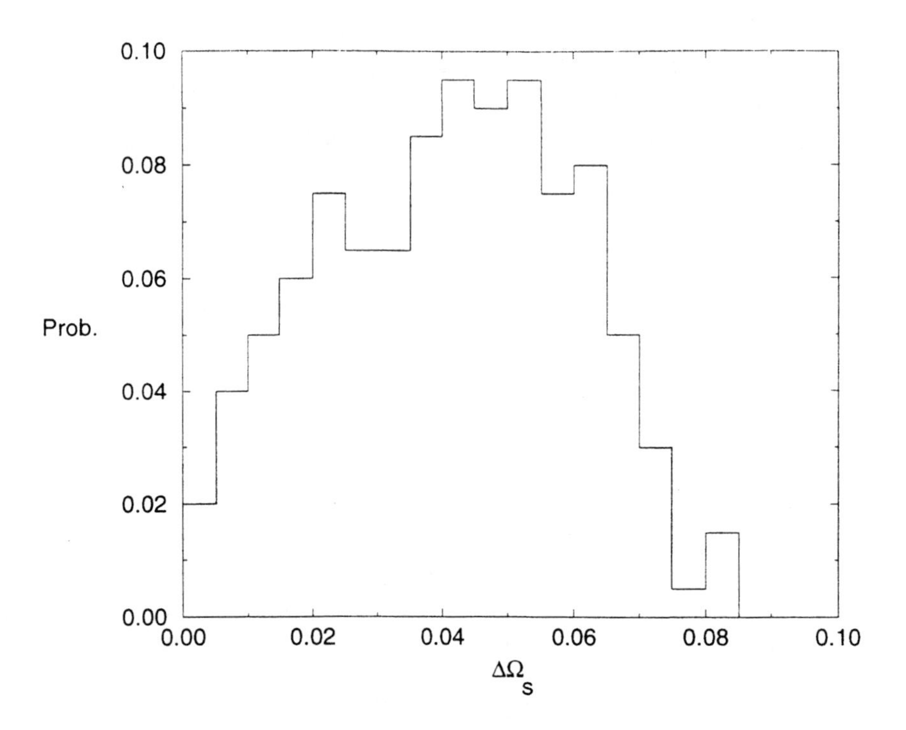

FIGURE 1. Histogram showing LISA's angular resolution $\Delta\Omega_S$ (steradians) for monchromatic sources with $f_{gw} = 10^{-4}$ Hz. The vertical axis gives the probability that a random detection has given resolution $\Delta\Omega_S$. We assumed that sources are uniformly distributed in space and orientation, and chose total $S/N = 10$ as the detection threshhold. The histogram is based on 200 random 'detections'.

sense that its signal-to-noise was above some threshhold. Here we set as a threshhold that the total signal-to-noise S/N be greater than 10. (Since LISA has three arms, it can be thought of as a pair of independent 2-arm interferometers; by 'total' S/N we mean the combined S/N for the pair [5].) All results are for one year of observation.

It is important to understand that the results in Figs. 1-3 depend only on the geoemtry of LISA's orbit–its motion around around the Sun, the precession of the plane containing the satellites, and the rotation of the satellites within that plane. The results do not depend at all on the size or shape of LISA's noise curve. Increasing the noise at some frequency would of course decrease the number of sources detected at that frequency, but (for fixed threshhold) would not affect the distribution of $\Delta\Omega_S$ for the sources that *are* detected. Similarly, although we started by saying that we had specified one particular binary chirp mass, it should now be clear that the same histogram would result for an arbitrary distribution of chirp masses–because any mass yields the same curve.

How one calculates $\Delta\Omega_S$ is explained in [5], and we refer the reader there for full details. Here the following short summary must suffice. For each detected source there are 7 physical parameters to be determined: the overall signal amplitude $\mathcal{A}$; an overall phase φ_0; the frequency f_{gw}; two angles θ_S, ϕ_S describing the source position; and two angles θ_L, ϕ_L describing the orientation of the binary's orbital plane (equivalently, the signal's polarization). We assume the detector noise is Gaussian. We calculate the Fisher matrix in the usual way; inverting the Fisher matrix yields the error variance-covariance matrix, up to correction terms of $(S/N)^{-1}$. Since typically S/N will be over order 10, these correction terms will not be negligible, but nevertheless we believe the zeroth-order piece we calculate should correctly determine LISA's angular resolution to within a factor of ~ 2, which is good enough for our present purposes. (There is one subtlety in our calculation worth mentioning. Since LISA has 3 arms, it effectively acts like two 2-arm interferometers, producing two datastreams. This decomposition into two outputs is not unique, but under the assumption that the noise is 'totally symmetric' in the 3 arms, it was shown in [5] that there is a particularly simple decomposition such

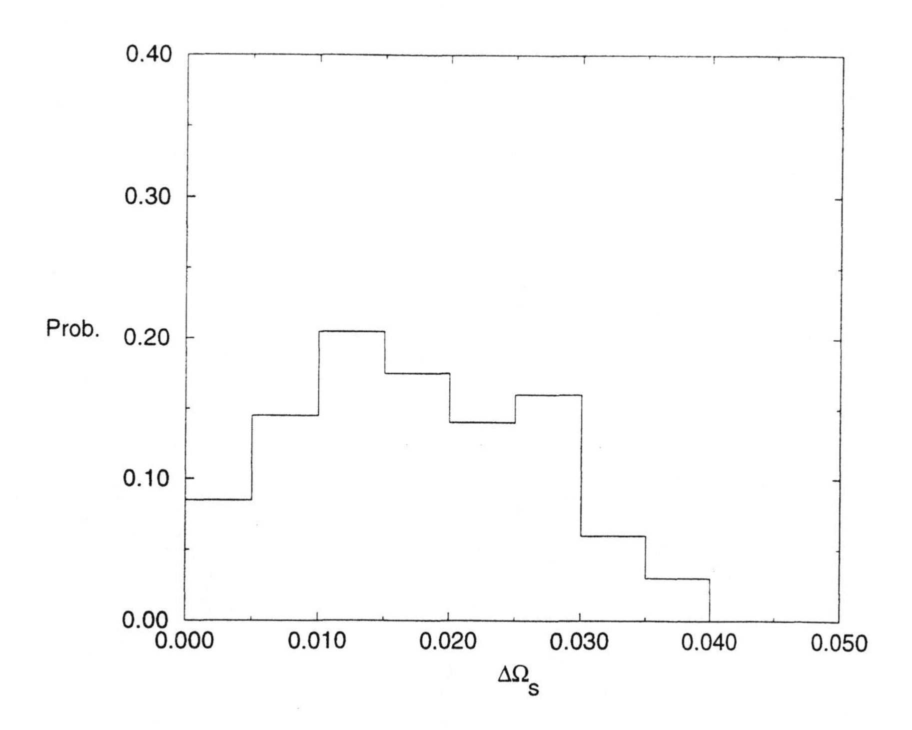

FIGURE 2. Same as Fig. 1, except $f_{gw} = 10^{-3}$ Hz.

that the two outputs have uncorrelated noise. Since we find that, in fact, having two outputs instead of one only decreases $\Delta\Omega_S$ by a factor ~ 2 for monochromatic sources, we think our results should depend only very weakly on our particular assumption about noise correlations in the two outputs.)

We define the solid angle $\Delta\Omega_S$ by

$$\Delta\Omega_S = 2\pi \left[\langle \Delta\mu_S \, \Delta\phi_S \rangle^2 - \langle \Delta\mu_S \, \Delta\phi_S \rangle^2 \right]^{1/2} \tag{1}$$

where $\mu_S \equiv \cos\theta_S$. The second term in brackets in Eq. (1) accounts for the fact that errors in μ_S and ϕ_S will in general be correlated, so that the error box on the sky is elliptical in general, not circular. We have chosen the normalization factor of 2π in our definition of $\Delta\Omega_S$ so that the probability that the source lies *outside* an (appropriately shaped) error ellipse enclosing solid angle $\Delta\Omega$ is simply $e^{-\Delta\Omega/\Delta\Omega_S}$. (For purposes of comparison, note that references [2-4] do not include the 2π normalization factor in their definition of the position error box.)

Fig. 1 is a histogram showing LISA's angular resolution $\Delta\Omega_S$ (steradians) for monchromatic sources with $f_{gw} = 10^{-4}$ Hz. The vertical scale gives the probability distribution of $\Delta\Omega_S$ for detectable sources, assuming a detection threshhold of total $S/N = 10$ and assuming that sources are uniformly distributed in space and orientation. Each histogram is based on 200 random 'detections', with results divided into 20 bins; the 'raggedness' of the plots is presumably due to small-number statistics. The vertical scale is normalized so that the sum of all the bin heights is one.

As one would expect, LISA's angular resolution improves at higher frequencies; it is roughly 70 times better (in solid angle) at 10^{-2} Hz than at 10^{-4} Hz. We see that for fixed f_{gw} the distribution in $\Delta\Omega_S$ is actually not very wide: it is rare for any detection to have position uncertainty $\Delta\Omega_S$ a factor of three worse than the mean value. The sources on the 'poor-resolution-tail' in Fig. 3 lie very close to the ecliptic, where the Doppler shift conveys little information.

We have also run our code for the case of only one detector output (the output available if one laser were to malfunction), though this is not displayed. We find this degrades $\Delta\Omega_S$ by only a factor $2 - 3$.

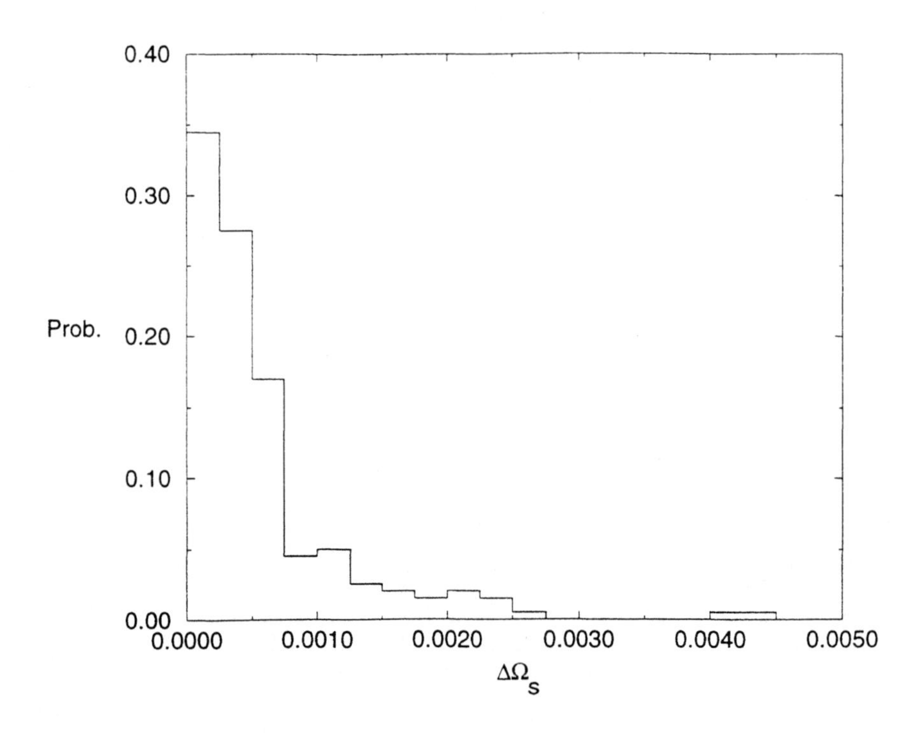

FIGURE 3. Same as Fig. 1, except $f_{gw} = 10^{-2}$ Hz.

We have run our code for $f_{gw} = 3$ mHz, to compare with results in [2-4]. We find that at this frequency (where almost all the position information comes from Doppler effects), the shortcomings of [2-4] mentioned in the Introduction lead to only modest changes in the expected value of $\Delta\Omega_S$, and these changes pull in opposite directions. At 3 mHz, correlations with the other parameters increase $\Delta\Omega_S$ by only a factor ~ 3. Using all the information provided by the 3 arms, for a fixed monochromatic source, decreases $\Delta\Omega_S$ by only a factor $\sim 2-3$. (Both the 'extra' parameter correlations and the 'extra' information from the third arm have a more drastic effect on $\Delta\Omega_S$ for the case of mergers of very massive black holes, where the effective integration time is much shorter than one year and most of the signal-to-noise is accumulated at frequencies well below 1 mHz [5].)

We turn now to a comparison of LISA's angular resolution and that of the proposed OMEGA mission. The OMEGA mission has a different orbit than LISA's: the OMEGA satellites orbit the Earth, lie approximately in the ecliptic plane, and the configuration rotates once around the z-axis every ~ 56 days. Actually, OMEGA's detector plane would deviate from the ecliptic by several degrees, and precess with period of ~ 7 years. But since the precession of OMEGA's detector plane is both small-amplitude and slow, as a first cut we have approximated the detector plane as fixed and lying in the ecliptic. In this approximation, essentially all information about the source position is encoded in the Doppler shift. This is in contrast to the case of LISA, where position information is also encoded in the signal's amplitude and phase modulation, resulting from the precession of the orbital plane.

Another way of saying this is, that in the limit of very low f_{gw}, (in which the Doppler effect becomes irrelevant), the Fisher matrix becomes degenerate for the case of OMEGA, but not for LISA. (Of course, this is strictly true only for an OMEGA whose detector plane is completely fixed.) One can understand this as follows. In the 2-D detector plane, the space of symmetric, trace-free tensors is only 2-dimensional, so in principle from a sinusoidal wave, OMEGA can measure (besides the frequency) only two complex numbers: the complex amplitudes for the two basis tensors. Now, one complex number just corresponds to an overall re-scaling of the distance and a shift in the zero of time. Therefore there is only one complex number (the complex ratio of the two amplitudes) that encodes all information about the source direction and polarization. But it is clearly impossible to determine four angles from one complex number: the problem is ill-determined. In contrast, LISA samples the gravitational wave in a one-parameter family of planes, which allows it in principle to extract both the position and orientation of the source, even in the limit where the Doppler information becomes negligible.

In Fig. 4 we plot OMEGA's angular resolution for $f_{gw} = 10^{-3}$ Hz. This histogram was constructed in precisely the same way as described above for LISA. Unlike for LISA, the results have a simple scaling with frequency; $\Delta\Omega_S \propto f^{-2}$. E.g., to find OMEGA's angular resolution for $f_{gw} = 10^{-2}$ Hz, simply multiply the horizontal axis of Fig. 4 by the factor 10^{-2}. (This scaling can be proved by a suitable change of variables [6].) The points on OMEGA's 'poor-angular-resolution' tail in Fig. 4 have the same explanation as for LISA in Fig. 3: they correspond to sources that lie very close to the ecliptic plane.

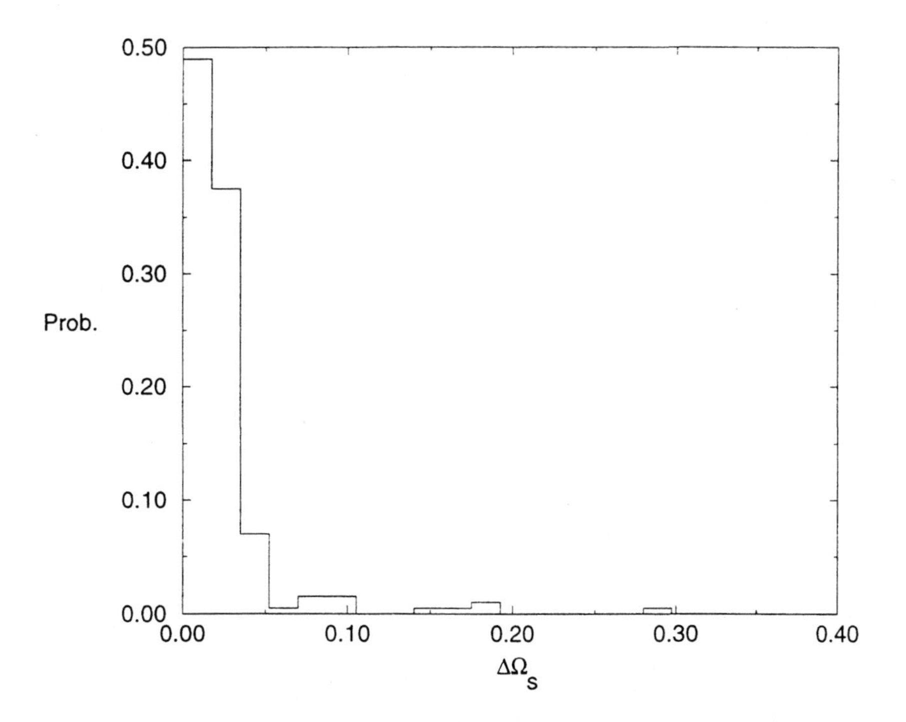

FIGURE 4. Same as Fig. 2, but for the OMEGA detector. Note, however, that the 'detectable' sources on which this figure is based have a different spatial distribution than than the detectable LISA sources of Fig. 2. Unlike for LISA, the distribution has a simple scaling with frequency: $\Delta\Omega_S \propto f_{gw}^{-2}$.

Comparing (appropriately scaled versions of) Fig. 4 with Figs. 1-3, we can see that, if we take the two missions to have exactly the same sensitivity, then OMEGA's angular resolution is roughly a factor 20 (in steradians) worse than LISA's at $f_{gw} = 10^{-4}$ Hz, but quite comparable at 10^{-3} and 10^{-2} Hz. In fact, we see that OMEGA's angular resolution is actually somewhat better than LISA's at 10^{-2} Hz. How can this be? It seems counterintuitive, since LISA clearly collects more 'information' about any given source, at fixed signal-to-noise? It turns out to be an artifact of the way we have made the comparison. Recall that we plotted the angular resolution of 'detectable' sources. Detectable sources are not evenly distributed on the sky; for OMEGA they crowd towards the ecliptic North and South poles and away from the ecliptic plane, much more so than for LISA. The distribution on the sky of detectable sources, for both LISA and OMEGA, is shown in Fig. 5. (The histograms for LISA and OMEGA in Fig. 5 are based on the same 200 detectable sources used in Figs. 1-3 and Fig. 4, respectively.) Thus LISA sources are more likely to live in regions of the sky where the Doppler shift carries relatively less information. (For a source located precisely at $\cos\theta_S = 0$, the change in magnitude of the Doppler shift is second order in the variation in $\cos\theta_S$; clearly then, the Doppler shift carries less information for sources living close to the ecliptic.) The main point, however, is that *for monochromatic sources*, for $f_{gw} \gtrsim 10^{-3}$ Hz, LISA and OMEGA have comparable angular resolution, assuming they have the same sensitivity. For $f_{gw} \sim 10^{-4}$ Hz, LISA's angular resolution is roughly an order of magnitude better.

We point out that the 'moral' of this LISA/OMEGA comparison is rather different for monochromatic sources than for the case of merging massive black hole binaries. In the latter case, where most of the S/N is accumulated on a timescale much shorter than a year, LISA's angular resolution is typically $\sim 10^{-4} - -10^{-3}$ srad, while OMEGA has essentially no angular resolution at all [1].

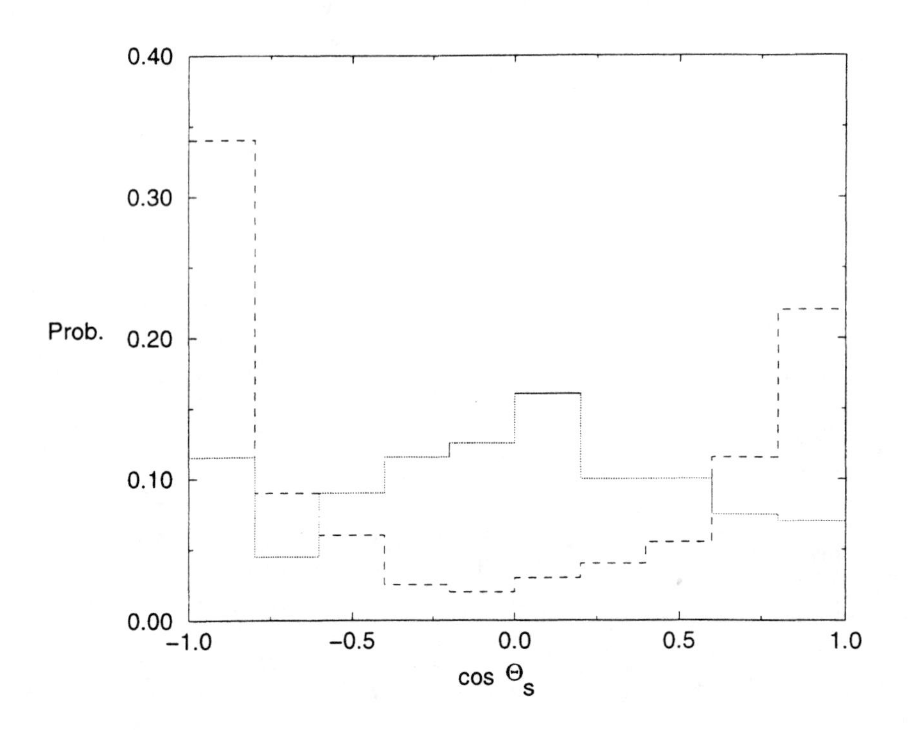

FIGURE 5. Shows distribution of $\cos\theta_S$ for detectable sources, for both LISA (solid line) and OMEGA (dashed line).

We have not discussed here LISA's ability to extract other information from monochromatic sources, such as the distance to the source and its polarization. That will be discussed in an expanded version of the work presented here, which is currently in preparation [6].

REFERENCES

1. A. Vecchio and C. Cutler, this proceedings (1998).
2. K. Danzmann et al., LISA Pre-Phase A Report, Max-Planck-Institut fur Quantenoptik, Report MPQ 209, Garching, Germany (1996).
3. M. Peterseim, O. Jennrich, and K. Danzmann, Class. Quant. Grav. **13**, A279 (1996).
4. M. Peterseim, O. Jennrich, K. Danzmann, and B. F. Schutz, Class. Quant. Grav. **114**, 1507 (1997).
5. C. Cutler, Phys. Rev. D **57**, 7089 (1998).
6. A. Vecchio and C. Cutler, in progress.

LISA: Parameter Estimation for Massive Black Hole Binaries

Alberto Vecchio and Curt Cutler

Max Plank Institut für Gravitationsphysik – Albert Einstein Institut
D-14473 Potsdam, Germany

Abstract. We calculate how accurately LISA and OMEGA can determine the parameters of massive black hole binary systems by observing the signal emitted during the in-spiral phase. In particular, we explore the role of the black hole spins and the induced effect of precession of the source orbital plane.

INTRODUCTION

The strongest sources of gravitational waves (GW) for the future generation of space-based detectors – LISA [1] and/or OMEGA [2] – are likely to be binary systems of massive black holes (MBH), as they would be detectable, in the final year of in-spiral, at a signal-to-noise ratio (SNR) $\sim 10^2 - 10^4$ for systems at cosmological distances (see Fig. 1). Such sources may consist of two MBH's (with mass in the range $\sim 10^4 - 10^7\, M_\odot$) or a low mass BH ($\sim 10\, M_\odot$) orbiting a massive one. If the event rate is reasonable (see [3–9] and references therein) gravitational wave observations would allow precise *direct measurements* of the BH fundamental parameters, could have strong implications on the estimation of the cosmological parameters and possibly the understanding of the scenario of structure formation in the Universe [1,10,11].

Here we report on some progress regarding the estimation of the error with which the distance, masses, spins, position of the source in the sky and instant of final collapse could be measured by the future generation of space-borne laser interferometers; indeed, this paper is complementary to [13] that deals with observations of monochromatic sources. The issue of the LISA angular resolution was first addressed by Peterseim et al. in [12] and, regarding massive black hole binaries (MBHB), by Cutler in [14]; here we extend Cutler's analysis in three respects: (a) for the case of binary systems with negligible spins (or, equivalently, whose contribution is accounted for only in the GW phase), we include in the GW phase the post^2-Newtonian terms; (b) we then take into account the role of the spins in affecting the orientation of the source orbital plane (the so-called "dragging of inertial frames"); (c) we apply this analysis not only to the LISA experiment but also to the OMEGA configuration.

THE SIGNAL AND ITS EXTRACTION

The gravitational radiation emitted during the in-spiral of a binary system is described by 17 independent parameters; indeed, the waveform can be very complex, in particular for general values of the orbit eccentricity and the BH spins. Here we shall assume that (i) the orbit is circular (which is probably quite realistic for binaries composed of two MBH's that have undergone a common evolution inside a galactic core, whereas is likely to be violated for solar mass compact objects and/or low mass BH's orbiting a massive one) and (ii) if spins are present, either the masses of the BH's are roughly equal, or one of the BH's has a negligible spin (which still describe a wide range of astrophysical situations); in this case the system undergoes the so-called *simple precession* [15].

We consider a binary source of masses m_1 and m_2 (we indicate with m, μ and $\mathcal{M}$ the total, reduced and chirp mass, respectively), spins $\mathbf{S}_1$ and $\mathbf{S}_2$ (we define $\mathbf{S} = \mathbf{S}_1 + \mathbf{S}_2$ and $\mathbf{J} = \mathbf{L} + \mathbf{S}$) at a luminosity distance D; with respect to an appropriate ecliptic reference frame, the location of the source in the sky is given by

CP456, *Laser Interferometer Space Antenna*
edited by William M. Folkner
 1-56396-848-7/98/$15.00

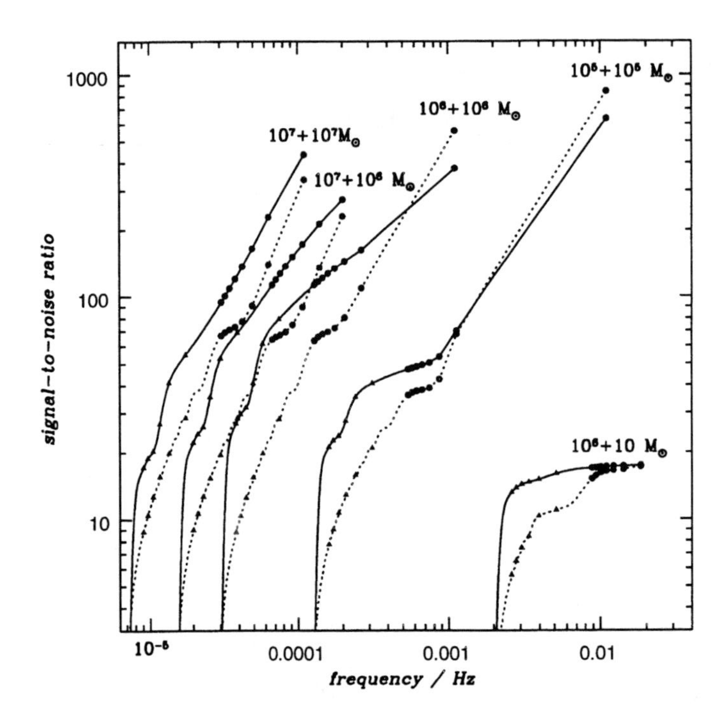

FIGURE 1. The signal-to-noise ratio at which massive black hole binaries can be detected by LISA and OMEGA. The plot shows the SNR as a function of the instantaneous frequency of emission during the final year of in-spiral of BH's of different masses (see labels); both instrumental and confusion noise are taken into account. The dots and triangles mark the final days and months, respectively, of in-spiral. The solid and dotted lines refer to LISA and OMEGA (one interferometer), respectively. The fiducial source is at $z = 1$ (where the cosmological parameters have been chosen according to $H_0 = 75\,\mathrm{km\,s^{-1}\,Mpc^{-1}}$, $\Omega_0 = 0$, $\Lambda_0 = 0$; indeed $D = z/H_0 = 3.997$ Gpc) with position and orientation in the sky given by: $\cos\theta_N = 0.3$, $\phi_N = 5$, $\cos\theta_L = 0.8$, $\phi_L = 2$. The black hole spins are assumed to be $\mathbf{S}_1 = \mathbf{S}_2 = 0$, so that the spin-orbit and spin-spin parameters, β and σ [18], respectively, are equal to zero.

$\hat{\mathbf{N}} = (\theta_N, \phi_N)$ and the direction of the orbital angular momentum is $\hat{\mathbf{L}} = (\theta_L, \phi_L)$, where θ and ϕ are polar angles. In the Fourier space, the signal $\tilde{h}^{(\iota)}(f)$ recorded at each of the two independent laser interferometers (labelled by ι) of a three arms 60° instrument (LISA or OMEGA) reads [14–17]:

$$\tilde{h}^{(\iota)}(f) \simeq \begin{cases} \mathcal{A}\, A_p^{(\iota)}[t(f)]\, f^{-7/6}\, e^{\{i\,[\Psi(f) - \varphi_p^{(\iota)}[t(f)] - \varphi_D[t(f)] - \delta_p\phi[t(f)]]\}} & 0 < f < f_{\mathrm{cut-off}} \\ 0 & f > f_{\mathrm{cut-off}} \end{cases}, \qquad (1)$$

where $\mathcal{A}$ and $\Psi(f)$ are the gravitational wave amplitude and phase, respectively; $A_p^{(\iota)}(t)$ and $\varphi_p^{(\iota)}(t)$ are the polarization amplitude (see Fig. 2) and phase, respectively; $\varphi_D(t)$ is the Doppler phase modulation and $\delta_p\phi(t)$ is the so-called Thomas precession phase [15]. As the signal must be cut off at some finite frequency, we set (conservatively) $f_{\mathrm{cut-off}} = f_{\mathrm{isco}} = 1/(6^{3/2}\,\pi\, m)$. We also recall that all the quantities in Eq. (1) are the observed one, as in general sources will be at cosmological distances. $\tilde{h}^{(\iota)}(f)$ is phase and amplitude modulated as result of three distinct effects: (i) the change of orientation of the detector; (ii) the orbital motion of the instrument around the Sun; (iii) the change of $\hat{\mathbf{L}}$ during the in-spiral, if the observed binary undergoes precession, which causes *intrinsical* modulations on the GW signal; furthermore $\delta_p\phi$ contributes to the phase of the wave.

The detector output is the sum of gravitational radiation $\tilde{h}^{(\iota)}(f;\mathbf{p})$ – where $\mathbf{p}$ is the vector of signal parameters – and noise $\tilde{n}(f)$, with spectral density $S_n(f)$, that we shall assume to be Gaussian and stationary. For space-borne laser interferometers, two main terms contribute to $S_n(f)$: the instrumental noise (whose expression

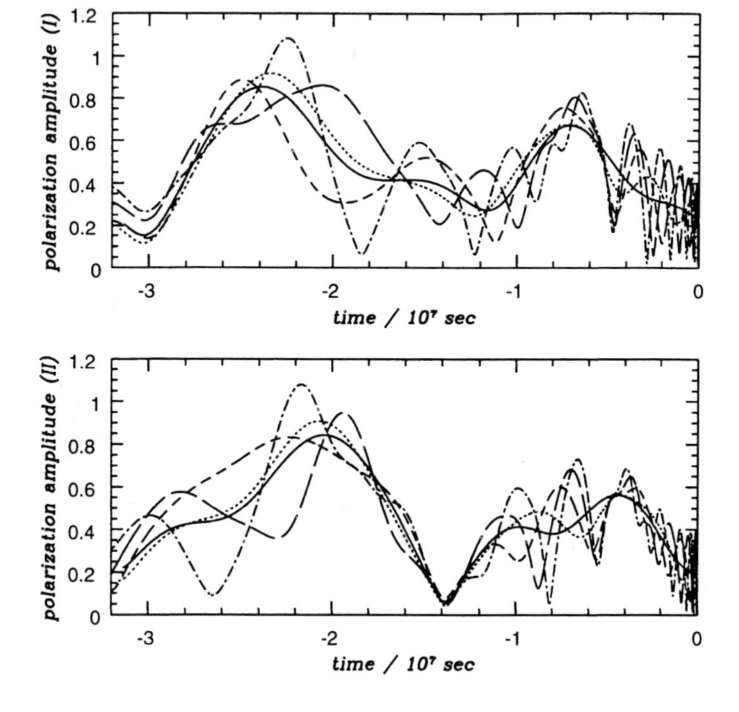

FIGURE 2. Polarization amplitudes in the final year of binary in-spiral in LISA observations. The plot shows the polarization amplitude $A_p^{(\iota)}$ as a function of time in the LISA detector $\iota = I$ and $\iota = II$ for a system of mass $m_1 = 10^7\, M_\odot$ and $m_2 = 10^6\, M_\odot$ and different values of the spin-orbit parameter β (solid line: $\beta = 0$; dotted line: $\beta = 1$; short dashed line: $\beta = 3$; long dashed line: $\beta = 5$; dotted-dashed line: $\beta = 7$). The source location in the sky is given by $\cos\theta_N = 0.3$ and $\phi_N = 5$, the constant direction of $\hat{\mathbf{J}}$ is $\cos\theta_J = 0.8$ and $\phi_J = 2$, and $\hat{\mathbf{S}}\cdot\hat{\mathbf{L}} = 0.9$. Similar or even stronger oscillations are observable in the the polarization phase $\varphi_p^{(\iota)}$.

is given in [14] and [19], for LISA and OMEGA, respectively) and the so-called confusion noise, due to the incoherent superposition of unresolved short-period stellar-mass binaries (a simple fit, based on the analysis reported in [22], can be found in [14]).

The optimal signal-to-noise ratio produced by a signal h is given by SNR $= (h|h)^{1/2}$, where $(.|.)$ is the usual inner product [16]. If $\mathbf{p}_0$ is the true value of the parameters, in general the presence of the noise will induce an error $\Delta\mathbf{p}$ in their determination; in the limit of high SNR [16,20,21] it can be easily estimated by computing the variance-covariance matrix:

$$\langle \Delta p^j\, \Delta p^k \rangle = \left[\left(\Gamma^{(\iota)}\right)^{-1}\right]^{jk} + \mathcal{O}(\mathrm{SNR})^{-1} \quad , \quad \Gamma^{(\iota)}_{jk} \equiv \left(\frac{\partial h^{(\iota)}}{\partial p^j} \middle| \frac{\partial h^{(\iota)}}{\partial p^k} \right) , \tag{2}$$

where $\Gamma^{(\iota)}_{jk}$ is the Fisher matrix; the diagonal elements of Γ^{-1} are the mean square errors associated to the parameter measurements, and we define the angular resolution of the instrument $\Delta\Omega^{(N)}_{(\iota)}$ according to [14]. For the case of the analysis with two or more detectors, whose noises are uncorrelated (for LISA and OMEGA this assumption is satisfied if the noise of the three arms is "totally symmetric" [14]), the Fisher matrix is simply: $\Gamma_{jk} = \sum_\iota \Gamma^{(\iota)}_{jk}$.

PARAMETER MEASUREMENTS: RESULTS

In this section we present some results regarding the expected errors associated with the measurements of the parameters that characterize a source within the signal model mentioned in the previous Section (a more general and thorough investigation is currently in progress [17]). The results refer to the combination of the data from the two independent interferometer outputs that one can construct out of the three arm signals [14], unless differently specified.

We assume first to know *a priori* that the observed source is composed by BH's with negligible spins or $\hat{\mathbf{S}}_1$, $\hat{\mathbf{S}}_2$ and $\hat{\mathbf{L}}$ are oriented in such a way that the direction of the angular momenta is constant; indeed the amplitude and phase of the signal are not intrinsically modulated by the precession of $\hat{\mathbf{L}}$ (equivalently, one could assume that, for simplicity, we take into account the spin contributions only to the GW phase $\Psi(f)$). In

TABLE 1. Signal-to-noise ratio, distance measurement error and angular resolution for LISA observations of the final year of MBHB in-spirals. The fiducial source is at $z = 1$ ($H_0 = 75\,\mathrm{km\,s^{-1}\,Mpc^{-1}}$, $\Omega_0 = 0$, $\Lambda_0 = 0$); the waveform is modelled through the post^2-Newtonian approximation and the spin effects are taken into account only into the GW phase $\Psi(f)$, with $\beta = 0$ and $\sigma = 0$. SNR, $(\Delta D/D)$ and $\Delta\Omega^{(N)}$ are given for the case of measurements with only one interferometer (labelled by I) and both instruments (labelled by $I + II$). $S_n(f)$ is given by the sum of the instrumental and confusion noise. The masses m_1 and m_2, $\Delta D/D$ and $\Delta\Omega^{(N)}$ are in units $M_\odot$, 10^{-2} and 10^{-5} srad, respectively.

mass	$\hat{\mathbf{N}}, \hat{\mathbf{L}}$ [a]	SNR_I	$(\Delta D/D)_I$	$\Delta\Omega_I^{(N)}$	SNR_{I+II}	$(\Delta D/D)_{I+II}$	$\Delta\Omega_{I+II}^{(N)}$
10^7 10^7	1	435	3.182	1.104	591	1.555	1.874
10^7 10^7	2	634	3.361	29.362	926	1.365	81.637
10^7 10^7	3	1358	10.503	34.324	2117	0.315	70.851
10^7 10^7	4	1083	2.664	23.281	1461	0.994	56.170
10^7 10^7	5	2024	215.027	23.810	2937	17.344	32.009
10^7 10^7	6	1031	5.844	55.368	1696	0.564	118.126
10^7 10^7	7	1484	1.864	77.097	1733	0.867	143.222
10^7 10^6	1	273	2.714	169.934	371	1.354	1.486
10^7 10^6	2	394	2.744	375.994	580	1.178	25.488
10^7 10^6	3	808	8.649	108.273	1259	0.338	24.410
10^7 10^6	4	648	2.473	195.042	878	0.908	17.670
10^7 10^6	5	1237	202.645	164.315	1787	15.280	18.126
10^7 10^6	6	616	4.763	118.747	1005	0.524	40.679
10^7 10^6	7	890	1.562	318.656	1043	0.711	52.702
10^6 10^6	1	377	2.313	113.560	516	1.178	0.766
10^6 10^6	2	549	2.298	255.911	802	1.030	17.927
10^6 10^6	3	1171	7.187	69.976	1823	0.273	17.683
10^6 10^6	4	934	2.038	130.227	1257	0.755	13.296
10^6 10^6	5	1742	129.453	102.271	2530	12.992	12.255
10^6 10^6	6	888	3.944	76.227	1461	0.427	27.735
10^6 10^6	7	1273	1.340	216.329	1490	0.593	37.714
10^5 10^5	1	634	4.217	299.699	868	2.505	1.874
10^5 10^5	2	933	3.990	558.823	1354	2.177	81.637
10^5 10^5	3	2047	13.030	216.564	3189	0.444	70.851
10^5 10^5	4	1628	4.202	365.391	2185	1.509	56.170
10^5 10^5	5	2995	218.400	206.663	4363	21.957	32.009
10^5 10^5	6	1551	6.594	263.980	2561	0.814	118.126
10^5 10^5	7	2217	2.415	635.150	2589	0.982	143.222

[a] The values of the seven combinations of $\hat{\mathbf{N}}$ and $\hat{\mathbf{L}}$ correspond to: (1) $\cos\theta_N = 0.3$, $\phi_N = 5$, $\cos\theta_L = 0.8$, $\phi_L = 2$; (2) $\cos\theta_N = -0.1$, $\phi_N = 2$, $\cos\theta_L = -0.2$, $\phi_L = 5$; (3) $\cos\theta_N = -0.8$, $\phi_N = 1$, $\cos\theta_L = 0.5$, $\phi_L = 3$; (4) $\cos\theta_N = -0.5$, $\phi_N = 3$, $\cos\theta_L = -0.6$, $\phi_L = -2$; (5) $\cos\theta_N = 0.9$, $\phi_N = 2$, $\cos\theta_L = -0.8$, $\phi_L = 5$; (6) $\cos\theta_N = -0.6$, $\phi_N = 1$, $\cos\theta_L = 0.2$, $\phi_L = 3$; (7) $\cos\theta_N = -0.1$, $\phi_N = 3$, $\cos\theta_L = -0.9$, $\phi_L = 6$.

TABLE 2. Timing accuracy and mass measurement errors for LISA observations of the final year of MBHB in-spirals. The signal and instrument parameters are as in Table 1. Δt_c is expressed in seconds; $\Delta\mathcal{M}/\mathcal{M}$ and $\Delta\mu/\mu$ are given in units 10^{-2}.

mass	$\hat{\mathbf{N}}$, $\hat{\mathbf{L}}$	$(\Delta t_c)_I$	$(\Delta\mathcal{M}/\mathcal{M})_I$	$(\Delta\mu/\mu)_I$	$(\Delta t_c)_{I+II}$	$(\Delta\mathcal{M}/\mathcal{M})_{I+II}$	$(\Delta\mu/\mu)_{I+II}$
10^7 10^7	1	1113.863	0.652	86.075	761.372	0.438	57.832
10^7 10^7	2	889.315	0.594	75.701	522.691	0.325	41.817
10^7 10^7	3	541.064	0.403	50.379	298.124	0.212	26.576
10^7 10^7	4	575.248	0.406	50.722	399.685	0.283	35.169
10^7 10^7	5	344.043	0.285	34.195	193.322	0.140	17.205
10^7 10^7	6	640.458	0.474	58.948	377.276	0.267	33.566
10^7 10^7	7	449.990	0.357	42.903	357.897	0.265	32.569
10^7 10^6	1	1041.898	0.398	31.603	717.697	0.269	21.361
10^7 10^6	2	817.420	0.351	26.988	496.570	0.200	15.533
10^7 10^6	3	523.466	0.255	19.172	285.986	0.131	9.905
10^7 10^6	4	550.462	0.248	18.785	377.215	0.171	12.859
10^7 10^6	5	337.748	0.178	12.904	184.119	0.085	6.350
10^7 10^6	6	596.168	0.281	21.112	359.884	0.164	12.433
10^7 10^6	7	424.832	0.212	15.496	341.956	0.162	12.066
10^6 10^6	1	36.705	0.057	11.351	24.718	0.038	7.591
10^6 10^6	2	28.311	0.050	9.674	16.753	0.028	5.464
10^6 10^6	3	15.727	0.030	5.737	9.106	0.017	3.273
10^6 10^6	4	17.777	0.034	6.297	12.290	0.024	4.437
10^6 10^6	5	9.536	0.024	4.372	5.874	0.012	2.238
10^6 10^6	6	19.851	0.038	7.239	11.850	0.022	4.106
10^6 10^6	7	13.580	0.030	5.409	10.721	0.022	4.083
10^5 10^5	1	17.989	0.011	3.158	1.707	0.006	1.864
10^5 10^5	2	13.095	0.008	2.336	3.039	0.005	1.327
10^5 10^5	3	8.624	0.004	1.182	4.869	0.003	0.750
10^5 10^5	4	5.593	0.006	1.441	2.439	0.004	0.993
10^5 10^5	5	8.880	0.004	1.001	3.048	0.002	0.508
10^5 10^5	6	8.883	0.006	1.529	5.924	0.003	0.914
10^5 10^5	7	3.412	0.005	1.314	1.405	0.004	0.932

this case we model the waveform through the post^2-Newtonian order, while the amplitude is retained at the lowest Newtonian quadrupolar approximation; the signal depends therefore on eleven independent parameters.

We start by discussing LISA observations of MBHB's for a time of integration corresponding to the final year of in-spiral. Table 1 and 2 show the SNR, $\Delta\Omega^{(N)}$, $\Delta D/D$, Δt_c, $\Delta\mathcal{M}/\mathcal{M}$ and $\Delta\mu/\mu$ for MBH's in the mass range $10^5 - 10^7\,M_\odot$ and a number of locations and orientations of the source in the sky; t_c corresponds to the time of BH coalescence. It is clear that the results depend strongly (they vary by orders of magnitude) on the actual values of the source parameters. We therefore performed a numerical simulation, keeping the source distance fixed and varying randomly $\hat{\mathbf{N}}$ and $\hat{\mathbf{L}}$ for the following mass combinations: $m_1 = m_2 = 10^7\,M_\odot$, $m_1 = 10^7\,M_\odot$ and $m_2 = 10^6\,M_\odot$, $m_1 = m_2 = 10^6\,M_\odot$ and $m_1 = m_2 = 10^5\,M_\odot$. The histograms of the distribution of SNR, $\Delta D/D$, Δt_c and $\Delta\Omega^{(N)}$ are given in Fig. 3 and 4: most of the events would be observable at a SNR $\sim 10^3$, with a distance determined with a relative error $\sim 1\%$, angular resolution $\simeq 10^{-4} - 10^{-3}$ srad and a timing accuracy that ranges from 10 to 10^3 seconds. However, long tails extend to both lower and higher values. Notice also, as was pointed out by Cutler in [14], the strong difference between the use of only one and both independent interferometers. From Table 4 it is also clear the main difference between the results reported here with respect to those given in [14], where the GW phase was computed only through the post$^{1.5}$-Newtonian approximation: $\Delta\Omega^{(N)}$ and $\Delta D/D$ hardly change at all, while $\Delta\mathcal{M}/\mathcal{M}$ and $\Delta\mu/\mu$ gets definitely worse, by roughly one order of magnitude, when the post^2-Newtonian contributions to $\Psi(f)$ are taken into account.

It is interesting to consider the case in which the time of observation is shorter that one year; in particular it could not be available the portion of the signal either at the beginning (the instrument might be "turned on" when some system is already closer to the final merger, or the low frequency cut-off of the detector prevents to access a full year of in-spiral) or the end (one would like to be able to *predict in advance* when and where a MBH merger takes place) of the in-spiral. Such situations are particularly relevant for the issue of the source

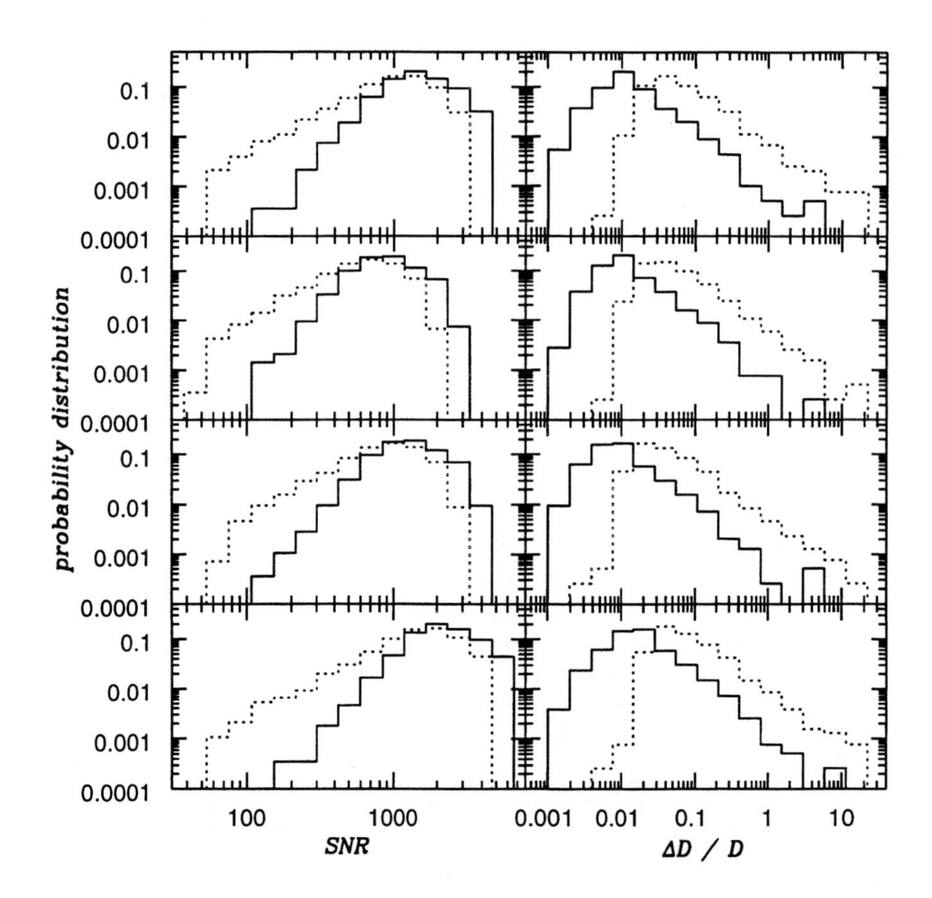

FIGURE 3. Probability distribution of signal-to-noise ratio and distance measurement error for observations of the final year of MBHB in-spiral by LISA. The histograms show the distribution of SNR and $\Delta D/D$ for 2000 random values of $\hat{\mathbf{N}}$ and $\hat{\mathbf{L}}$ (dotted line: one interferometer; solid line: two interferometers). The fiducial sources are all at $z = 1$ ($H_0 = 75\,\mathrm{km\,s^{-1}\,Mpc^{-1}}$, $\Omega_0 = 0$, $\Lambda_0 = 0$) and we take $\beta = 0$ and $\sigma = 0$ as true values of the spin parameters. The panels refer to the following four combinations of masses, from top to bottom, respectively: $m_1 = m_2 = 10^7\,M_\odot$, $m_1 = 10^7\,M_\odot$ and $m_2 = 10^6\,M_\odot$, $m_1 = m_2 = 10^6\,M_\odot$ and $m_1 = m_2 = 10^5\,M_\odot$. The GW model is the same as in Table 1, 2 and 3.

identification and the observation of MBH mergers by other (electro-magnetic) telescopes. Regarding the first case (lack of signal at the beginning of the in-spiral) we checked that the SNR is basically unaffected if the final week of signal is recorded (cfr. Fig. 1) and that the final month is sufficient to preserve (within a factor of two) the distance and location determination. On the contrary, the determination of parameters such as t_c, $\mathcal{M}$ and μ is strongly degraded (by a factor of 10 or more), as most of the wave cycles would be lost. For the second case (lack of signal at the end of the in-spiral), one day before the plunge of the two BH's into the common horizon, $\Delta\Omega^{(N)}$ is degraded by a factor 2-to-10 and even more is Δt_c, while $\Delta D/D$ is weakly affected. So one could know with a precision of better than one hour when a merger would take place and with an angular resolution that in some favorable case could still be of order of one square degree. A much stronger degradation would occur when only one detector is available.

It has been long emphasized that the in-spiral of a low mass BH into a massive one would be of great interest (see [10,23] an references therein). We therefore considered observations of a binary system with the following

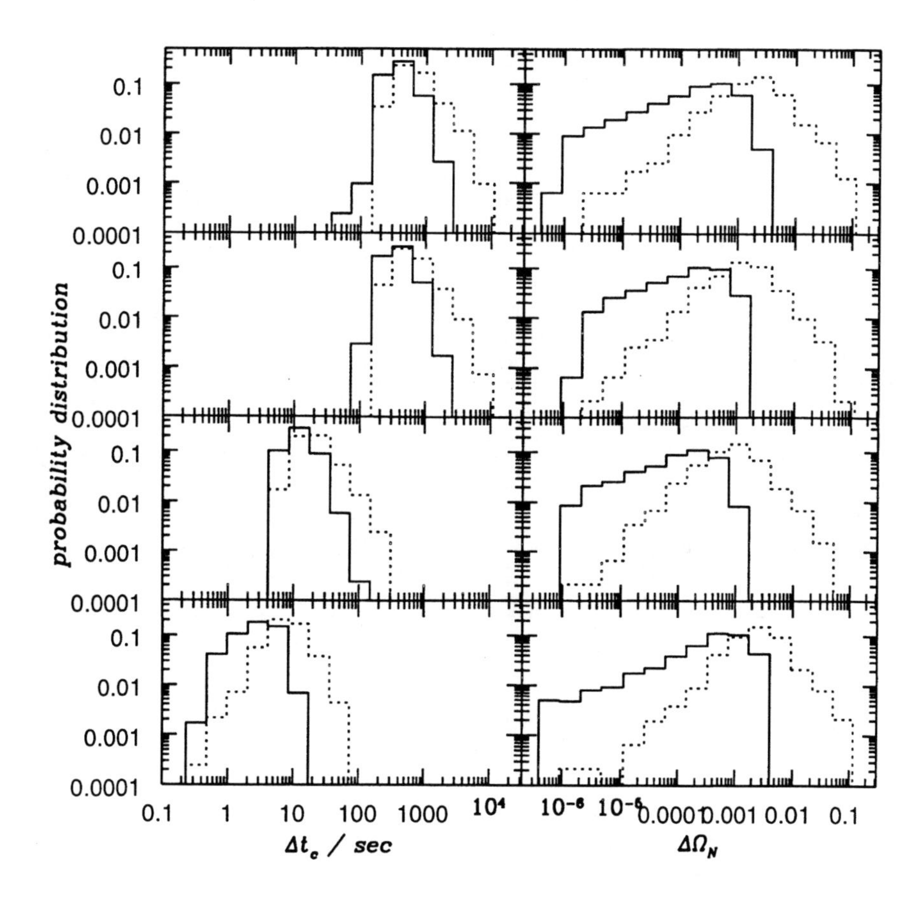

FIGURE 4. Probability distribution of time of coalescence and angular resolution for observations of the final year of MBHB in-spiral by LISA. Same as Fig. 3 but now for Δt_c and $\Delta\Omega^{(N)}$.

extreme mass ratio: $m_1 = 10^6\, M_\odot$ and $m_2 = 10\, M_\odot$. For such kind of source, the restrictions to circular orbit and negligible spins are severe. However, this analysis can be still considered an interesting one, in setting lower-limits to the parameter measurement errors. In Table 3 we show results for a fiducial source at $z = 1$. Here, the angular resolution and the distance determination are much worse than with respect to the MBHB case, due to the lower (~ 10) SNR at which the radiation can be detected. Notice however the extremely good accuracy in the determination of the chirp and reduced mass.

OMEGA [2] is a space interferometric mission rather similar to LISA, although with a larger instrumental noise spectral density (we recall that at frequencies below ~ 1 mHz, LISA and OMEGA share the same $S_n(f)$, as in this region the largest noise contribution comes from unresolved short-period stellar-mass binaries). However a crucial difference is the time-dependent orientation of the interferometer, whose arms basically remain on the plane of the ecliptic. Indeed, OMEGA does not sweep the sky in a LISA-like peculiar way. Such motion is crucial for the reconstruction of the geometry of the source through the characteristic modulation of the signal phase and amplitude. For OMEGA, we performed the same analysis that was carried out for the LISA configuration. The main outcomes are the following: OMEGA would unambiguously detect radiation emitted by MBHB's, with a SNR comparable to the LISA one (cfr. Fig. 1); due to the radically different (and poor) "orientation motion", the instrument has essentially no angular and distance resolution; regarding the other parameters, in particular t_c, $\mathcal{M}$ and μ, the OMEGA measurement errors are worse by a factor typically 5-to-10

TABLE 3. Signal-to-noise ratio and parameter measurement errors for LISA observation of the final year of in-spiral of a $10\,M_\odot$ BH orbiting a $10^6\,M_\odot$ MBH at distance $z = 1$. Signal and instrument parameters as well as units are the same as in Table 1 and 2.

mass	system	SNR_I	$(\Delta D/D)_I$	$(\Delta\Omega)_I$	SNR_{I+II}	$(\Delta D/D)_{I+II}$	$(\Delta\Omega)_{I+II}$
10^6 10	1	17.432	55.080	1112.348	28.677	34.229	195.338
10^6 10	2	22.204	364.459	2178.673	32.917	225.149	390.212
10^6 10	3	52.071	10.726	4992.853	77.747	6.574	840.650
10^6 10	4	38.172	136.496	3149.102	47.132	88.755	505.139
10^6 10	5	64.371	164.476	7721.325	97.202	90.142	442.833
10^6 10	6	39.240	16.909	8009.717	64.800	11.882	1720.845
10^6 10	7	41.297	1354.567	4731.088	52.938	835.288	1021.153
		$(\Delta t_c)_I$	$(\Delta\mathcal{M}/\mathcal{M})_I$	$(\Delta\mu/\mu)_I$	$(\Delta t_c)_{I+II}$	$(\Delta\mathcal{M}/\mathcal{M})_{I+II}$	$(\Delta\mu/\mu)_{I+II}$
10^6 10	1	1078.440	0.057	2.919	195.338	0.029	1.560
10^6 10	2	1227.801	0.050	3.382	390.212	0.024	1.801
10^6 10	3	378.978	0.031	1.011	840.650	0.015	0.655
10^6 10	4	675.429	0.028	1.799	505.139	0.017	1.099
10^6 10	5	232.000	0.037	0.619	442.833	0.013	0.393
10^6 10	6	569.537	0.032	1.515	1720.845	0.018	1.009
10^6 10	7	1310.555	0.031	3.317	1021.153	0.017	1.741

TABLE 4. Angular resolution, distance and mass measurement errors for LISA observations of the final year of a MBHB in-spiral using different gravitational wave models. The fiducial source, with $m_1 = 10^7\,M_\odot$ and $m_2 = 10^6\,M_\odot$, is at $z = 1$ ($H_0 = 75\,\mathrm{km\,s^{-1}\,Mpc^{-1}}$, $\Omega_0 = 0$, $\Lambda_0 = 0$). The waveform is modelled in the three following ways: (A) the GW phase $\Psi(f)$ is computed through the post^2-Newtonian approximation, and the spin-orbit and spin-spin interactions are taken into account only in $\Psi(f)$ (no effects on the polarization amplitude and phase); σ is set to 0; the number of independent parameters is 11; (B) same as case (A) but with $\Psi(f)$ computed through the post$^{1.5}$-Newtonian approximation; the number of independent parameters is 10; (C) the GW phase $\Psi(f)$ is modelled through the post$^{1.5}$-Newtonian approximation, but we do include the change of orientation of $\hat{\mathbf{L}}$ and $\hat{\mathbf{S}}$, induced by the spin-orbit coupling, that intrinsically modulates the phase and amplitude of the wave; $\tilde{h}^{(\iota)}(f)$ depends on 12 independent parameters. $\hat{\mathbf{N}}$, $\hat{\mathbf{J}}$ and $\hat{\mathbf{S}}$ are the same as in Fig. 2. The labels I and $I+II$ refer to observations with one and two independent interferometers, respectively. $\Delta\Omega^{(N)}$ is given in units 10^{-5} srad, whereas $\Delta D/D$, $\Delta\mathcal{M}/\mathcal{M}$ and $\Delta\mu/\mu$ are in units 10^{-2}

parameter	GW model	$\beta = 3$	$\beta = 3$	$\beta = 5$	$\beta = 5$	$\beta = 7$	$\beta = 7$
		I	I+II	I	I+II	I	I+II
	A	159.15	1.50	152.56	1.50	146.40	1.50
$\Delta\Omega^{(N)}$	B	150.91	1.48	144.94	1.48	139.39	1.49
	C	116.25	1.51	83.76	1.55	76.44	1.52
	A	2.64	1.31	2.59	1.28	2.54	1.25
$\Delta D/D$	B	2.64	1.32	2.60	1.29	2.56	1.26
	C	1.88	0.84	1.70	0.71	1.58	0.62
	A	0.40	0.27	0.40	0.27	0.40	0.27
$\Delta\mathcal{M}/\mathcal{M}$	B	0.07	0.04	0.07	0.04	0.07	0.04
	C	0.007	0.005	0.007	0.005	0.006	0.004
	A	31.86	21.56	31.99	21.65	32.11	21.71
$\Delta\mu/\mu$	B	2.78	1.79	2.79	1.80	2.79	1.80
	C	0.25	0.18	0.21	0.15	0.20	0.14

with respect to the corresponding LISA values, due to the larger noise contribution.

In the discussion that we have carried out so far, we have assumed negligible BH spins in order to simplify the model of the waveform. However, several observations suggest that MBH's do carry a substantial intrinsic angular momentum. Indeed, a more realistic model of the GW signal should contain such information; according to it, we consider binary systems that undergo simple precession [15], and we investigate how the results presented so far are affected. $\tilde{h}^{(\iota)}(f)$ depends now on 12 independent parameters, and naively one could expect a degradation of the measurement accuracy as a larger number of parameters has to be extracted from the same data stream. However, when the precession of the orbital plane is included, the parameters leave even more peculiar "finger-prints" on the recorded signal. A feeling for the richness of the structure of $\tilde{h}^{(\iota)}(f)$ can be simply derived by looking at the behavior of the polarization phase A_p shown in Fig. 2. In order to explore this issue we have computed the parameter errors using different models for the GW signal. We report some results in Table 4; as the parameter space is very large, a more complete investigation is required and is presently in progress [17]. The most striking features of the results, particularly if compared to the no-precession case, are the following: although the number of parameters is greater, $\Delta\Omega_{I+II}^{(N)}$ and $(\Delta D/D)_{I+II}$ are basically unaffected or slightly decrease, depending on the actual value of β; however, $\Delta\Omega_I^{(N)}$ and $(\Delta D/D)_I$ show a larger improvement, by an amount between 30% and a factor ~ 2, due to the fact that the geometry of the source can be reconstructed through the intrinsic GW modulation caused by the precession of the source orbital plane. The errors of the mass determination are also strongly affected when precession is included into the model, decreasing by a factor ~ 100; in fact $\mathcal{M}$ and μ regulate not only the rate of binary in-spiral, but also the rate of precession of $\mathbf{S}$ and $\mathbf{L}$, leaving therefore a trace in A_p and φ_p.

We conclude that according to our preliminary results, the BH spins do play a crucial role for the parameter estimation with space-based observations and are likely to provide, in a wide range of astrophysical situations, extra information that could improve the measurements; however, in order to address with more confidence the performances of LISA and OMEGA , a thorough exploration of the parameter space and more accurate GW models should be implemented (this work in currently in progress [17]). It is fair to say that any reference value characterizing the estimation of the source parameters should be really considered *cum grano salis*, as it characterizes the behavior of an instrument only with respect to the particular source the observations refer to.

REFERENCES

1. Bender, P. et al., *LISA Pre-Phase A Report; Second Edition*, MPQ 233 (1998).
2. Hellings, R. W., *Ω Orbiting Medium Explorer for Gravitational Astrophysics*, Mixed Proposal (1995).
3. Begelman, M.C., Blandford, R.D., and Rees, M.J., Nature **287**, 307 (1980).
4. Haehnelt, M.G., Mont. Not. Roy. Astron. Soc. **269**, 199 (1994).
5. Rees, M. J., in *Black holes and relativity*, edited by Wald, R. (University of Chicago Press, Chicago, 1997).
6. Vecchio, A. , Class. and Quantum Grav. **14**, 1431 (1997).
7. Sigurdsson, S., and Rees, M. J., Mont. Not. Royal Astron. Soc. **284**, 318 (1996)
8. Sigurdsson, S. , Class. and Quantum Grav. **14**, 1425 (1997).
9. See also the following contributions to the Proceedings of the *Second LISA Symposium*: Blandford, R.D., *Binary black holes in galactic nuclei*; Richstone, D., *Very massive black holes, then and now*; Haehnelt, M.G., *Supermassive black holes as sources for LISA*; Sigurdsson, S., *High mass ratio sources of low frequency gravitational radiation.*
10. Thorne, K.S., in *Black holes and relativity*, edited by Wald, R. (University of Chicago Press, Chicago, 1997).
11. Schutz, B.F., Nature **323**, 310 (1986).
12. Peterseim, M, Jennrich, O., and Danzmann, K., Class. and Quantum Grav. **13**, 279 (1997).
13. Cutler, C., *Observing binary systems with LISA and OMEGA: Angular resolution and astrophysical parameter measurements. I. Periodic Sources*, to be published in the Proceeding of the *Second LISA Symposium.*
14. Cutler, C., Phys. Rev. D **57**, 7089 (1998).
15. Apostolatos, T. A., Cutler, C., Sussman, G. S., and Thorne, K. S., Phys. Rev. D **49**, 6274 (1994).
16. Cutler, C., and Flanagan, E.E., Phys. Rev. D **49**, 2658 (1994).
17. Cutler, C. and Vecchio, A., in preparation.
18. Blanchet, L., Damour, T., Iyer, B.R., Will, C.M. and Wiseman, A.G., Phys. Rev. Lett. **74**, 3515 (1995).
19. Hellings, R.W., private communication.
20. Balasubramanian, R., Sathyaprakash, B.S., Dhurandhar, S.V., Phys. Rev. D **53**, 3033 (1996).
21. Nicholson, D., and Vecchio, A., Phys. Rev. D **57**, 4588, (1998).
22. Bender,P. L., and Hils, D., Class. and Quantum Grav. **14**, 1439 (1997).
23. Poisson, E., Phys. Rev. D **54**, 5939 (1996). See also Poisson, E., *Measuring black hole parameters using gravitational wave data from space-based interferometers*, to be published in the Proceeding of the *Second LISA Symposium.*

Filtering Gravitational Waves from Supermassive Black Hole Binaries

B.S. Sathyaprakash
Cardiff University, 5, The Parade, Cardiff, CF2 3YB, U.K.
B.Sathyaprakash@astro.cf.ac.uk

Abstract. The effectualness of post-Newtonian wave forms and recently proposed P-approximants in extracting gravitational waves from inspiralling binaries of supermassive black holes is explored. Based on the extent of the overlaps of these approximate wave forms with the exact ones, we conclude that P-approximants are essential to reliably measure the parameters of supermassive binaries. The wave form from such a source depends on several parameters: The masses of the component bodies, their spins, eccentricity of the orbit, angles describing the orientation of the binary relative to the LISA detector, direction to the source, etc. Making a choice of search templates in such a multi-dimensional parameter space could be pretty complicated. We suggest a method based on the *principal component analysis* which could potentially simplify the search to a great extent by enabling us to choose templates in a parameter space of much smaller dimension than the physical parameter space.

I INTRODUCTION

Inspiralling and merging binaries, consisting of component objects both of which are supermassive black holes (SMBH) or one that is a SMBH and the other a stellar mass compact object (a neutron star or a black hole), are candidate sources [1] for the Laser Interferometer Sapce Antenna (LISA) [2]. LISA will be able to detect gravitational waves emitted during the spiralling in of these objects out to cosmological distances. There is mounting evidence that the nucleus of every galaxy (whether active or not), including the Milky Way, contains a massive or a supermassive (10^6-$10^9 M_\odot$) black hole [3]. If this is true then galaxy collisions observed in high red-shift surveys are accompanied by the inspiral and merger of the black holes at their centres. Order of magnitude estimates suggest that LISA should see SMBH-SMBH binary inspiral at roughly once a year [4]. The capture and inspiral of compact bodies by a SMBH might be much more frequent, making observation of such events possibly quite routine [5].

These sources have a lot of very interesting physics and offer an opportunity to test some of the predictions of general relativity in the strongly non-linear regime, such as the tails of gravitational waves, spin-orbit coupling induced precession, non-linear tails, hereditary effects, etc. [6–8]. They are also good test beds to constrain other theories of gravity. Gravitational waves emitted either during the inspiral and merger of rotating supermassive black holes or when a stellar mass compact object falls into a SMBH, can be used to map the structure of spacetime and test uniqueness theorems on rotating black holes [6]. LISA will be able to see the formation of massive black holes at cosmological distances by detecting the waves emitted in the process [6]. These are but a very modest list of physics that will be borne out of observation of gravitational waves from these sources. (See the article by Thorne in this volume for a comprehensive treatment of the subject).

Doing all this exciting physics is possible if only we can measure the source parameters accurately and without any systematic bias. Gravitational wave measurements will be made by matched filtering the detector output in a multi-dimensional parameter space of signals, using search templates that are essentially copies of the expected signals. However, presently we only know the inspiral wave form approximately (computed using post-Newtonian expansion of Einstein's equations) [9–14], and this situation is not likely to change significantly when LISA is likely to fly. That being the case, there is a need to assess the reliability of the approximate wave forms in extracting and measuring the true wave forms.

CP456, *Laser Interferometer Space Antenna*
edited by William M. Folkner
 1-56396-848-7/98/$15.00

To aid us in this assessment we shall use as true signals those constructed from the exact solution to the approximate two-body equations obtained in the limiting case where one of the bodies is assumed to be a test particle of negligibly small mass and the other a Schwarzschild black hole [16,17]. The test mass is assumed not to alter the structure of the spacetime. There have also been computations of the post-Newtonian wave forms for such a source [17–19]. These, and their improved versions called P-approximants [20], will serve as test signals whose reliability we shall explore. The results obtained here are indicative of what may happen in the more realistic and interesting case of the inspiral and merger of two black holes of comparable masses.

Having chosen an approximate wave form that will serve as a search template the problem is to set up a bank of filters corresponding to a discrete set of values of the a priori unknown signal parameters. Templates must be so chosen that every inspiral wave form of strength larger than a certain minimal strength is detected by passing the detector output through the bank of templates. Doing this by a brute force approach might prove very complex and not tractable as the number of signal parameters for binaries consisting of a SMBH and a stellar mass compact object, or another SMBH, is quite large: The masses, spins, direction to the source, Keplerian and higher order eccentricity parameters, orientation of the binary orbit with respect to the line of sight, etc., to name a few. It is, therefore, necessary to devise algorithms that can simplify the search yet covering all possible signal shapes. The *principal component analysis,* well-known in metrology and other areas where the number of parameters in the problem is quite large, is an excellent tool to track the true degrees of freedom in the possible shapes of a complicated wave form. We shall demonstrate the effectualness of this method by applying it to inspiral signals from binaries consisting of spinless, point mass, stars in circularised orbits. Its application to a generic inspiral signal expected in LISA will be addressed in a future work.

The rest of this paper is organised as follows. In Sec. II we shall briefly review the Post-Newtonian and P-approximant models and discuss their reliability in extracting and measuring SMBH-SMBH inspiral signals. In Sec. III we shall discuss the parameter space of search for SMBH binaries and introduce the method of principal components. We shall illustrate how the method works by applying it to the case of gravitational waves emitted by a system of two point masses in orbit around each other. In Sec. IV we discuss the scope for further research in this area. In this article we use a system of units in which $G = c = 1$.

II MODELLING INSPIRAL SIGNALS FROM A COMPACT BINARY

Understanding the evolution of a pair of spinning, inspiralling and merging black holes is an outstanding problem in general relativity. A solution to this problem has to basically address two related issues: The motion of the two bodies and the associated gravitational wave flux. Unfortunately, the problem is too hard to solve exactly and therefore theorists have resorted to numerical and perturbative methods to tackle these issues. These methods are far from providing an accurate solution to the interesting case of two rotating black holes in orbit around each other but they have yielded a solution to the case where one of the bodies is assumed to be a non-spinning test particle in a quasi-circular orbit around a Schwarzschild hole. In this case we have an exact analytical expression for the two-body energy $E(v)$ and an exact numerical solution [16,17] to the gravitational wave flux $F(v)$, as well as a post-Newtonian expansion of the flux to order v^{11} beyond the standard quadrupole formula [17–19]. Here v is an invariantly defined "velocity" related to the instantaneous *gravitational wave* frequency f (= twice the *orbital* frequency) by $v = (\pi m f)^{\frac{1}{3}}$, where $m \equiv m_1 + m_2$ is the total mass of the binary.

There have been attempts to perturbatively solve the two-body problem using a post-Newtonian (Taylor) expansion of the relevant physical quantities. This programme has so far yielded analytical expressions for the gravitational wave flux and the two-body energy to order v^5 beyond the well-known quadrupole approximation [10–14] and theorists are close to extending these results by another v^2 order [15].

In the following three Sections we shall summarise the post-Newtonian wave forms and their improvements called P-approximants, followed by a discussion of their effectualness.

A Post-Newtonian Expansion

Cutler et al. [21] pointed out that for the purpose of signal detection only post-Newtonian corrections in the phase of the inspiral wave form are important and that the amplitude corrections can be neglected. We shall, therefore, work with the restricted post-Newtonian approximation in which the signal is given by $h(t) = a(t)\cos\phi(t)$, where the gravitational wave phase $\phi(t)$ is essentially twice the orbital phase $\Phi(t)$. Starting from the energy and flux functions, $E(v)$ and $F(v)$ respectively, one can compute the "phasing formula" giving the evolution of the gravitational wave phase $\phi(t;\boldsymbol{\alpha}) = 2\Phi(t;\boldsymbol{\alpha})$. This formula involves a set of parameters $\boldsymbol{\alpha} \equiv \{\alpha_i\}$ of the binary, such as the masses of the component stars, their spins, direction to the source, etc., carrying information about the emitting binary. Binaries with different sets of values of the parameters will have distinctly different phase developments that can be discriminated by cross correlating the two wave forms, as we shall do below. The standard energy-balance equation $dE_{\rm tot}/dt = -F$, where $E_{\rm tot}$ is the total relativistic energy of the system related to the dimensionless energy function $E(v)$ via $E(v) = m(1 + E_{\rm tot}(v))$, gives the following parametric representation of the phasing formula:

$$t(v) = t_c + m \int_v^{v_{\rm lso}} dv \frac{E'(v)}{F(v)}, \quad \Phi(v) = \Phi_c + \int_v^{v_{\rm lso}} dv v^3 \frac{E'(v)}{F(v)}. \tag{1}$$

where, for the sake of clarity, we have not shown the dependence on signal parameters $\boldsymbol{\alpha}$. In the above equation t_c and Φ_c are integration constants, $E'(v) \equiv dE/dv$ and $v_{\rm lso}$ is a reference velocity which can be conveniently set equal to the velocity of the system at the last stable circular orbit defined by $E'(v_{\rm lso}) = 0$.

The standard post-Newtonian expansions of the energy and flux to order v^n, $E_{T_n} \equiv \sum_{k=0}^{n} E_k(\eta) v^k$ and $F_{T_n} \equiv \sum_{k=0}^{n} F_k(\eta) v^k$, where $\eta \equiv m_1 m_2/m^2$ is the symmetric mass ratio, can be used to construct a sequence of approximate wave forms $h_n^T(t;\boldsymbol{\alpha})$. In formal terms, any such construction defines a *map* from the set of Taylor coefficients (E_k, F_k) into the (functional) space of wave forms. Up to now, the literature has only considered the standard map, say T,

$$(E_{T_n}, F_{T_n}) \xrightarrow{T} h_n^T(t;\boldsymbol{\alpha}), \tag{2}$$

obtained by inserting the successive Taylor approximants into the phasing formula [21,16].

B P-Approximants

Damour et al. [20] proposed a new map, called "P", based on two essential ingredients: (i) The introduction of two new, supposedly more basic, energy-type and flux-type functions, say $e(v;\eta)$ and $f(v;\eta)$, and (ii) the systematic use of Padé approximants [22] (instead of straightforward Taylor expansions) when constructing successive approximants of the intermediate functions $e(v;\eta)$, $f(v;\eta)$. Here we have explicitly shown the dependence on the parameter η since in the following we would like to distinguish between binaries consisting of a test mass (i.e., $\eta \to 0$) and a Schwarzschild black hole from those containing two Schwarzschild black holes of comparable masses (i.e., $\eta \neq 0$). The motivation for using Padé approximants comes from the fact that the physical quantities exhibit pole singularities on the real axis which cannot be captured by a Taylor series while the corresponding Padé approximation, which is a rational polynomial, is potentially capable of capturing the pole. Schematically, the procedure is [20]:

$$(E_{T_n}, F_{T_n}) \to (e_{T_n}, f_{T_n}) \to (e_{P_n}, f_{P_n}) \to (E[e_{P_n}], F[e_{P_n}, f_{P_n}]) \to h_n^P(t;\boldsymbol{\alpha}), \tag{3}$$

where e_P and f_P are the Padé approximants of the truncated Taylor series e_T and f_T, respectively. The new energy function $e(x)$ is symmetric in the two masses and in the limit $\eta \to 0$ it is given by

$$e(x;\eta=0) = -x\frac{1-4x}{1-3x} \tag{4}$$

which has a simple pole singularity at $x \equiv v^2 = 1/3$. The function $E(x)$ entering the phasing formulas is given in terms of $e(x)$ by

$$E(x) = \left[1 + 2\eta\left(\sqrt{1+e(x)} - 1\right)\right]^{1/2} - 1. \tag{5}$$

On the grounds of mathematical continuity between the cases $\eta \to 0$ and $\eta \neq 0$ one can expect the exact function $e(x, \eta \neq 0)$ to admit a simple pole singularity on the real axis $\propto (x - x_{\rm pole})^{-1}$. We do not know the location of this pole. However, *Padé approximants* are excellent tools for obtaining accurate representations of functions having such singularities [22]. Indeed, it turns out that the Padé approximant of the second post-Newtonian expansion of $e(x)$ gives the exact energy function [20]. The use of $E(x)$, constructed from a Padé approximant to $e(x)$, can therefore be expected to greatly improve the accuracy of the phasing formula.

It has been pointed out [18] that the flux function $F(v;\eta=0)$ too has a simple pole at the light ring $v^2 = 1/3$. This motivates the introduction of the following "factored" flux function $f(v;\eta)$, Padé approximants f_{P_n}, and the corresponding flux function F_{P_n} entering the phasing formula:

$$f(v;\eta) \equiv (1 - v/v_{\rm pole})\, F(v;\eta), \quad F_{P_n}(v;\eta) \equiv (1 - v/v_{\rm pole})^{-1}\, f_{P_n}(v;\eta). \tag{6}$$

In the next Section we shall see that the mapping defined by (3) greatly improves the inspiral signal model over that defined by (2).

C Model Wave Forms Vis-A-Vis Exact Wave Forms

Several people have assessed the accuracy of the post-Newtonian wave forms using different yard sticks, such as the number of cycles that arise at different post-Newtonian orders [21,10], convergence of the post-Newtonian expansion [23], the overlap integrals of approximate wave forms with the exact wave forms where the latter is known [24,16,20], Cauchy convergence of the model wave forms [20], etc. What is relevant to detection and estimation is really the overlap integral of two wave forms weighted down by the detector spectral noise density, which we shall discuss next.

The matched filtering technique employed in detecting and measuring compact binary inspiral wave forms leads naturally to the definition of the scalar product. Given wave forms $a(t)$ and $b(t)$ their scalar product is defined as:

$$\langle a, b\rangle \equiv 2\int_0^\infty \frac{df}{S_h(f)}\left[\tilde{a}(f)\tilde{b}^*(f) + \tilde{a}^*(f)\tilde{b}(f)\right], \tag{7}$$

where $S_h(f)$ is the one-sided detector noise spectral density, $\tilde{a}(f) \equiv \int_{-\infty}^{\infty} dt\; a(t)\exp(2\pi i f t)$ is the Fourier transform of the signal and a * denotes complex conjugation. In our study, we shall deal mainly with wave forms of unit norm; i.e. those that obey $\langle a, a\rangle = 1$. The scalar product of wave forms whose shapes are not the same, or whose parameters are mismatched, or in general both, is less than 1. This motivates the definition of the ambiguity function $\mathcal{A}$ [25]:

$$\mathcal{A} \equiv \langle a(\boldsymbol{\alpha}_1), b(\boldsymbol{\alpha}_2)\rangle, \tag{8}$$

where a and b are both of unit norm. Though the parameter vectors $\boldsymbol{\alpha}_1$ and $\boldsymbol{\alpha}_2$ need not, in general, be of the same dimension, we shall assume that to be the case but shall allow the wave form shapes to be different. For instance, a could be a model wave form, a h_T or a h_P, and b could be the exact wave form h_X. In that case $\mathcal{A}$, maximised over wave forms' kinematic parameters, such as constant time- and phase-offsets, gives a measure of how good is our signal model h_T or h_P in measuring the exact wave form h_X, and we shall refer to this number as the *overlap* of the two wave forms. (A measure of a signal model's effectiveness in detection can be obtained by maximising the overlap over the template's dynamical parameters such as the masses, spins, etc.)

TABLE 1. Overlap of model wave forms h_T (second row) and h_P (third row) with the exact wave form h_X, computed in the limit $\eta \to 0$, at post-Newtonian orders v^n (first row). Standard Taylor expansion yields overlaps less than 85% even at a very high (v^{11}) post-Newtonian order while P-approximants yield overlaps in excess of 95% at post-Newtonian orders v^4 and higher. There is no entry for h_P at v^{10} order since (diagonal) P-approximants, it turns out, have a singularity in the region of interest (i.e. for $v < v_{\rm lso}$) at that order.

n	4	5	6	7	8	9	10	11
$\langle h_T, h_X\rangle$	0.67	0.54	0.82	0.79	0.78	0.76	0.80	0.77
$\langle h_P, h_X\rangle$	0.96	0.97	0.99	0.99	0.99	0.99	—	1.00

TABLE 2. Signal-to-noise ratios for inspiral waves from binaries consisting of pairs of supermassive black holes, or a supermassive black hole and a stellar mass compact object, at a distance of 3 Gpc. The observation is assumed to last for a year starting at an initial frequency f_i and ending at or before the frequency $f_{\rm lso} = 1/(\pi 6^{3/2} m)$ corresponding to the last stable circular orbit. Also listed are the rough estimates of the kind of models necessary to detect signals and estimate their parameters. See the discussion before Eq. (9) and following Eq. (12) for a practical definition of *chirping* binaries.

Nature of the signal	Component masses $m_1, m_2 (M_\odot)$	f_i (Hz)	Typical SNR	Model wave form for: Detection	Estimation
1. Non-chirping binaries	$1.4 \le m_1 \le 5$ $10^5 \le m_2 \le 10^8$	$< 10^{-4}$	10–100	None	Ambiguous
2. Moderately chirping binaries	$1.4 \le m_1 \le 10$ $10^5 \le m_2 \le 10^8$	$> 10^{-4}$	50–200	P-wave forms needed	P-wave forms essential
3. Strongly chirping binaries	$10^3 \le m_1 \le 10^8$ $10^3 \le m_2 \le 10^8$	Any	10^3–10^4	Approximate models sufficient	P-wave forms or better essential

In Table 1 we have listed the overlaps of the standard post-Newtonian signals and their P-approximants with the exact wave form, computed in the limit $\eta \to 0$. We have some what inconsistently used this limiting case to construct a wave form for a system consisting of two black holes of comparable masses. We believe, however, that this is good enough to gauge the effectualness of the different signal models. From Table 1 we observe that the agreement between standard post-Newtonian models and the exact wave forms is rather poor (overlaps ≤ 0.85) even at an order as high as v^{11}. On the other hand P-approximants constructed out of these models are in excellent agreement (overlaps ≥ 0.95) with the exact wave forms, at all orders beyond v^4. This is possibly true even for P-approximants constructed out of standard post-Newtonian series for a binary consisting of two objects of comparable masses.

The question then is: When are the P-approximants essential? LISA will see several types of sources for some of which the signal-to-noise ratio is so high that the signal might be visible in the time-series itself without the aid of sophisticated data processing. There are cases, that we shall discuss below, for which the change in signal's intrinsic frequency due to inspiral, during the entire period of observation, is unimportant; there are others whose frequency change is significant but the signal amplitude is very large. Clearly, detecting such signals will not require good signal models. However, an accurate determination of the signal parameters, in the latter case, will only be possible by fitting the detector output with an accurate model of the wave form. It is for this second stage of measurement that P-approximants are essential in most cases. However, there are also a class of sources for which P-approximants are essential even for detection.

Based on the results quoted in Table 1, in Table 2 we list various systems whose inspiral wave LISA is expected to observe, together with the estimated signal-to-noise ratios and the model wave form needed to detect the signal and accurately measure its parameters. Of course, the actual signal that we will observe is characterised by many other parameters and this Table needs to be updated after a study of the effect of those parameters on signal detection and estimation.

III A SEARCH ALGORITHM TO DETECT SMBH BINARIES

A Parameter Space for SMBH Binaries

The inspiral waves observed by ground-based detectors are quite easy to parametrise. In most cases, it is only the masses of the component stars, that is important. The eccentricity of the orbit would decay well before a binary enters the detector band of a ground-based interferometer; the spins are unimportant [10] except in the case of maximally spinning black holes and that too if the spin angular momentum is largely misaligned with the orbital angular momentum [26]. Since the waves are visible in the detector for less than about 15 mins, the direction to the source, or its orientation in the plane of the sky, is unimportant in tracking the emitted wave form except when the spins are so large as to cause a modulation via a change in the polarisation pattern that the detector observes [27]. Moreover, the systems we observe will inspiral and coalesce during the observation period so that we don't have to keep track of which part of the inspiral wave we observe.

FIGURE 1. The parameter space (m, η, f_i) for inspiral signal searches from supermassive black hole binaries. Masses of component stars is taken to be in the range $1M_\odot \leq m_1, m_2 \leq 10^6 M_\odot$. (Note that not all sources in the shaded area are observable by LISA but only a fraction depending on their amplitude and frequency.) For a given initial frequency f_i of the source search templates are needed for those sources above the curve corresponding to that frequency. The actual parameter space is much larger involving spins of the bodies, direction to and orientation of the source, eccentricity of the orbit, etc.

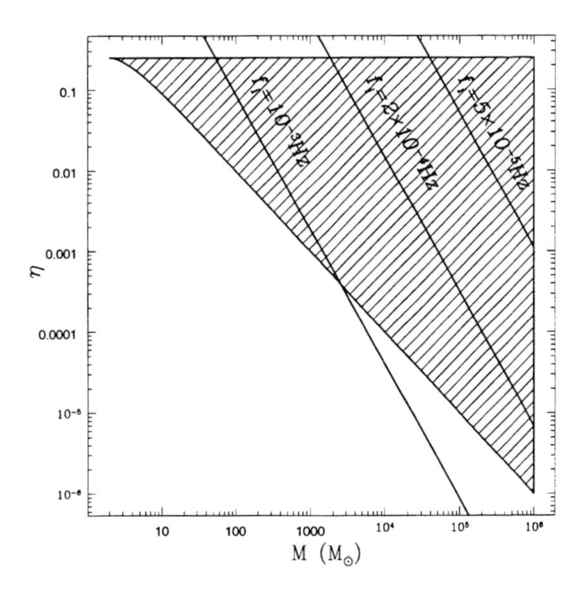

However, all these effects will be pretty much important for supermassive black hole binaries: One cannot neglect eccentricity of the orbits and spins of the component stars since there won't be enough time for the system to lose all its eccentricity nor for the spins to align with the orbit. The Doppler modulation caused by LISA's orbit around Sun will mean the direction to the source and its orientation in the plane of the sky, will have to be taken into account in our search algorithms. In addition to the masses of the component stars, one has to also bear in mind that at the start of the search the system could have any frequency whatsoever in the LISA band, and yet not sweep the whole band even after a year's observation.

To get an idea of the complexity of the search problem, let us take a look at the parameter space of masses and the frequency of the system when the observation begins, ignoring spins, angles describing direction to source and its orientation, etc. More precisely, we shall use the parameter set (m, η, f_i), where m is the total mass of the system, $\eta = m_1 m_2/m^2$ is the symmetric mass ratio and f_i is the initial frequency. Let us suppose we integrate for a year and want to search for systems whose primary is a massive black hole and the companion is either another massive black hole or a stellar mass compact object. The parameter space of masses for such a system is shown in Fig. 1. Search templates will not be needed over the entire parameter space since there are systems whose frequency will not change significantly during a year's observation. Therefore, our search algorithms will have to distinguish between sources whose frequency does *not* change more than the frequency resolution defined by the duration of observation—we shall call these *non-chirping* binaries—from those whose frequency does change more than the resolution—we shall call these *chirping* binaries.

Beware that some physicists refer to that phase of the binary wave form, in which the adiabatic approximation $\dot{f}/f^2 < 1$ fails, as the *chirp wave form.* The adiabatic condition states that the change in frequency during one cycle is smaller than the instantaneous frequency. In the case of binaries relevant to our discussion adiabatic approximation is almost always valid; nevertheless the frequency change is large enough to worry about. We use the word *chirp* in a more practical sense.

In the quadrupole approximation, the frequency of the orbit changes according to

$$f(t) = f_i \left(1 - \frac{t}{\tau_0}\right)^{-3/8}, \tag{9}$$

where f_i is the frequency of the system at the beginning of the observation (i.e., at $t = 0$) and τ_0, called the *chirp time*, is a function of the masses of the two bodies given by

$$\tau_0 = \frac{5}{256}\eta^{-1}m^{-5/3}(\pi f_i)^{-8/3}. \tag{10}$$

The chirp time is the time left till the two bodies would coalesce, starting from a time when the frequency was f_i. Note, however, that post-Newtonian calculations become increasingly inaccurate close to the last stable orbit which occurs at a frequency $f_{\rm lso} = 1/(6^{3/2}\pi m)$ and therefore our templates will have to stop well before the corresponding time $t_{\rm lso}$, which can be inferred by inverting Eq. (9) or its post-Newtonian generalisation. It is true that this distinction between τ_0 and $t_{\rm lso}$ is insignificant for binaries observed by ground-based detectors in the sense that $\tau_0 - t_{\rm lso}$ is much less than the duration of observation. However, for sources that LISA will see, this difference could be several years and should therefore be taken into account in computing the time available for observation starting from a certain initial frequency.

After a time T the change in frequency of the system is

$$\Delta_c f = f_i\left[\left(1-\frac{T}{\tau_0}\right)^{-3/8} - 1\right]. \tag{11}$$

If $\tau_0 \gg T$ then we can approximate

$$\Delta_c f \simeq \frac{3f_iT}{8\tau_0}. \tag{12}$$

Equating this to the frequency resolution $1/T$ achieved after an observation time T, we can arrive at a relation between m, η, f_i and T. Setting T to be, say one year, and substituting for τ_0 from Eq. (10) we get

$$\eta = 8.7\times 10^{-5}\left(\frac{m}{10^6\,M_\odot}\right)^{-5/3}\left(\frac{f_i}{10^{-4}\ \mathrm{Hz}}\right)^{-11/3}\left(\frac{T}{1\ \mathrm{Year}}\right)^{-2}. \tag{13}$$

For a given total mass and initial frequency, the maximum value of the symmetric mass ratio before the signal starts chirping is given by the above equation. We have plotted in Fig. 1 curves in the m–η plane corresponding to different starting frequencies f_i which separate sources for which chirping is important from those for which it is not. For a given initial frequency f_i, sources above the corresponding curve chirp, some moderately and others strongly, and search templates in the m–η space are needed in that region, but not below the curve, since those sources won't chirp during our observation of them. Thus, the parameter space that warrants templates is pretty complicated.

When we count all the relevant parameters, a brute force search problem looks overwhelmingly complex. At the same time computational costs are unlikely to be formidable, at least by the standards that we can expect in a decade from now, especially because the sampling rate will be quite low (less than a sample every ten seconds). Therefore, our aim in designing search algorithms should be simplicity, perhaps even at the expense of increasing computational costs, rather than complexity in favour of minimising computational costs. This is unlike the present trend in designing search algorithms for ground-based detectors.

The method of principal components has the flexibility to offer not only simplicity in our search algorithms but due to the way it works it automatically minimises the computational costs and lends itself naturally to hierarchical search methods. We shall discuss this method next.

B Covariance Matrix

The ambiguity function introduced in Sec. II C can be used to compute the variances in the measurement of the signal's parameters. This is done by expanding the ambiguity function of two wave forms, of identical shapes but of slightly different parameters, around its maximum and retaining only terms quadratic in the differences of the parameters. Then in the limit of high signal-to-noise ratio (i.e., $\rho^2 \equiv \langle h, h\rangle \gg 1$) the information matrix $\Gamma^{\boldsymbol{\alpha}}_{ij}$ for the parameter set $\boldsymbol{\alpha}$ is given by [25]

$$\Gamma^{\boldsymbol{\alpha}}_{ij} = \left\langle \frac{\partial h}{\partial \alpha_i}, \frac{\partial h}{\partial \alpha_j}\right\rangle. \tag{14}$$

The covariance matrix $C^{\boldsymbol{\alpha}}$ is the inverse of the Fisher information matrix [25]:

$$\mathcal{C}^{\boldsymbol{\alpha}} \equiv \overline{(\boldsymbol{\alpha}^{\mathrm{t}} - \boldsymbol{\alpha}^{\mathrm{m}})(\boldsymbol{\alpha}^{\mathrm{t}} - \boldsymbol{\alpha}^{\mathrm{m}})^T} = (\Gamma^{\boldsymbol{\alpha}})^{-1}, \tag{15}$$

where $\boldsymbol{\alpha}^{\mathrm{t}}$ and $\boldsymbol{\alpha}^{\mathrm{m}}$ are the true and measured parameters, respectively, $(\boldsymbol{\alpha}^{\mathrm{t}} - \boldsymbol{\alpha}^{\mathrm{m}})^T$ denotes the transpose of the vector $(\boldsymbol{\alpha}^{\mathrm{t}} - \boldsymbol{\alpha}^{\mathrm{m}})$ and an overline denotes the average over an ensemble of measurements of the same signal by identical detectors, at the same location, but with independent noise realisations. The covariance matrix is a real, symmetric matrix and has, in general, non-zero off-diagonal components. The diagonal elements of the covariance matrix are, as clear from the above equation, variances in the measurement of the various parameters. The off-diagonal elements $\mathcal{C}_{ij}$, $i \neq j$, termed covariances, measure the correlation between the ith and jth parameters. It is often convenient to deal with correlation-coefficients $c_{ij} \equiv \mathcal{C}_{ij}/(\mathcal{C}_{ii}\mathcal{C}_{jj})^{1/2}$, $i \neq j$, and they take values in the range $[-1, 1]$.

A particular correlation coefficient, say c_{12}, being equal to $+1$ or -1, indicates that one of the parameters α_1 or α_2, is redundant. In that case a systematic variation of the two parameters, for instance their sum, will account for the full variation of the signal achieved by varying the two parameters independently. Indeed, the variances, expressed as fractions of the trace of the covariance matrix, give a measure of the variations caused in the wave form shape due to different parameters. However, if correlations exist between two parameters then varying them both independently does not give rise to distinct wave form shapes.

C Principal Component Analysis

The method of principal components [28] aims at removing the redundancy that exists in a given parameter set by transforming to a new set and using this new set, of possibly unphysical parameters, to devise search algorithms. Those parameters, after diagonalisation, which account for most of the variation in the shape of the wave form, are called *principal components:* The parameter accounting for the maximum variation is the *first principal component,* the one that accounts for the next highest variation is the *second principal component,* and so on.

Let us define a new set of signal parameters $\boldsymbol{\beta}$ related to the old set $\boldsymbol{\alpha}$ via the orthogonal transformation $\boldsymbol{Q}$:

$$\boldsymbol{\alpha} = \boldsymbol{Q}\boldsymbol{\beta}, \quad \boldsymbol{\beta} = \boldsymbol{Q}^T\boldsymbol{\alpha}. \tag{16}$$

The purpose of transforming to a new parameter set is to achive diagonalisation of the new covariance matrix. In other words, we want the new parameters to satisfy:

$$\mathcal{C}^{\boldsymbol{\beta}} \equiv \overline{(\boldsymbol{\beta}^{\mathrm{t}} - \boldsymbol{\beta}^{\mathrm{m}})(\boldsymbol{\beta}^{\mathrm{t}} - \boldsymbol{\beta}^{\mathrm{m}})^T} = \boldsymbol{\Lambda} \tag{17}$$

where $\boldsymbol{\Lambda}$ is a diagonal matrix. The covariance matrix for the new parameter set can also be worked out starting from Eq.(16).

$$\mathcal{C}^{\boldsymbol{\beta}} \equiv \overline{(\boldsymbol{\beta}^{\mathrm{t}} - \boldsymbol{\beta}^{\mathrm{m}})(\boldsymbol{\beta}^{\mathrm{t}} - \boldsymbol{\beta}^{\mathrm{m}})^T} = \boldsymbol{Q}^T\,\overline{(\boldsymbol{\alpha}^{\mathrm{t}} - \boldsymbol{\alpha}^{\mathrm{m}})(\boldsymbol{\alpha}^{\mathrm{t}} - \boldsymbol{\alpha}^{\mathrm{m}})^T}\,\boldsymbol{Q} = \boldsymbol{Q}^T\mathcal{C}^{\boldsymbol{\alpha}}\boldsymbol{Q}. \tag{18}$$

Using Eq. (17) in Eq. (18) we get

$$\boldsymbol{Q}^T\mathcal{C}^{\boldsymbol{\alpha}}\boldsymbol{Q} = \boldsymbol{\Lambda} \Rightarrow \mathcal{C}^{\boldsymbol{\alpha}}\boldsymbol{Q} = \boldsymbol{\Lambda}\boldsymbol{Q}. \tag{19}$$

The last identity above is simply the eigen-value equation for the covariance matrix. Thus, we see that (i) the new variances are simply the eigen-values of the old covarinace matrix and (ii) the transformation matrix required consists of columns that are eigen-vectors of the old covariance matrix. Therefore, all that is needed, the new variances as well as the matrix to transform to the new parameter space, is contained in the covariance matrix that we begin with.

The rest of the job is easy: One can set up search templates in the new parameter space starting from the parameter with the largest variance (the first principal component) followed by the parameter with the next largest variance (the second principal component), and so on, until a desired variation of the signal shapes is covered. If it is required to carry out a hierarchical search one has to simply choose the first few parameters of largest variances. In this way, one can greatly simplify the search algorithm.

The method has its drawbacks too. And the major one among them is that the desired transformation (16) is unlikely to be integrable, for an arbitrary signal over the entire parameter space. The problem is not different from geometers trying to errect a Cartesian coordinate system in a curved space. A single Cartesian coordinate system cannot, of course, be constructed. But we do know how to circumvent this problem by using small coordinate patches, each of which is locally Euclidean, that neatly mesh up to span the entire space. Analogous operations will have to performed in dealing with a complicated signal space but this is not perhaps as complicated as choosing templates in the original parameter space that one started with.

Owen [29] has discovered a differential geometric method similar to principal component analysis in many respects and has applied it choosing search templates for ground-based detectors (also see Owen and Sathyaprakash [30]).

D An Illustrative Example

Let us consider binary inspiral waves emitted by a system of two point masses in orbit around each other. Working with the parameter space of instant of coalescence t_C, phase at the instant of coalescence φ_C, the Newtonian chirp time τ_0 and the first post-Newtonian chirp time τ_1, Balasubramanian et al. [31] have derived the following covariance matrix

$$C_{ij} = 10^2 \times \begin{pmatrix} 0.41 & 0.14 & -0.89 & 0.48 \\ & 0.04 & -0.29 & 0.15 \\ & & 2.03 & -1.15 \\ & & & 0.67 \end{pmatrix}. \qquad (20)$$

The various correlation coefficients have magnitudes quite close to unity: $c_{12} = 0.997$, $c_{13} = -0.972$, $c_{14} = 0.911$, $c_{23} = -0.954$, $c_{24} = 0.881$ and $c_{34} = -0.982$. This indicates that the parameter set $(t_C, \varphi_C, \tau_0, \tau_1)$ is possibly too redundant. On diagonalising the above matrix we get

$$C'_{ij} = 10^2 \times \begin{pmatrix} 0.001 & & & \\ & 0.003 & & \\ & & 0.051 & \\ & & & 31.2 \end{pmatrix}. \qquad (21)$$

Thus, most of the variation in the wave form shape caused by varying all the four parameters independently, can be accounted for by varying just one of the new parameters, corresponding to the last entry in the above matrix.

IV FUTURE PROSPECTS

In this talk I have discussed two related problems in detecting and measuring inspiral signals from super-massive black hole binaries.

The first problem is the accuracy of the model wave forms needed in detection and measurement. Here we have seen that accurate signal models, such as the recently proposed and amply tested P-approximants, are necessary, in particular, for a precise measurement of the astrophysical parameters. But our signal models have been too idealised and have not involved many other parameters, such as the spins of the component stars, eccentricity of the orbit, orientation of and direction to the source, etc. It is particularly important to assess the biases incurred in the estimation of parameters of a wave form with an approximate signal model.

Our confidence in signal detection lies in predicting catastrophic astrophysical events before they can be observed by other means. We should be able to tell our optical and radio astronomer friends where and when an event is likely to occur, what the source parameters are, etc., with tolerable uncertainty and little bias, so that they can point their telescopes in the right direction and be sure their instruments are sensitive enough to catch possible electromagnetic signals from the source that we detect and measure. To do so, we need accurate signal models.

The second problem is the choice of templates that need be employed in our search algorithms. Here we have argued that working with the original physical parameter space may unnecessarily complicate search algorithms since they don't take into account possible correlations that may exist between different parameters. However, the method of principal components allows us to transform to a new set parameters, only a small number of which may account for the full variation of the signal. Working with a small number of, albeit unphysical, parameters could potentially simplify the search to a great extent. However, the transformation to the new set may not be integrable requiring a different transformation matrix at different points in the parameter space. This latter problem deserves to be studied in detail in the context of supermassive black hole binaries. It is also necessary to explore a bigger parameter space than what is treated in this exploratory article.

Acknowledgements

Thanks are due to Thibault Damour and Bala Iyer for many enlightening discussions on the post-Newtonian expansion and P-approximants. I am indebted to Bernard Schutz for bringing to my notice the method of principal components. I have benefited from conversations with Sreeram Valluri on principal components.

REFERENCES

1. K.S. Thorne and V.B. Braginsky, *Astrophys. J.,* **204,** L1 (1976).
2. Bender, P. *et al.* LISA*: Pre-Phase A Report,* MPQ 208 (Max-Planck-Institut für Quantenoptik, Garching, Germany). (Also see the Second Edition, July 1998.)
3. A. Sillanpää, S. Haarala, M.J. Valtonen, B. Sundelius and G.G. Byrd, Astrophys. J., **325,** 628 (1988); L. Valtoja, M.J. Valtonen, G.G. Byrd, Astrophys. J., **343,** 47 (1989); N. Roos, J.S. Kastra and C.A. Hummel, Astrophys. J., **409,** 130, 1993; J. Kormendy and D. Richstone, Ann. Rev. Astron. & Astrophys., **33,** 581, (1995); R. Bender, J. Kormendy and W. Dehnen, Astrophys. J. Lett., **464,** L123 (1996); A. Eckart and R. Genzel, Nature, **383,** 415 (1996). C.M. Gaskell, Astrophys. J., **646,** L107 (1996). R.P. van der Marel, T. de Zeeuw, H.W. Rix and G.D. Quinlan, Nature, **385,** 610 (1996); M.J. Rees, Class. Quantum Grav., **14,** 1411 (1997).
4. D. Hils and P.L. Bender, Astrophys. J. Lett., **445,** L7 (1995); S. Sigurdsson and M.J. Rees, Mon. Not. R. Astron. Soc., **284,** 318 (1996).
5. T. Fukushige, T. Ebisuzaki and J. Makino, Astrophys. J. **396,** L61; M.G. Haehnelt, Mon. Not. R. Astron. Soc., **269,** 199 (1994); M. Rajagopal and R.W. Romani, Astrophys. J., **446,** 543 (1995); A. Vecchio, Class. Quantum. Gr., **14,** 1431, (1997).
6. K.S. Thorne, in *Proceedings of Snowmass 1994 Summer Study on Particle and Nuclear Astrophysics and Cosmology,* E.W. Kolb and R.D. Peccei, Eds., (World Scientific, Singapore, 1995) pp 160-184; B.F. Schutz, gr-qc/9710080, gr-qc/9710079.
7. L. Blanchet and B.S. Sathyaprakash, Phys. Rev. Lett., **74,** 1067 (1995)
8. B.S. Sathyaprakash and B.F. Schutz, submitted to Living Reviews in Relativity (1998).
9. T. Damour and N. Deruelle, Phys. Lett. **87A**, 81 (1981); and C.R. Acad. Sci. Paris **293** (II) 537 (1981); T. Damour, C.R. Acad. Sci. Paris **294** (II) 1355 (1982); and in *Gravitational Radiation,* ed. N. Deruelle and T. Piran, pp 59-144 (North-Holland, Amsterdam, 1983).
10. L. Blanchet, T. Damour, B.R. Iyer, C.M. Will and A.G. Wiseman, Phys. Rev. Lett. **74**, 3515 (1995).
11. L. Blanchet, T. Damour and B.R. Iyer, Phys. Rev. **D51**, 5360 (1995).
12. C.M. Will and A.G. Wiseman, Phys. Rev. **D54**, 4813 (1996).
13. L. Blanchet, B.R. Iyer, C.M. Will and A.G. Wiseman, Class. Quantum. Gr. **13**, 575, (1996).
14. L. Blanchet, Phys. Rev. **D54**, 1417 (1996).
15. L. Blanchet, B. R. Iyer and B. Joguet, paper in preparation; L. Blanchet, G. Faye, B. R. Iyer, B. Joguet and B. Ponsot, work in progress; C. M. Will and M. E. Pati, work in progress; G. Schäfer and P. Jaranowski, work in progress.
16. E. Poisson, Phys. Rev. **D52**, 5719 (1995).
17. T.Tanaka, H.Tagoshi and M.Sasaki, Prog.Theor.Phys. **96,** 1087 (1996).
18. C. Cutler, L.S. Finn, E. Poisson and G.J. Sussmann, Phys. Rev. **D47**, 1511 (1993).
19. E. Poisson, Phys. Rev. **D47**, 1497 (1993); H. Tagoshi and T. Nakamura, Phys. Rev. **D49**, 4016 (1994); M. Sasaki, Prog. Theor. Phys. **92**, 17 (1994); H. Tagoshi and M. Sasaki, Prog. Theor.Phys. **92**, 745 (1994).
20. T. Damour, B.R. Iyer, B.S. Sathyaprakash, Phys. Rev. **D 57**, 885 (1998).
21. C. Cutler et al., Phys. Rev. Lett. **70**, 2984 (1993).
22. C.M. Bender and S.A. Orszag, *Advanced mathematical methods for scientists and engineers* (McGraw Hill, Singapore, 1984).
23. L. E. Simone, S. W. Leonard, E. Poisson and C. M. Will, Class. Quantum Grav. **14**, 237 (1997).
24. T. A. Apostolatos, Phys. Rev. **D52**, 605 (1995).
25. C. W. Helstrom, *Statistical Theory of Signal Detection*, 2nd edition (Pergamon Press, London, 1968).
26. T. A. Apostolatos, Phys. Rev. **D54**, 2421 (1996).

27. A. Vecchio, private communication.
28. There are many texts on principal component analysis; see e.g., F. Murtagh and A. Heck, Multivariate Analysis (Reidel, Dortrecht, 1987); I.J. Jolliffe, Principal Component Analysis (Springer-Verlag, New York, 1986). For an application of the method in gravitational wave data analysis see, A. Ortolan, G. Vedovato, M. Cerdonio and S. Vitale, Phys. Rev. **D 50,** 4737 (1994).
29. B.J. Owen, Phys. Rev. **D53**, 6749 (1996).
30. B.J. Owen and B.S. Sathyaprakash, gr-qc/980876, submitted to Phys. Rev. **D**.
31. R. Balasubramanian, B.S. Sathyaprakash and S.V. Dhurandhar, Phys. Rev. **D 53,** 3033 (1996)

LISA as a Xylophone Interferometer Detector of Gravitational Radiation

Massimo Tinto

Jet Propulsion Laboratory,
California Institute of Technology,
4800 Oak Grove Drive, MS. 161-260,
Pasadena, California, 91109

Abstract. A filtering technique, for reducing the frequency fluctuations of the laser entering into the two-way Doppler tracking data measured with two spacecraft, is discussed. This method takes advantage of the sinusoidal behavior of the transfer function of this noise source to the Doppler observable, which displays sharp nulls at selected Fourier components. The non-zero gravitational wave signal remaining at these frequencies makes spacecraft to spacecraft laser Doppler tracking the equivalent of a *xylophone interferometer* detector of gravitational radiation.

The data analysis technique presented in this paper could be implemented with the LISA mission at the Fourier frequencies where the algorithm for unequal-arm interferometers fails to work, or as a backup option in case of failure of one of the three spacecraft.

INTRODUCTION

Non-resonant detectors of gravitational radiation (with frequency content $0 < f < f_H$) are essentially interferometers, in which a coherent train of electromagnetic waves (of nominal frequency $\nu_0 \gg f_H$) is folded into several beams, and at points where these intersect relative fluctuations of frequency or phase are monitored (homodyne detection). The observed low frequency signal is due to frequency variations of the source of the beams about ν_0, to relative motions of the source and the mirrors (or amplifying transponders) that do the folding, to temporal variations of the index of refraction along the beams, and, according to general relativity, to any time-variable gravitational fields present, such as the transverse traceless metric curvature of a passing plane gravitational wave train. To observe these gravitational fields in this way, it is thus necessary to control, or monitor, the other sources of relative frequency fluctuations, and, in the data analysis, to optimally use algorithms based on the different characteristic interferometer responses to gravitational waves (the signal) and to the other sources (the noise) [1].

Space-based interferometers, such as the coherent microwave tracking of interplanetary spacecraft [2] and proposed Michelson optical interferometers in planetary orbits [3], are most sensitive to milliHertz gravitational waves and have arm lengths ranging from 10^6 to 10^8 kilometers.

In present single-spacecraft Doppler tracking observations many of the noise sources can be either reduced or calibrated by implementing appropriate microwave frequency links and by using specialized electronics, so the fundamental limitation is imposed by the frequency (time-keeping) fluctuations inherent to the reference clocks that control the microwave system. Hydrogen maser clocks, currently used in Doppler tracking experiments, achieve their best performance at about 1000 seconds integration time, with a fractional frequency stability of a few parts in 10^{-16}. This is the reason why these one-arm interferometers in space are most sensitive to milliHertz gravitational waves. This integration time is also comparable to the microwave propagation (or "storage") time $2L/c$ to spacecraft en route to the outer solar system ($L \simeq 3AU$) [4].

By comparing phases of split beams propagated along non-parallel arms, source frequency fluctuations can be removed and gravitational wave signals at levels many orders of magnitude lower can be detected. Especially for space-based interferometers such as LISA, that use lasers with a frequency stability of a few parts in 10^{-13},

CP456, *Laser Interferometer Space Antenna*
edited by William M. Folkner
 1-56396-848-7/98/$15.00

it is essential to be able to remove these fluctuations when searching for gravitational waves of dimensionless amplitudes less than 10^{-20} in the milliHertz band.

Since the armlengths of these space-based interferometers can be different by several percent, the direct recombination of the two beams at a photo detector will not remove completely the laser noise. This is because the frequency fluctuations of the laser will be delayed by a different amount of time into the two different-length arms. In order to solve for this problem, a technique involving heterodyne interferometry with unequal arm lengths and independent readouts was proposed [5], which yielded data from which source frequency fluctuations were removed by many orders of magnitude. The technique discussed in [5], however, is not effective at frequencies equal to integer multiples of the inverse of the round trip light times in the two arms. This is because the transfer functions of the laser fluctuations to the Doppler tracking responses have sharp nulls at these frequencies. This implies that the information about the laser fluctuations provided by one of the two Doppler responses at these Fourier frequencies is lost, and the calibration of these fluctuations at these frequencies can not be performed [5].

In this paper we will show, however, that it is still possible to make measurements of gravitational radiation at these frequencies by using only the data generated by a single pair of spacecraft. In this sense we can regard spacecraft to spacecraft coherent laser tracking as a *xylophone interferometer* detector of gravitational radiation [6]. An outline of the paper is presented below.

In section II, after deriving the transfer functions of the noise sources affecting the two-way tracking data set, we show that there exist selected Fourier frequencies at which the transfer function of the laser frequency fluctuations into the Doppler observable is essentially null. These are what we will refer to as the frequencies of the xylophone.

Since the xylophone frequencies are equal to integer multiples of the inverse of the round trip light time, any variation in the distance between the spacecraft implies changes in these frequencies. For the selected LISA trajectory, we can successfully implement the xylophone technique by integrating the data for time intervals during which the variations of the xylophone frequencies are smaller than the frequency resolution bin. Estimates of the maximum integration times allowed by each of the three pairs of spacecraft are presented.

The noise levels, achievable with this technique with two LISA spacecraft, are presented in section III. We find that a strain sensitivity of 2.5×10^{-21} at the frequency 3×10^{-2} Hz can be reached when searching for sinusoids and by integrating the data for 10 days. At this sensitivity level, gravitational radiation from galactic binary systems should be observable. Our comments and conclusions are then presented in section IV.

SPACECRAFT TO SPACECRAFT COHERENT LASER TRACKING AS A XYLOPHONE INTERFEROMETER

Let us consider two of the three LISA spacecraft, each acting as a free falling test particle, and continuously tracking each other via coherent laser light. One spacecraft, which we will refer to as spacecraft a, transmits a laser beam of nominal frequency ν_0 to the other spacecraft (spacecraft b). The phase of the light received at spacecraft b is used by a laser on board spacecraft b for coherent transmission back to spacecraft a. The relative two-way frequency (or phase) changes $\Delta\nu/\nu_0$, as functions of time, are then measured at a photo detector. When a gravitational wave crossing the solar system propagates through this electromagnetic link, it causes small perturbations in $\Delta\nu/\nu_0$, which are replicated three times in the Doppler data with maximum spacing given by the two-way light propagation time between the two spacecraft.

Let us introduce a set of Cartesian orthogonal coordinates (X, Y, Z) centered on one of the two spacecraft, say spacecraft a. The Z axis is assumed to be oriented along the direction of propagation of a gravitational wave pulse, and (X, Y) are two orthogonal axes in the plane of the wave (see Figure 1). In this coordinate system we can write the two-way Doppler response, measured by spacecraft a at time t, as follows

$$\left(\frac{\Delta\nu(t)}{\nu_0}\right)_a \equiv y(t) = -\frac{(1-\mu)}{2}\, h(t) \; - \; \mu\, h(t-(1+\mu)L) + \frac{(1+\mu)}{2}\, h(t-2L) \\ + C_a(t-2L) \; - \; C_a(t) \; + \; 2B_b(t-L) \; + \; B_a(t-2L) \; + \; B_a(t) \\ + TR_b(t-L) \; + \; N_{2a}(t) \; , \tag{1}$$

where $h(t)$ is equal to

$$h(t) = h_+(t)\cos(2\phi) + h_\times(t)\sin(2\phi) \; . \tag{2}$$

Here $h_+(t)$, $h_\times(t)$ are the wave's two amplitudes with respect to the (X, Y) axis, (θ, ϕ) are the polar angles describing the location of spacecraft b with respect to the (X, Y, Z) coordinates, μ is equal to $\cos\theta$, and L is the distance between the two spacecraft (units in which the speed of light $c = 1$).

We have denoted with $C_a(t)$ the random process associated with the frequency fluctuations of the laser on board spacecraft a; $B_a(t)$, $B_b(t)$ are the joint effects of the noises from buffeting by non gravitational forces on the test masses on board spacecraft a and b respectively, $TR_b(t)$ is the noise due to the optical transponder on board spacecraft b, and $N_{2a}(t)$ is the noise from the photo detector on board spacecraft a where two-way phase changes are measured.

From Eq. (1) we deduce that gravitational wave pulses of duration longer than the round trip light time $2L$ have a Doppler response $y(t)$ that, to first order, tends to zero. The tracking system essentially acts as a pass-band device, in which the low-frequency limit f_l is roughly equal to $(2L)^{-1}$ Hz, and the high-frequency limit f_H is set by the shot noise at the photo detector [2,4].

In Eq. (1) it is also important to note the characteristic time signatures of the random processes $C_a(t)$, $B_a(t)$, and $B_b(t)$. The time signature of the noise $C_a(t)$ can be understood by observing that the frequency of the signal received at time t contains laser frequency fluctuations transmitted $2L$ seconds earlier. By subtracting from the frequency of the received signal the frequency of the signal transmitted at time t, we also subtract the frequency fluctuations $C_a(t)$ with the net result shown in Eq. (1). As far as the fluctuations due to buffeting of the test-mass on board spacecraft a are concerned, the frequency of the received signal is affected at the moment of reception as well as $2L$ seconds earlier. Since the frequency of the signal generated at time t does not contain yet any of these fluctuations, we conclude that $B_a(t)$ is positive-correlated at the round trip light time $2L$. The time signature of the noise $B_b(t)$ in Eq. (1) can be understood through similar considerations [6].

Among all the noise sources included in Eq. (1), the frequency fluctuations due to the lasers are expected to be the largest. A space-qualified single-mode laser, such as a diode-pumped Nd:YAG ring laser of frequency $\nu_0 = 3.0 \times 10^{14}$ Hz and phase-locked to a Fabry-Perot optical cavity, is expected to have a spectral level of frequency fluctuations equal to about $1.0 \times 10^{-13}/\sqrt{Hz}$ in the milliHertz band. Frequency stability measurements performed on such a laser by McNamara *et al.* [7] indicate that a stability of about $1.0 \times 10^{-14}/\sqrt{Hz}$ might be achievable in the same frequency band.

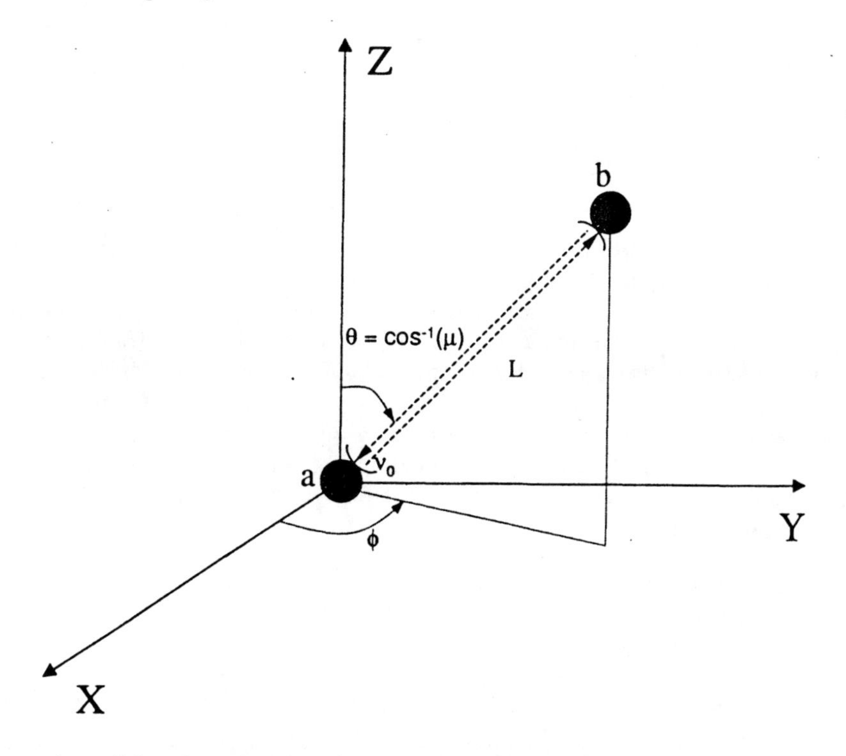

FIGURE 1. Coherent laser light of nominal frequency ν_0 is transmitted from spacecraft a to spacecraft b, and coherently transponded back. The gravitational wave train propagates along the Z direction, and the cosine of the angle between its direction of propagation and the laser beam is denoted by μ.

For the moment we will not make any assumptions on the frequency stability of the onboard laser, and return to this point later. We will focus instead on its transfer function and on the transfer function of the gravitational wave signal as shown in Eq. (1). If we denote with $\widetilde{y}(f)$ the Fourier transform of the time series $y(t)$, defined as

$$\widetilde{y}(f) \equiv \int_{-\infty}^{+\infty} y(t)\ e^{2\pi i f t}\ dt\ , \tag{3}$$

then Eq. (1) can be rewritten in the Fourier domain as follows

$$\begin{aligned} \widetilde{y}(f) = & \left[\frac{(\mu-1)}{2} - \mu\ e^{2\pi i f(1+\mu)L} + \frac{(1+\mu)}{2}\ e^{4\pi i f L}\right] \widetilde{h}(f) \\ & + \widetilde{C_a}(f)\ \left[e^{4\pi i f L} - 1\right] + \widetilde{B_a}(f)\ \left[e^{4\pi i f L} + 1\right] + 2\ \widetilde{B_b}(f)\ e^{2\pi i f L} \\ & + \widetilde{TR_b}(f)\ e^{2\pi i f L} + \widetilde{N_a}(f)\ . \end{aligned} \tag{4}$$

Note that the transfer function of the noise C_a is equal to zero at frequencies that are integer multiples of the inverse of the round trip light time, while the transfer function of the gravitational wave signal is in general different from zero. By making coherent laser tracking measurements at these frequencies, we are in fact making *xylophone interferometric measurements* of gravitational waves.

If we define Δf to be the frequency resolution of our data set (equal to the inverse of the integration time τ), to first order in $(\Delta f\ L)$ and at the xylophone frequencies $f_k = k/2L$, the response $\widetilde{y}(f_k)$ can be approximated by the following expression [6]

$$\begin{aligned} \widetilde{y}(f_k) \simeq & \ \mu\ \left[1 + (-1)^{k+1} e^{\pi i k \mu}\right] \widetilde{h}(f_k)\ \pm\ \widetilde{C_a}(f_k)\ (2\Delta f L) \\ & + 2\left[\widetilde{B_a}(f_+) + \widetilde{B_b}(f_+)\right] + (-1)^k\ \widetilde{TR_b}(f) + \widetilde{N_a}(f_k)\ , \end{aligned} \tag{5}$$

If we take $L = 5\ \times 10^6$ km, and assume an integration time τ of three months as a numerical example, we find that the amplitudes of the frequency fluctuations due to the laser are reduced at the xylophone frequencies by a factor of

$$\frac{2\ \Delta f\ L}{c} = 4.1 \times 10^{-6}\ . \tag{6}$$

Eq. (5) shows some interesting, and somewhat peculiar, properties of the remaining gravitational wave signal at the xylophone frequencies. The response to a gravitational wave pulse goes to zero not only when the wave propagates along the line of sight between the spacecraft ($\mu = \pm 1$), but also for directions orthogonal to it ($\mu = 0$). This is consequence of the fact that for $\mu = 0$ the Doppler response y to a gravitational wave becomes a *two-pulse* response, identical to the response of the laser noise, and therefore it cancels out at the xylophone frequencies.

For sources randomly distributed in the sky, as in the case of a stochastic background of gravitational waves, we can assume the angles (θ, ϕ) to be random variables uniformly distributed over the sphere. Since the average over (θ, ϕ) of the response given in Eq. (4) is equal to zero, it follows that the variance (denoted with $\Sigma^2(f)$) of the antenna pattern is equal to [2]

$$\Sigma^2(f) = \frac{(2\pi f L)^2 - 3}{(2\pi f L)^2} - \frac{(2\pi f L)^2 + 3}{3(2\pi f L)^2}\ \cos(4\pi f L) + \frac{2}{(2\pi f L)^3}\ \sin(4\pi f L)\ . \tag{7}$$

At the xylophone frequencies f_k, Σ^2 becomes the following monotonically increasing function of the integer k

$$\Sigma^2(k) = \frac{2}{3} - \frac{4}{(\pi k)^2}\ , \tag{8}$$

ESTIMATED SENSITIVITIES

In order to take advantage of the xylophone technique it is necessary to integrate over a sufficiently long period of time. This is because we want to reduce the noise due to the laser to a level as close as possible to that identified by the remaining noise sources at the xylophone frequencies (see Eq. (5)). Since the xylophone frequencies change in time as the distance between the spacecraft varies, we can not coherently integrate our data indefinitely. Coherent integration can be performed only on a time scale τ during which the variations of the xylophone frequencies are smaller than the frequency resolution $\Delta f = 1/\tau$.

In order to identify the maximum time of coherent integration for our xylophone interferometer detector, we need to identify the time dependence of the separation $L(t)$ between two of the LISA spacecraft. From the definition of the frequencies f_k, we can then derive the variation of the xylophone frequencies, δf_k, in terms of the relative change in the distance between the spacecraft, $\delta L(t)/L(0)$. Since δf_k is related to $\delta L(t)$ by the following equation [6]

$$\delta f_k = \frac{k}{2L} \times \frac{\delta L(t)}{L(0)} , \tag{9}$$

it follows that the maximum time of integration can be computed by requiring δf_k to be smaller than the frequency resolution $\Delta f = 1/\tau$.

It has been calculated by Folkner *et al.* [8] that the relative longitudinal speeds between the three pairs of spacecraft, during approximately the first year of the mission, can be written in the following approximated form

$$V_{i,j}(t) = V_{i,j}^{(0)} \sin\left(\frac{2\pi t}{T_{i,j}}\right) \qquad (i,j) = (a,b) \; ; \; (a,c) \; ; \; (b,c) \; , \tag{10}$$

where we have denoted with $(a,b), (a,c), (b,c)$ the three possible spacecraft pairs, $V_{i,j}^{(0)}$ is a constant velocity, and $T_{i,j}$ is the period for the pair (i,j). In reference [8] it has also been shown that the LISA trajectory can be selected in such a way that two of the three arms' rates of change are essentially equal during the first year of the mission. This configuration is particularly attractive because it implies an almost null variation in differential armlength for one of the three interferometers. Following reference [8], we will assume $V_{a,b}^{(0)} = V_{a,c}^{(0)} \neq V_{b,c}^{(0)}$, with $V_{a,b}^{(0)} = 1$ m/s, $V_{b,c}^{(0)} = 13$ m/s, $T_{a,b} = T_{a,c} \approx 4$ months, and $T_{b,c} \approx 1$ year. With these numerical values we can calculate the maximum integration times for different xylophone frequencies. The calculation is straightforward, and we will not go through it here [6]. The results of this analysis, however, indicate that the data from the two pairs of spacecraft, $(a,b), (a,c)$, can be integrated coherently at the frequency $f_1 = 3 \times 10^{-2}$ Hz for about 10 days. A shorter integration time of about 3 days is needed instead to make xylophone measurements at the frequency 1.5 Hz. For the remaining pair of spacecraft, due to their larger relative speed, we have found that coherent integration at f_1 can be performed for about 6 days, while at 1.5 Hz the maximum integration time goes down to about 2 days.

The numerical values of the maximum integration times derived above allow us to estimate, at the xylophone frequencies, the one-sided power spectral density of the noise affecting the data set y. In what follows we will consider two spacecraft with identical optical and mechanical payloads, and equal to those described in [3]. We will also assume the random processes associated with the noise sources affecting the stability of the coherent one-way tracking data to be uncorrelated with each other, and their one-sided power spectral densities to be consistent with those given in reference [3]. Since our xylophone will be sensitive to gravitational radiation at frequencies equal to or larger than the inverse of the round trip light time ($c/2L \approx 3 \times 10^{-2}$ Hz), the dominant noise sources determining its strain sensitivity will be the photon-shot noise, and the frequency fluctuations of the laser [3].

After taking into account Eq. (5), and the expressions of the one-sided power spectral density for the shot-noise and the frequency fluctuations of the laser given in reference [3], the one-sided power spectral density $S_y(f)$ of the noise in the response y, estimated in the frequency band of the xylophone, can be written as follows

$$S_y(f) = 10^{-38} f^2 + \left[10^{-28} f^{-2/3} + 6.3 \times 10^{-37} f^{-3.4}\right] \sin^2(2\pi f L) , \tag{11}$$

In Figure 2 we have plotted this function by assuming an integration time of 3 days. Note that, with such an integration time, the one-sided power spectral density of the laser noise is reduced, at the xylophone frequencies, by a factor of $(2\Delta f\ L)^2 = 1.6 \times 10^{-8}$. Since the function $S_y(f)$ plotted in Figure 2 is monotonically decreasing at the xylophone frequencies, we conclude that with such an integration time the noise due to the laser is still the dominant one.

The 3 days time interval implies a variation of the largest of the xylophone frequencies smaller than the frequency resolution bin. At smaller xylophone frequencies, however, the one-sided power spectral densities should be rescaled according to the appropriate maximum integration times. For instance, since at the frequency 3×10^{-2} Hz we can coherently integrate the data from two of the three pairs of spacecraft for 10 days, the one-sided power spectral density at this frequency would be smaller than the value shown in Figure 2 by a factor $(3/10)^2$. From the estimate of the one-sided power spectral density of the noise given in Figure 2, it is possible to calculate the root-mean-squared (r.m.s.) noise level $\sigma(f_k)$ of the frequency fluctuations in the bins of width Δf, around the frequencies f_k ($k = 1, 2, 3, ...$). This is given by the following expression

$$\sigma(f_k) \equiv [S_y(f_k)\ \Delta f]^{1/2} \ , \ k = 1, 2, 3, ... \tag{12}$$

This measure of the Doppler sensitivity is appropriate for sinusoidal signals, such as those generated during the coalescence of a binary system, while it over estimates the sensitivity to bursts and to a stochastic background of gravitational radiation. The quantitative results implied by the formula given in Eq. (12) should therefore be considered by keeping in mind this observation. For a detailed and quantitative analysis covering bursts and stochastic waveforms instead, the reader is referred to [6].

From Eqs. (5, 8) we derive that, at 3×10^{-2} Hz, the xylophone will have an r.m.s. strain sensitivity to sinusoids, averaged over an isotropic distribution of source directions, of about 2.6×10^{-21}. A binary system containing two $1.0\ M_\odot$ stars, for instance, could radiate sinusoidally at $f = 3 \times 10^{-2}$ Hz, and during a period of 17 days the frequency of the radiation would change by an amount smaller than the frequency resolution of the data. Since the Doppler data can be integrated coherently for about 10 days at the xylophone's fundamental, we find that such a binary system could be observed out to a distance of about 3 kpc.

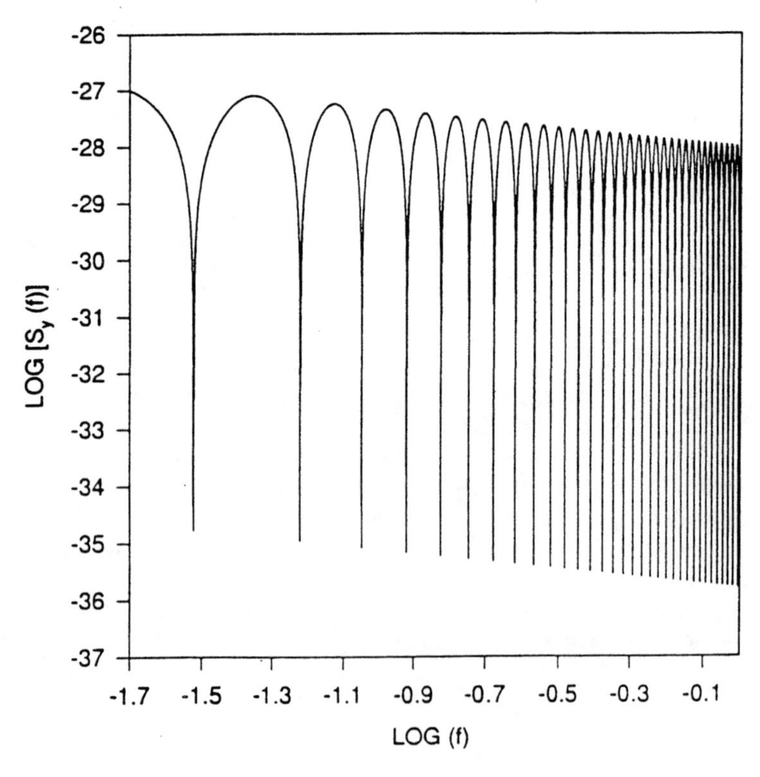

FIGURE 2. The one-sided power spectral density of the noise, $S_y(f)$, entering into the response y. The minima of S_y have been estimated by assuming an integration time of 3 days. See text for explanation.

CONCLUSIONS

We have discussed a data analysis technique for performing searches of gravitational radiation in space with two spacecraft tracking each other via coherent laser light. The main result of our analysis, deduced in Eq. (5), shows that we can reduce, by several orders of magnitude, the frequency fluctuations introduced in the Doppler data by the laser. This is achieved by making measurements at the Fourier frequencies where the transfer function of the laser fluctuations to the Doppler observable has sharp minima. In this respect spacecraft to spacecraft coherent laser tracking can be regarded as a *xylophone interferometer* detector of gravitational radiation.

When searching for sinusoids, we have found that a strain sensitivity of 2.6×10^{-21} at the frequency 3×10^{-2} Hz can be obtained after coherently integrating the two-way Doppler data for 10 days. At this sensitivity level, gravitational radiation from galactic coalescing binary systems should be observable.

Spacecraft to spacecraft xylophone interferometric measurements of gravitational radiation could be implemented with the LISA mission at the Fourier frequencies where the algorithm for unequal-arm interferometers fails to work, or as backup option in case of failure of one of the three spacecraft.

ACKNOWLEDGEMENTS

It is a pleasure to thank Frank B. Estabrook and John W. Armstrong for several useful discussions, and their encouragement during this work. This research was performed at the Jet Propulsion Laboratory, California Institute of Technology, under contract with the National Aeronautics and Space Administration.

REFERENCES

1. M. Tinto, and F.B. Estabrook, *Phys. Rev. D*, **52**, 1749, (1995).
2. F.B. Estabrook and H.D. Wahlquist, *Gen. Relativ. Gravit.* **6**, 439 (1975).
3. LISA: (Laser Interferometer Space Antenna) *A Cornerstone Project in ESA's long term space science program "Horison 2000 Plus"*. **MPQ 208**, (Max-Planck-Institute für Quantenoptic, Garching bei München, 1995).
4. J.W. Armstrong. In these proceedings
5. G. Giampieri, R. Hellings, M. Tinto, J.E. Faller, *Optics Communications*, **123**, 669, (1996).
6. M. Tinto, *Phys. Rev. D*, October 15, 1998.
7. P.W. McNamara, H. Ward, J. Hough, and D. Robertson, *Clas. Quantum Grav.*, **14**, 1543, (1997).
8. W.M. Folkner, F. Hechler, T.H. Sweetser, M.A. Vincent, and P.L. Bender, *Clas. Quantum Grav.*, **14**, 1543, (1997).

Detection of continuous gravitational wave signals: pattern tracking with the Hough Transform

M.Alessandra Papa*, Bernard F. Schutz*,
Sergio Frasca‡† and Pia Astone†

* *Max-Planck-Institut für Gravitationsphysik*
Albert Einstein Institut
Schlaatzweg 1, D-14473 Potsdam, Germany

‡ *Dip. di Fisica, Univ. di Roma "La Sapienza"*
P.le A. Moro 2, 00185 Roma, Italy

† *INFN Sez. di Roma I*
P.le A. Moro 2, 00185 Roma, Italy

Abstract.
Searching for patterns originating from continuous signals in time-frequency diagrams – such diagrams are produced by the very first states of a hierarchical procedure which are described in [2] – is the issue that we shall address in this presentation. We shall outline the main features of a strategy based on the use of the Hough transform by presenting the concept with an easy example and then illustrating the way we will apply it to the specific continuous gravitational wave signal search.

INTRODUCTION

The aim of the procedure we shall outline here is that of identifying candidate continuous signals in the data from a gravitational wave detector by means of a procedure acting on spectra computed on a suitable time stretch. The motivation for this incoherent analysis is that computational time constraints make it necessary to investigate alternative strategies to coherent matched filtering for all-sky all-frequencies searches [1].

The basic idea is that a low signal to noise (snr hereafter) signal will not show up as a significant peak in one of the short spectra. But still, if a signal is there, the occurence of the same peak in many spectra may acquire significance and help the detection. So, the strategy consists of two steps: the first step when one selects peaks and produces time-frequency diagrams [2], the second when one finds in these diagrams patterns which are consistent with the expected time evolution of the istantaneous frequencies of the searched signals. The latter is the issue addressed in this presentation.

PATTERN TRACKING

Fig. 1 shows a typical time frequency diagram. In it, a few signal points are hidden, but, due their "low density", it is not possibile to tell which they are by simple visual inspection. A specific pattern tracking technique must thus be implemented which is capable of operating in such low snr conditions. A promising one is based on the use of the Hough transform (HT, hereafter): "*the Hough transform is recognized as being a powerful tool in shape analysis which gives good results even in the presence of noise and occlusion*"([4]). This is a transformation from the space of the data points to the space of the parameters describing the signals one is looking for. Therefore, is relies on the a-priori knowledge of the searched shape. The Hough transform was invented by Paul V.C. Hough in the early sixties ([3]) in order to identify particle tracks in bubble chambers

CP456, *Laser Interferometer Space Antenna*
edited by William M. Folkner

of high energy physics experiments ([5]). Since then, a great deal of work has been done on the subject – see review paper by Leavers ([4]) – with the scope of improving the performance of the Hough transform at detecting complex shapes embebbed in noise while overcoming the growing computational and storage cost that must be paid for growing number of parameters.

The basic idea of the HT is rather simple and we shall explain it by means of an easy example. Suppose one knows that in a set of *(x,y)* points, such as those shown in the left panel of fig. 1, a subset that follows a linear law ($y{=}ax{+}b$) is hidden and that one wants to estimate the parameters a and b. A way to do this is to set a grid on one of the parameters, say b_j, and then, for each data point (x_i, y_i), by inversion, find the corresponding a_{ij}. One can then set a grid also on the parameter a and tell how many counts fall in each pixel[1] of the resulting (a, b) plane: $x(a_i, b_j)$. The claim is that all the signal points will add up coherently in the same pixel, corresponding to the right value of the parameters, whereas noise points will be randomly scattered in the plane[2]. So, by studying the clustering properties of the maps in the space of parameters it is possibile to give an estimate of the right parameter values and of the corresponding false alarm probability. In fig.2 the panel on the right shows the map obtained by Hough transforming the data set of fig.1 for the searched linear behaviour. The maximum clustering takes place in the pixel $a = 1, b = 0$ which actually corresponds to the correct values of the parameters of the signal. The inversion does not, though, produce a uniform histogram: there will be pixels that are more likely to have higher counts than others just by chance and this appears clearly in the left plot of fig. 2 where the expected value of the count in each pixel for noise only (estimated over 50 trials) is colour coded. In order to be able to compare the outcomes in different pixels, we can "normalize" this variable to its expected value and standard deviation in order to make it indipendent of the particular pixel one is looking at and to define a measure of the contrast with respect to the expected mean. Thus we define a new variable

$$t(a_i, b_j) = \frac{x(a_i, b_j) - \mu(a_i, b_j)}{\sigma(a_i, b_j)},$$

where $\sigma(a_i, b_j)$ can be estimated from $\mu(a_i, b_j)$ by noting that $x(a_i, b_j)$ is a variable that follows a binomial distribution with probability $p \simeq \frac{\mu}{N}$, where N is the number of trials performed to estimate μ. Figure 3 shows the map for this new variable. The most significant pixel found is that corresponding to the signal, namely the $(a = 1, b = 0)$ pixel with $t = 4.59$, even though not at a very high overall significance level, which is about 30% (false alarm probability to have the value of $t = 4.59$ in any pixel of the map). Improvements can be obtained by using a finer grid on a and b. In the absence of errors in the position of the data points, in fact, using a finer grid in the parameter space enhances the significance of the clustering in the signal pixel simply because the expected noise in the pixels decreases with pixel area. The grid that was used in the example

---

1) In the language of HTs pixels are usually refered to as "accumulator cells".

2) And also signal points inverted for "wrong" values of the parameters.

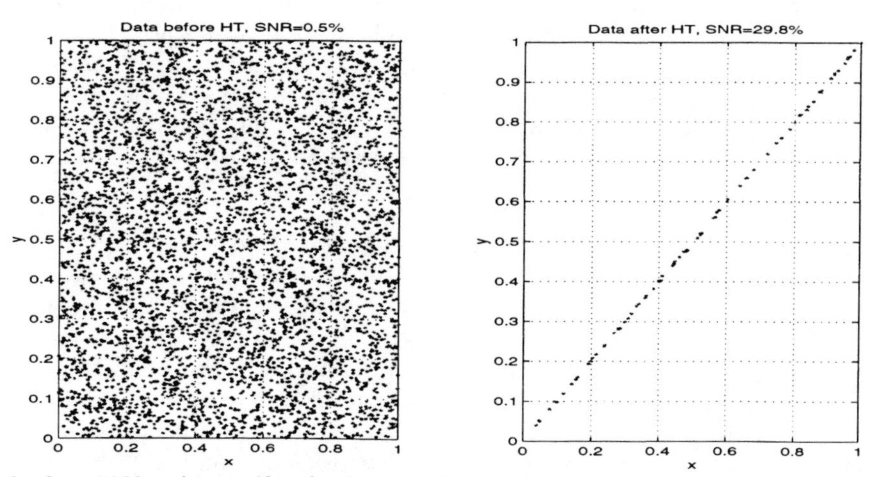

FIGURE 1. Left plot: 5000 points uniformly distributed on the *(x,y)* axis plus a 0.5% of couples that follow un unknown linear law. The density of signal points is so low that it is not possible to guess where they are by visual inspection. Right plot: the selected data after Hough filtering cointain a much higher fraction of signal, round 30%.

was just about the coarsest one could set to barely detect a signal at SNR $= 0.5\%$. If one were actually doing an analysis, one would then isolate the candidates points contributing to the most significant pixels (in this case a threshold could be placed, say, around $t = 4$) and then repeat the HT on a finer grid around the pixel parameter values and for the selected points only. With such a hierarchy of HTs one would improve SNR and significance of a putative signal. In practice, there is a limit to how well one can perform because usually there is an uncertainty in the position of signal points so, if the grid in the parameters becomes much finer than the uncertainty induced on these by the data error, this will cause the signal in the HT plane to be smeared out in different pixels thus decreasing SNR. Understanding optimal grid setting is a crucial step in optimizing HT performance.

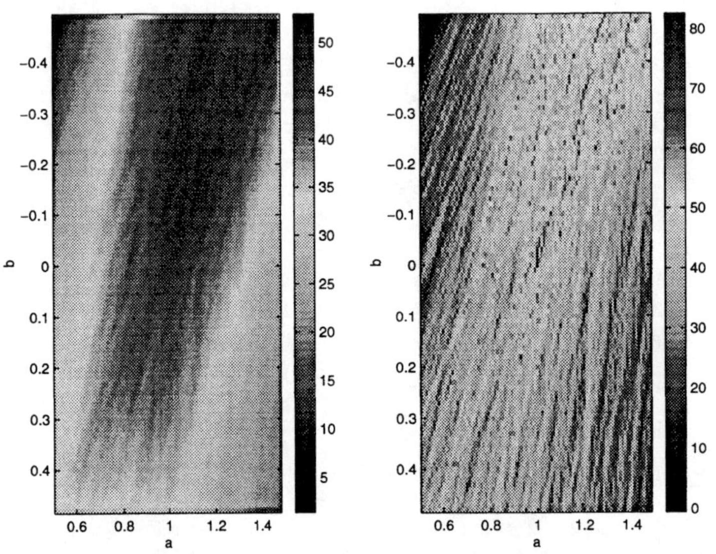

FIGURE 2. Left plot: expected Hough transform histogram for noise only (μ) in a bounded (a, b) plane. Right plot: Hough transform histogram, x, color coded in the plane of the parameters (a, b).

In the context of tracking patterns in time-frequency diagrams, the HT that must be performed is more complex than the example above: even in the simplified source case that we are studying now, the signal shape is more complicated than a linear law and the problem is not static. To start with, we have restricted our analysis to the case that the istantaneous intrinsic frequency of the source may be considered constant and that its apparent variation is due to the relative motion between the source and the detector on Earth. Under these assumptions, three parameters completely describe a signal: two celestial coordinates for the position of the source – say its right ascension α and declination δ – and its intrinsic frequency, f_0. The scheme that we are following to implement the HT is the following: set a grid of trial intrinsic frequencies and a grid on δ. For each point in the diagram – f_{ij} at time t_i and in the j-th frequency bin – the inversion yields two possibile values of α. As a matter of fact, one gets an equation in α and δ that describes a curve in the (α, δ) plane, thus what we really do, is to increment the count in the corresponding pixels. It turns out that it is most convenient to work in ecliptic coordinates and in these coordinates one finds that the curves are ellipses, (λ, β), with the center at latitude $\simeq 0$. The time t_i determines the position of the center of this ellipse and, for a given time, the "radius" depends on the distance between f_{ij} and f_0. At different times, the f_{ij}s originating from the same source will generate different circles all intersecting in a point, which identifies the position of the source. This way a map of the clustering over the whole sky is produced. Its features depend on the time scale covered by the data. For the circles to be spread with maximum uniformity, a year must be considered, but uniformity is not necessary for the analysis to work. Also note that *not* 1 year of effective observation is necessary, but only that the FFTs that we are using are spread during one year. The chance probability of the clustering count in every bin is computed as follows: the expected average value of the clustering count, μ_{ij}, for a given observation time, can easily be estimated, e.g. numerically. Then, this is used to compute the chance probability of the outcomes of the observations because the number of counts in each pixel is a random variable that follows a Poisson distribution with expected value μ_{ij}. For a year of observation, for example, μ_{ij} does not depend on the longitude of the source and we have verified that the number count follows a Poisson distribution, at least for latitutes $|\beta| < 75^o$. So it is possible to associate to each pixel its Poisson probability thus constructing such probability maps for each trial f_0. A threshold must then be set in order to select candidate sources, i.e. triplets (β, λ, f_0), that will undergo further processing.

We have implemented this HT algorithm and we are testing it on simulated spectral data (figs. 4 and 5 show the count map and the probability map for 200 spectra over 1 year). Optimal performance, as defined in [2], has not yet been reached as the implementation still needs some "fine tuning". For example, as already mentioned above, it is crucial to establish optimal gridding. In our problem this is related to the uncertainty in the definition of the frequencies f_{ij} and f_0, due to the finite resolution of the spectra. In fact, while performing the HT, we are not inverting for the actual istantaneous frequency of the signal, but for its discretized counterpart. This produces errors in the resulting position ellipse: it is as though the ellipse gained "a width" in the (λ, β) plane. Moreover, this width varies with time and position, making it difficult to set an a priori optimal grid. The HT scheme that we have outlined here seems particularly suited to take into account these effects quite naturally: for example, for every f_{ij}, a weight can be given to pixels neighbouring the ellipse according to the expected error on the position, at each time, induced by frequency discretization in the time-frequency diagrams. This should largely cure degradation of snr due to the choice of a non optimal gridding with a minor additional processing cost. Moreover, clustering properties in the sky maps seem to be little sensitive to small (of order 5 frequency frequency bins) deviations from the correct value of f_0, thus making it possible to search, at first, on a coarse grid in f_0 space and save computational time. Finally, points of ellipses of different radius could be weighted differently in order to construct more uniform count maps and make the statistical evaluation of the outcomes of the analysis more straightforward to interpret.

These are the issues that we are currently investigating in order to optimizate the HT scheme which we think is a promising approach for incoherent searches of continuous gravitational wave signals.

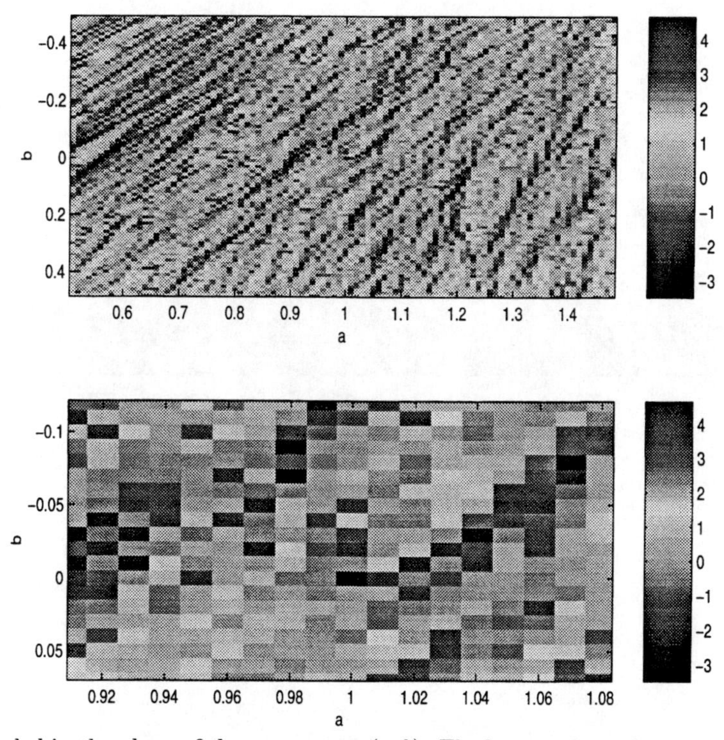

FIGURE 3. t, color coded in the plane of the parameters (a, b). The bottom figure is a zoom - near the correct values - of the map above. The most significant value for t actually occurs in the correct $a = 1, b = 0$ pixel with a value of $t = 4.59$.

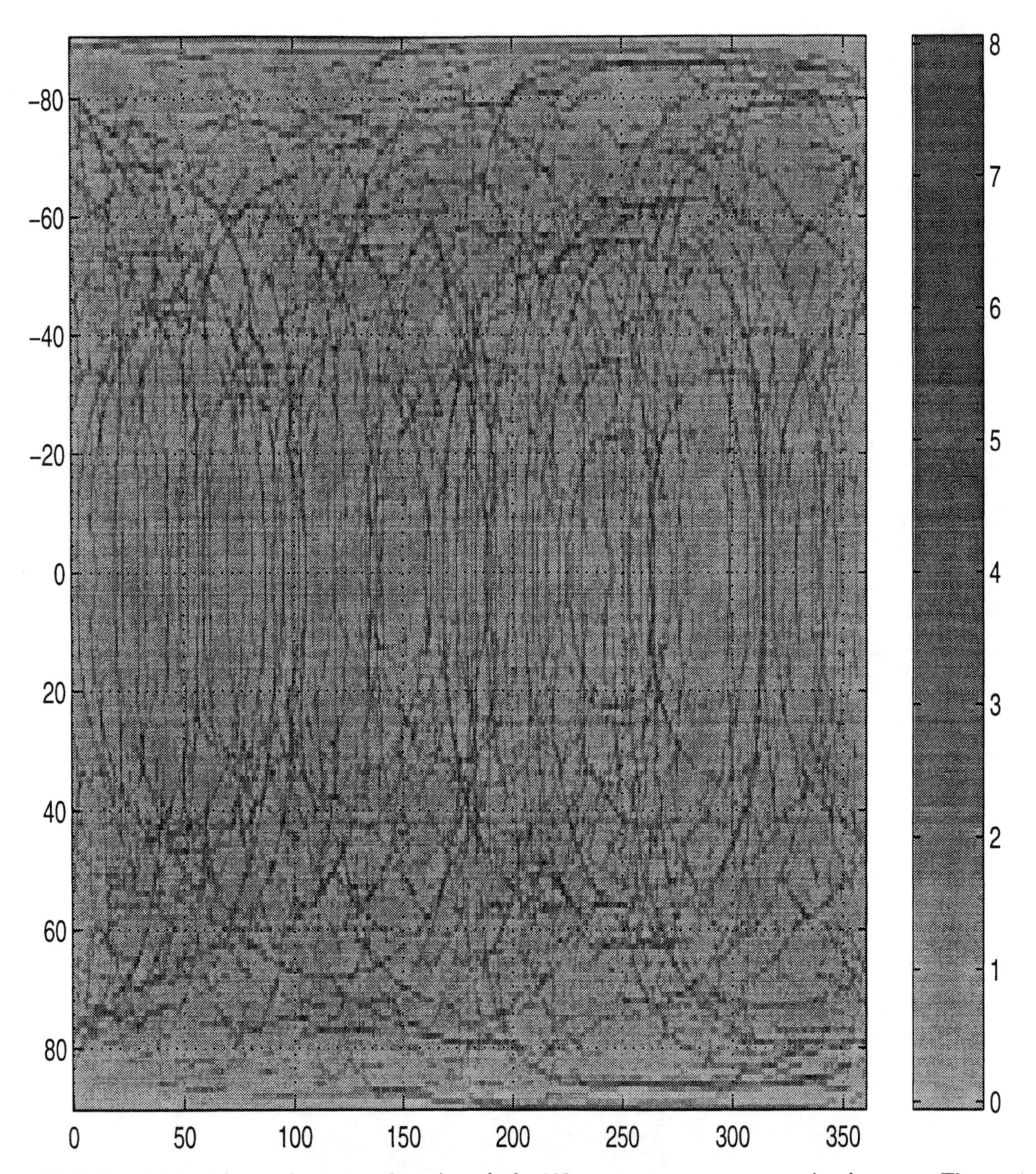

FIGURE 4. Map of the number counts for noise only for 200 spectra over one year, no signal present. The maximum clustering count is 8.

ACKNOWLEDGMENTS

We thank Prof. Lucia Zanello for having suggested to investigate the Hough transform method.

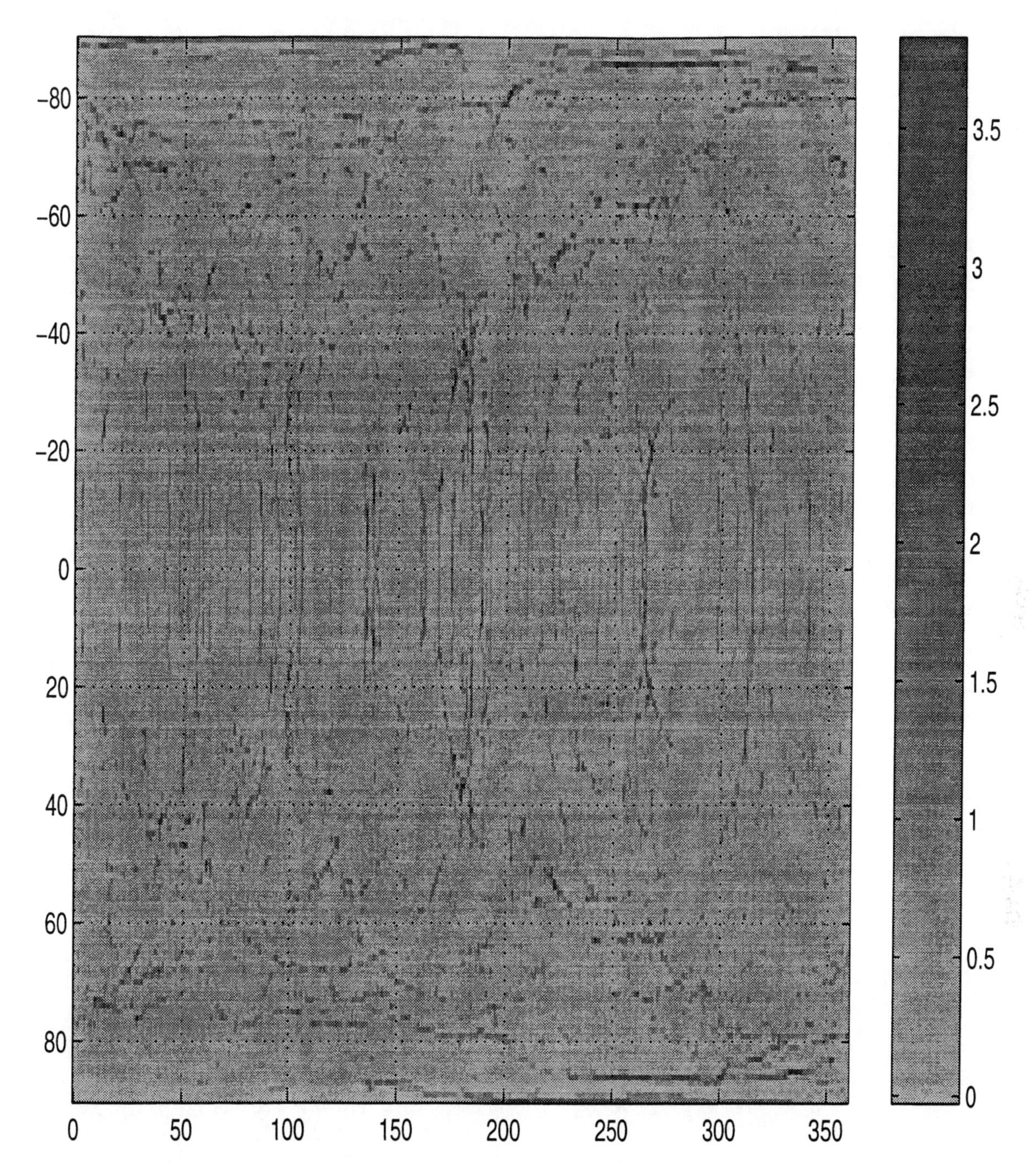

FIGURE 5. Probability map (-log10 color coded)corresponding to the clustering count map of the previous figure (no signal). The most significant value of the probability is $10^{-3.8}$ and it stands in the pixel $\lambda = 40.5, \beta = 42.5$.

REFERENCES

1. Brady P.R., Creighton T., Cutler C. and Schutz B.F "searching for periodic sources with LIGO", *Phys. Rev. D*, **57**, 4, 2101-2115, (1998).
2. Papa M.A., Astone P., Frasca S. and Schutz B.F. "Searching for continuous waves by line identification " *Gravitational Wave Data Analysis Workshop II*, Orsay, November 1997 also in *Contributions from the AEI to the GWDA2 Workshop* **AEI-057**, February 1998.

3. Hough P.V.C., "Methods and means for recognizing complex patterns", *U.S. Patent No. 3069654*, 1962.
4. Leavers, V.E., "Which Hough Transform ? ", *CVGIP: Image Understanding* **58**,2, 250-264, (1993).
5. Ermolin, Y.,and Ljuslin, C. *Nucl. Instrum. Methods Phys. Res. Sect A* **A289(3)**, 592-596 (1990).

Removal of interference from external coherent signals

Alicia M. Sintes and Bernard F. Schutz

Max-Planck-Institut für Gravitationsphysik (Albert-Einstein-Institut), Schlaatzweg 1, D-14473 Potsdam, Germany

Abstract. We present a technique that we call coherent line removal, for removing external coherent interference from gravitational wave interferometer data. We illustrate the usefulness of this technique applying it to the the data produced by the Glasgow laser interferometer in 1996 and removing all those lines corresponding to the electricity supply frequency and its harmonics. We also find that this method seems to reduce the level of non-Gaussian noise present in the interferometer and therefore, it can raise the sensitivity and duty cycle of the detectors.

INTRODUCTION

In the measured noise spectrum of the different gravitational wave interferometer prototypes [1–3], one observes peaks due external interference, where the amplitudes are not stochastic in contrast to the stochastic noise. The most numerous are powerline frequency harmonics. In this paper we review how to remove these very effectively using a technique we call coherent line removal (CLR) [4,5].

CLR is an algorithm able to remove interference present in the data while preserving the stochastic detector noise. CLR works when the interference is present in many harmonics, as long as they remain coherent with one another. Unlike other existing methods for removing single interference lines [6,7], CLR can remove the external interference without removing any 'single line' signal buried by the harmonics. The algorithm works even when the interference frequency changes. CLR can be used to remove all harmonics of periodic or broad-band signals (e.g., those which change frequency in time), even when there is no external reference source. CLR requires little or no a priori knowledge of the signals we want to remove. This is a characteristic that distinguishes it from other methods such as adaptive noise cancelling [8]. It is 'safe' to apply this technique to gravitational wave data because we expect that coherent gravitational wave signals will appear with at most the fundamental and one harmonic [9]. Lines with multiple harmonics must be of terrestrial origin.

In this paper, we illustrate the usefulness of this new technique by applying it to the data produced by the Glasgow laser interferometer in March 1996 and removing all those lines corresponding to the electricity supply frequency and its harmonics. As a result the interference is attenuated or eliminated by cancellation in the time domain and the power spectrum appears completely clean allowing the detection of signals that were buried in the interference. Therefore, this new method appears to be good news as far as searching for continuous waves (as those ones produced by pulsars [9,10]) is concerned. The removal improves the data in the time-domain as well. Strong interference produces a significant non-Gaussian component to the noise. Removing it therefore improves the sensitivity of the detector to short bursts of gravitational waves [11].

COHERENT LINE REMOVAL

In this section, we summarize the principle of CLR. For further details we refer the reader to [5].

We assume that the interference has the form

$$y(t) = \sum_n a_n m(t)^n + (a_n m(t)^n)^* \ , \tag{1}$$

where a_n are complex amplitudes and $m(t)$ is a nearly monochromatic function near a frequency f_0. The idea is to use the information in the different harmonics of the interference to construct a function $M(t)$ that is

CP456, *Laser Interferometer Space Antenna*
edited by William M. Folkner
 1-56396-848-7/98/$15.00

as close a replica as possible of $m(t)$ and then construct a function close to $y(t)$ which is subtracted from the output of the system cancelling the interference. The key is that real gravitational wave signals will not be present with multiple harmonics and that $M(t)$ is constructed from many frequency bands with independent noise. Hence, CLR will little affect the statistics of the noise in any one band and any gravitational wave signal masked by the interference can be recovered without any disturbance.

We assume that the data produced by the system is just the sum of the interference plus noise

$$x(t) = y(t) + n(t) , \tag{2}$$

where $y(t)$ is given by Eq. (1) and the noise $n(t)$ in the detector is a zero-mean stationary stochastic process. The procedure consists in defining a set of functions $\tilde{z}_k(\nu)$ in the frequency domain as

$$\tilde{z}_k(\nu) \equiv \begin{cases} \tilde{x}(\nu) & \nu_{ik} < \nu < \nu_{fk} \\ 0 & \text{elsewhere} , \end{cases} \tag{3}$$

where (ν_{ik}, ν_{fk}) correspond to the upper and lower frequency limits of the harmonics of the interference and k denotes the harmonic considered. These functions are equivalent to

$$\tilde{z}_k(\nu) = a_k \widetilde{m^k}(\nu) + \tilde{n}_k(\nu) , \tag{4}$$

where $\tilde{n}_k(\nu)$ is the noise in the frequency band of the harmonic considered. Their inverse Fourier transforms yield

$$z_k(t) = a_k m(t)^k + n_k(t) . \tag{5}$$

Since $m(t)$ is supposed to be a narrow-band function near a frequency f_0, each $z_k(t)$ is a narrow-band function near kf_0. Then, we define

$$B_k(t) \equiv [z_k(t)]^{1/k} , \tag{6}$$

that can be rewritten as

$$B_k(t) = (a_k)^{1/k} m(t) \beta_k(t) , \qquad \beta_k(t) = \left[1 + \frac{n_k(t)}{a_k m(t)^k}\right]^{1/k} . \tag{7}$$

All these functions, $\{B_k(t)\}$, are almost monochromatic around the fundamental frequency, f_0, but they differ basically by a certain complex amplitude. These factors, Γ_k, can easily be calculated, and we can construct a set of functions $\{b_k(t)\}$

$$b_k(t) = \Gamma_k B_k(t) , \tag{8}$$

such that, they all have the same mean value. Then, $M(t)$ can be constructed as a function of all $\{b_k(t)\}$ in such a way that it has the same mean and minimum variance. If $M(t)$ is linear with $\{b_k(t)\}$, the statistically the best is

$$M(t) = \left(\sum_k \frac{b_k(t)}{\mathrm{Var}[\beta_k(t)]}\right) \Big/ \left(\sum_k \frac{1}{\mathrm{Var}[\beta_k(t)]}\right) , \tag{9}$$

where

$$\mathrm{Var}[\beta_k(t)] = \frac{\langle n_k(t) n_k(t)^* \rangle}{k^2 |a_k m(t)^k|^2} + \text{corrections} . \tag{10}$$

In practice, we approximate

$$|a_k m(t)^k|^2 \approx |z_k(t)|^2 , \tag{11}$$

and we assume stationary noise. Therefore,

$$\langle n_k(t) n_k(t)^* \rangle = \int_{\nu_{ik}}^{\nu_{fk}} S(\nu) d\nu , \tag{12}$$

where $S(\nu)$ is the power spectral density of the noise.

Finally, it only remains to determine the amplitude of the different harmonics, which can be obtained applying a least square method.

REMOVAL OF 50 HZ HARMONICS

In this section, we present experimental results that demonstrate the performance of the CLR algorithm and show its potential value. We apply this method to the data produced by the Glasgow laser interferometer in March 1996 and the electrical interference is successfully removed.

In the study of the Glasgow data, we observe in the power spectrum many lines. Some of them are due to thermal noise (which we will not consider here) and many others at multiples of 50 Hz due external interference, where the amplitudes are not stochastic. In long-term Fourier transforms, the lines at multiples of 50 Hz are broad, and the structure of different lines is similar apart from an overall scaling proportional to the frequency. In smaller length Fourier transforms, the lines are narrow, with central frequencies that change with time, again in proportion to one another. It thus appears that all these lines are harmonics of a single source (e.g., the electricity supply) and that their broad shape is due to the wandering of the incoming electricity frequency.

In the Glasgow data, those lines at 1 kHz have a width of 5 Hz. Therefore, we can ignore these sections of the power spectrum or we can try to remove this interference in order to be able to detect gravitational waves signals masked by them.

In order to remove the electrical interference, we separate the data into groups of 2^{19} points (approximately two minutes) and, for each of them, the coherent line removal algorithm is applied. A detailed description can

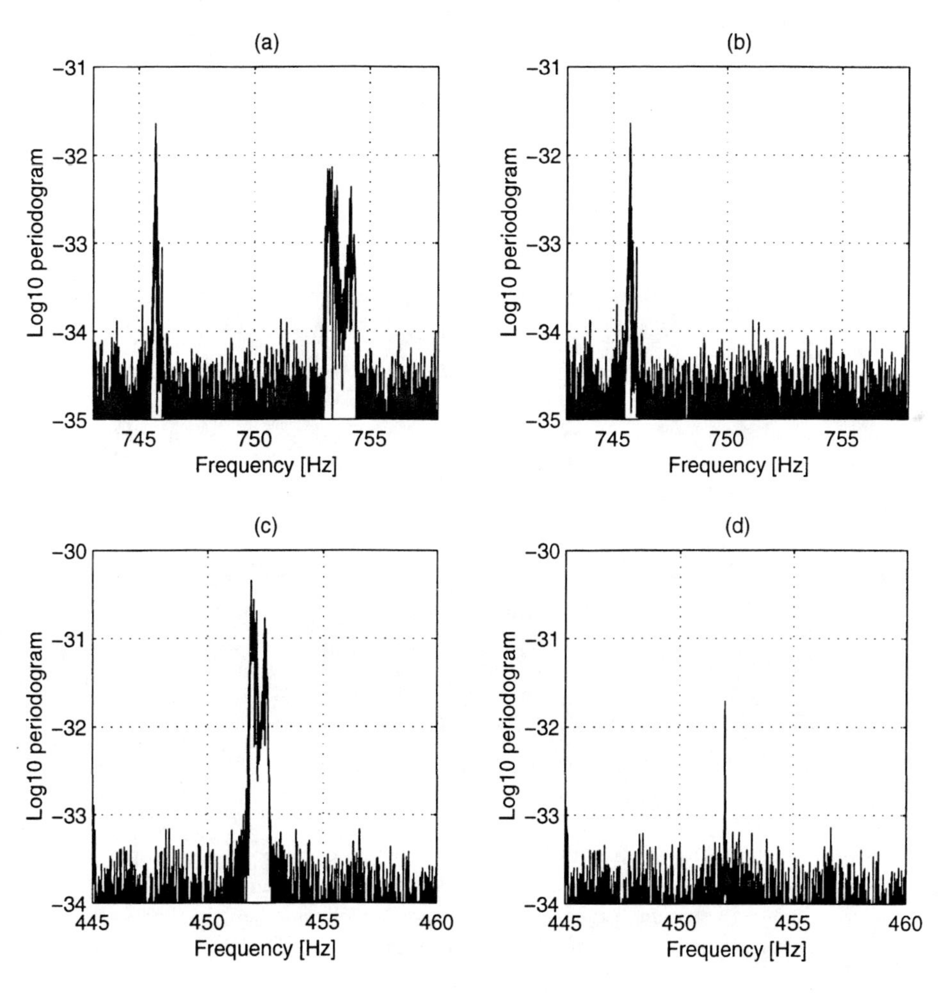

FIGURE 1. Decimal logarithm of the periodogram of 2^{19} points (approximately 2 minutes) of the Glasgow data. (a) One of the harmonics near 754 Hz. (b) The same data after the removal of the interference as described in the text. (c) The same experimental data with an artificial signal added at 452 Hz. (d) The data in (c) after the removal of the interference, revealing that the signal remains detectable. Its amplitude is hardly changed by removing the interference.

be found in [5].

In Fig. 1, we show the performance of CLR on two minutes of data. We can see how CLR leaves the spectrum completely clean of the electrical interference and keeps the intrinsic detector noise. CLR is also applied to the true experimental data with an external simulated signal at 452 Hz, that is initially hidden due to its weakness and we succeed in removing the electrical interference while keeping the signal present in the data, obtaining a clear outstanding peak over the noise level.

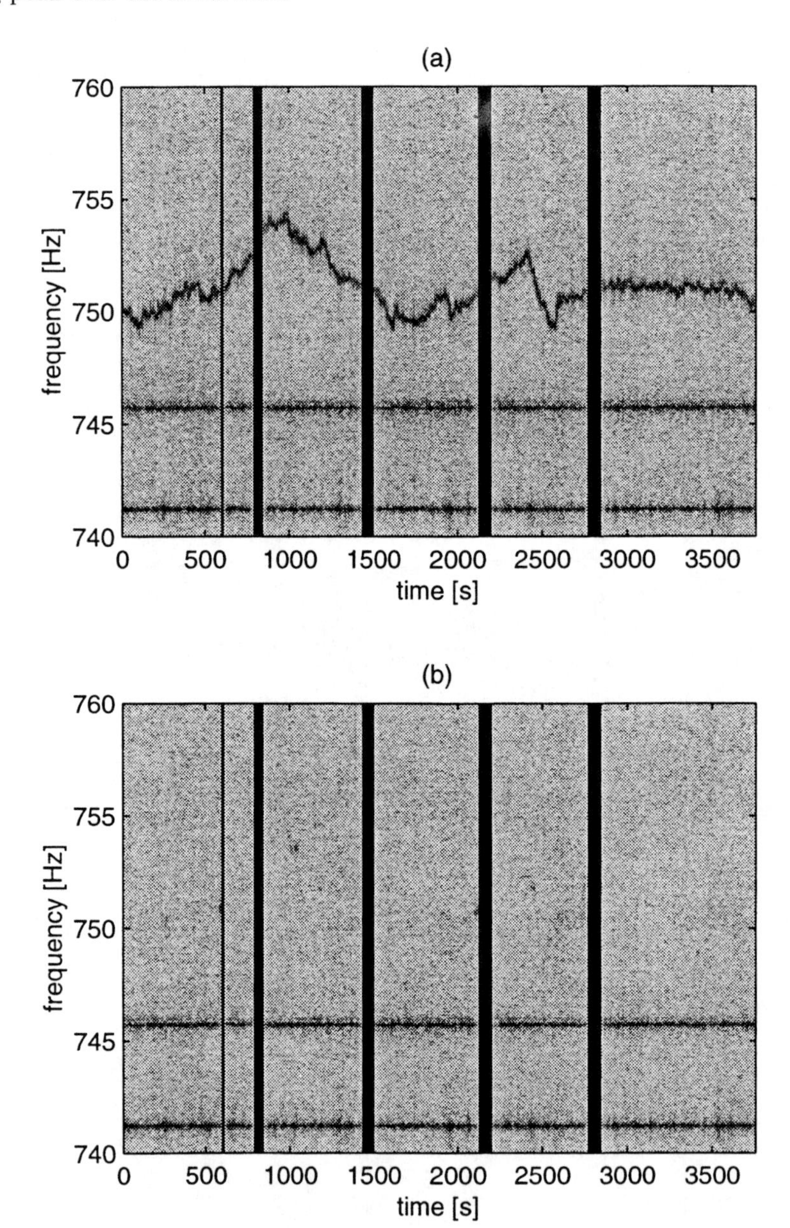

FIGURE 2. Comparison of a zoom of the spectrogram. The dark areas correspond to the periods in which the detector is out of lock. (a) is obtained from the Glasgow data. We can observe the wandering of the incoming electrical signal. The other two remaining lines at constant frequency correspond to violin modes. (b) The same spectrogram as in (a) after applying coherent line removal, showing how the electrical interference is completely removed.

In Fig. 2, we compare a zoom of the spectrogram for the frequency range between 740 and 760 Hz. There we can see the performance of the algorithm on the whole data stream. We show how a line due to an harmonic of the electrical interference in the initial data is removed.

We are interested in studying possible side effects of the line removal on the statistics of the noise in the time domain. We observe that the mean value is hardly changed. By contrast, a big difference is obtained for the standard deviation. For the Glasgow data, its value is around 1.50 Volts. After the line removal, the standard deviation is reduced, obtaining a value around 1.05 Volts. This indicates that a huge amount of power has been removed.

Further analysis reveals that values of skewness and kurtosis are getting closer to zero after the line removal. Values of skewness and kurtosis near zero suggest a Gaussian nature. Therefore, we are interested in studying the possible reduction of the level of non-Gaussian noise. To this end, we take a piece of data and we study their histogram, calculating the number of events that lie between different equal intervals. If we plot the logarithm of the number of events versus $(x-\mu)^2$, where x is the central position of the interval and μ is the mean, in case of a Gaussian distribution, all points should fit on a straight line of slope $-1/2\sigma^2$, where σ is the standard deviation. We observe that this is not the case (see figure 3). Although, both distributions seem to have a linear regime, they present a break and then a very heavy tail. The two distributions are very different. This is mainly due to the change of the standard deviation.

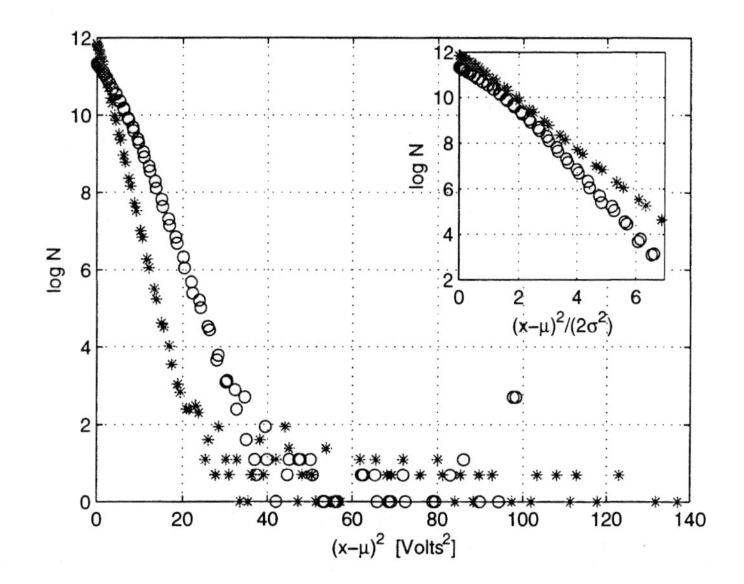

FIGURE 3. Comparison of the logarithm plot of the histogram for 3.2×2^{19} points as a function of $(x-\mu)^2$. The circles correspond to the Glasgow data and the stars to the same data after removing the electrical interference. The Glasgow data is characterized by $\mu = -0.0182$ Volts and $\sigma = 1.5151$ Volts. After the line removal, we obtain the values of $\mu = -0.0182$ Volts and $\sigma = 1.0449$ Volts. In the right-hand corner, there is zoom of the original figure, but rescaled so that the abscissa corresponds to $(x-\mu)^2/2\sigma^2$. If the data resembles a Gaussian distribution, we will expect a single straight line of slope -1. This is not the case for the Glasgow data, but it seems to be satisfied for the clean data up to 4σ. The large number of points in the highest bin of the Glasgow data is an effect of saturating the ADC. These points are spread to higher and lower voltages by line removal.

We can zoom the 'linear' regime and change the scale in the abscissa to $(x-\mu)^2/(2\sigma^2)$. Then, any Gaussian distribution should fit into a straight line of slope -1. We observe that after removing the interference, it follows a Gaussian distribution quite well up to 4σ. The original Glasgow data does not fit a straight line anywhere.

In order study the Gaussian character, we have also applied two statistical tests to the data: the chi-square test that measures the discrepancies between binned distributions, and the one-dimensional Kolmogorov-Smirnov test that measures the differences between cumulative distributions of a continuous data.

We computed the significance probability for every 2^{12} points of the data using both tests and we checked whether the distribution are Gaussian or not. The two tests are not equivalent but in any case, the values of the significance probability would be close to unity for distributions resembling a Gaussian distribution. In both tests, the significance probability increased after removing the electrical interference, showing that this procedure suppresses some non-Gaussian noise, although, generally speaking, the distribution was still non-Gaussian in character, presumably because of the heavy tails, which are not affected by line removal. See [5] for details.

ACKNOWLEDGMENTS

We would like to thank C. Cutler, A. Królak, M.A. Papa and A. Vecchio for helpful discussions, and J. Hough and the gravitational waves group at Glasgow University for providing their gravitational wave interferometer data for analysis. This work was partially supported by the European Union, TMR Contract No. ERBFMBICT972771.

REFERENCES

1. Abramovici A., Althouse W., Camp J., Durance D., Giaime J.A., Gillespie A., Kawamura S., Kuhnert A., Lyons T., Raab F.J., Savage Jr. R.L., Shoemaker D., Sievers L., Spero R., Vogt R., Weiss R., Whitcomb S., Zucker M., *Physics Letters A* **218**, 157 (1996).
2. Maischberger K., Ruediger A., Schilling R., Schnupp L., Winkler W., Leuchs G., "Status of the Garching 30 meter prototype for a large gravitational wave detector", eds. Michelson P.F., En-Ke Hu & Pizzella G., in *Experimental Gravitational Physics.* World Scientific, Singapore, 1991, pp. 316-21.
3. Jones G.S., 1996, *Fourier analysis of the data produced by the Glasgow laser interferometer in March 1996*, internal report, Cardiff University, Dept. of Physics and Astronomy.
4. Sintes A.M., Schutz B.F., "Removal of interference from gravitational wave spectrum", in *proceedings of the 2nd workshop on Gravitational Wave Data Analysis.* Orsay, France, 1998.
5. Sintes A.M., Schutz B.F., "Coherent Line Removal: A new technique to remove interference from the gravitational wave spectrum", 1998, to be published.
6. Thomson D.J., 1982, "Spectrum Estimation and Harmonic Analysis", in *Proceedings of the IEEE*, **70**, 1055-96 (1982).
7. Percival D.B., Walden A.T., *Spectral analysis for physical applications*, first edition, Cambridge University Press, 1993.
8. Widrow B., Glover J.R., McCool J.M., Kaunitz J., Williamns C.S., Hearn R.H., Zeidler J.R., Dong E., Goodlin R.C., "Adaptive Noise Cancelling: Principles and Applications", in *Proceedings of the IEEE*, **63**, 1692-1716 (1975).
9. Schutz B.F., "Detection of Gravitational Waves", in *Relativistic Gravitation and Gravitational Radiation*, Marck J.A., Lasota J.P., eds., Cambridge University Press, 1997.
10. Thorne K.S., eds. Kolb E.W. & Peccei R., in *Proceedings of Snowmass 1994 Summer Study on Particle and Nuclear Astrophysics and Cosmology.* World Scientific, Singapore, 1995, p. 398.
11. Nicholson D., Dickson C.A., Watkins W.J., Schutz B.F., Shuttleworth J., Jones G.S., Robertson D.I., Mackenzie N.L., Strain K.A., Meers B.J., Newton G.P., Ward H., Cantley C.A., Robertson N.A., Hough J., Danzmann K., Niebauer T.M., Rüdiger A., Schilling R., Schnupp L., Winkler W., *Physics Letters A* **218**, 175-180 (1996).

TECHNOLOGY FOR

GRAVITATIONAL-WAVE DETECTORS

IN SPACE

Laser Phase-Locking Techniques for LISA : Experimental Status

P. W. McNamara, H. Ward, J. Hough

Department of Physics and Astronomy, Kelvin Building, University of Glasgow, Glasgow G12 8QQ, Scotland, UK

Abstract. Phase-locking of a slave laser to weak incoming light is an essential ingredient of the laser transponders to be used in LISA. Here we report on an experiment to evaluate such a weak-light phase-locking scheme, in which two independent diode-pumped Nd:YAG non-planar ring oscillators are locked together with a 15MHz frequency offset. With a weak-light intensity of 17 nW we observe close to shot-noise limited performance of the phase locking system over a frequency range from 800 Hz down to 10 Hz. Excess noise is observed at lower frequencies – reaching two orders of magnitude above the shot-noise limit at 10^{-4} Hz – most probably due to seismic and temperature effects on the optical layout.

The level of the unlocked relative phase fluctuations is observed to be anomalously low, compared with that deduced from separate experiments in which one of the lasers was frequency stabilised to a reference cavity [1,2]. The possibility of partial injection locking of the lasers due to leakage light cannot be excluded. Caution in interpretation of the results is therefore warranted.

INTRODUCTION

LISA, the Laser Interferometer Space Antenna, is a proposed spaceborne gravitational wave detector, designed to detect gravitational waves in the frequency range of 10^{-4} to 10^{-1} Hz. LISA consists of three widely-separated spacecraft providing three arms whose relative lengths are monitored using laser light. In essence each pair of arms operates as a simple one-bounce Michelson interferometer, with the laser beams transmitted and received *via* 30 cm diameter Cassegrain telescopes. There are, however, two significant differences from a conventional interferometer. Firstly there is no beam splitter; instead, each spacecraft has two independent lasers that are phase-locked to provide the two beams that are projected along adjacent arms. A further difference arises from the fact that the very long arm length of LISA (5×10^9 m) results in the emitted beam diverging to a diameter of over 25 km by the time it propagates to the spacecraft at the end of an arm. Direct reflection of this light by a 30 cm diameter optic would lead to only $\sim$ 200 photons per hour being detected back at the originating spacecraft. The LISA target sensitivity can only be achieved with the less significant photon counting fluctuations associated with a much larger detected power. The solution is to phase-lock the local laser in a spacecraft to the weak incoming light and to send the higher power, phase-locked, light back to the originating craft. The detected power is thereby increased to approximately 10^8 photons per second ($\sim$ 70 pW) back at the originating craft. With this scheme, the arrangement at the far spacecraft can be thought of as an "amplifying mirror".

The sensitivity goals for LISA, coupled with the unavoidable differences in the arm lengths, set a relatively stringent requirement on the tolerable laser frequency fluctuations. Active frequency stabilisation of one laser in the LISA system to a resonance of a stable reference cavity is therefore planned. Initial [1] and current [2] results of frequency stabilisation experiments are reported elsewhere.

EXPERIMENTAL DEMONSTRATION OF PHASE-LOCKING

The ultimate goal of the current experiment is to demonstrate phase locking of two LISA-style lasers using relative powers comparable to those planned for LISA. Two separate diode-pumped Nd:YAG non-planar ring oscillators (NPROs) [3] manufactured at the Laser Zentrum, Hannover, are used as the light sources. In our

CP456, *Laser Interferometer Space Antenna*
edited by William M. Folkner
 1-56396-848-7/98/$15.00

experiment the lasers are operated at output powers of order one hundred milliwatts. Most of the light is split off for frequency stabilisation purposes, leaving ~ 10 mW for the phase locking experiments.

The general approach is to use an attenuated fraction of light from the 'master' laser to simulate the weak light received by a LISA spacecraft. The other 'slave' laser is then phase locked to this weak light with an offset frequency provided by a stable signal generator. An independent out-of-loop measurement of the resulting relative phase fluctuations is then made using a separate – and higher power – optical mixing of beams from the two lasers.

Optics

The optical arrangement used is shown in Figure 1. Polarisation techniques are used to obtain the two optical recombination paths and to produce the necessary effective optical attenuation of the master laser.

The polarisation of the master laser output beam is linearised and oriented using a combination of quarter and half wave plates. The light is then passed through a polariser, ensuring linear polarisation in the s-plane propagates to the main polarising beam splitter.

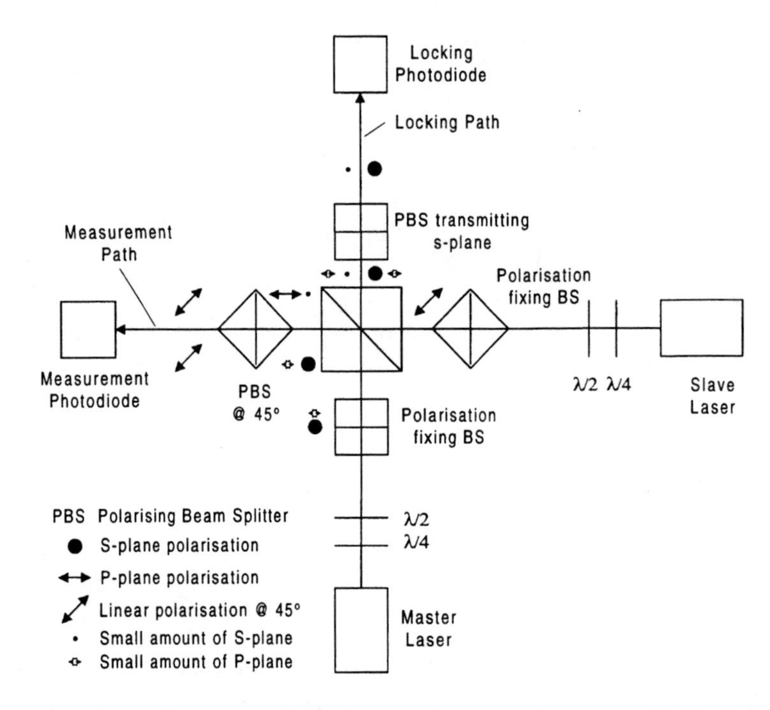

FIGURE 1. Optical layout for phase locking the two independent Nd:YAG NPROs.

The polarisation of the light from the slave laser is also corrected, but this time it is set at 45° to that of the master laser. The slave laser light then strikes an orthogonal port of the main polarising beam splitter. This splitter directs the incoming light along two paths – the locking path and the measurement path. The interference signal from the locking path is used to drive the feedback system for the phase-locking, and the signal from the measurement path is used to measure the residual relative phase noise of the slave laser with respect to the master.

The two beams along the locking path are *a*) approximately half of the slave laser light, and *b*) a tiny fraction of the master light. The light from the master laser along the locking path, at the output of the main beamsplitter, is composed of two orthogonal polarisation components – the desired small amount of light ($\sim 10^{-5}$ of the input master laser intensity) in the s-plane leaking through the polariser, and an unwanted component in the p-plane. This unwanted component arises from the p-component of the elliptically polarised light coming from the master laser, imperfectly rejected by the polariser following the wave plates at the output of the laser. The slave light along the locking path after the main beamsplitter will also be composed of two components, although the s-plane component will be very much larger than the component in the p-plane. After the main polarising splitter, the light in the locking path passes through another polariser, oriented to transmit s-polarisation. Without this final polariser the unwanted p-components from the master and slave would give rise to a beat note at the locking photodiode that was not insignificant compared with the wanted signal due to interference of the s-components. Any residual p-component that survives to the locking photodiode is orthogonal to the s-components and hence does not interfere; the extra intensity of such a component is negligible and does not affect the shot noise limit which is dominated by the weak s-component arriving from the master laser.

The second output port of the main polarising beamsplitter is used for the independent phase noise measurement. Most of the master laser light is sent along this path (s-polarisation), along with approximately half of the slave light (p-polarisation). Interference is produced by passing the light through a polarising splitter oriented at 45° to the two polarisations. The intensity of the master laser at the photo-detector in this path is much higher than that in the locking path, and hence the shot noise limited sensitivity to relative phase fluctuations is correspondingly lower.

Phase-locking Servo

As in the case of LISA, frequency offset locking is adopted, with an offset frequency of 15 MHz chosen for this demonstration. The signal from the locking photo-detector is bandpass filtered around 15 MHz, and the demodulated signal is used as the error point for the electronic feedback servo.

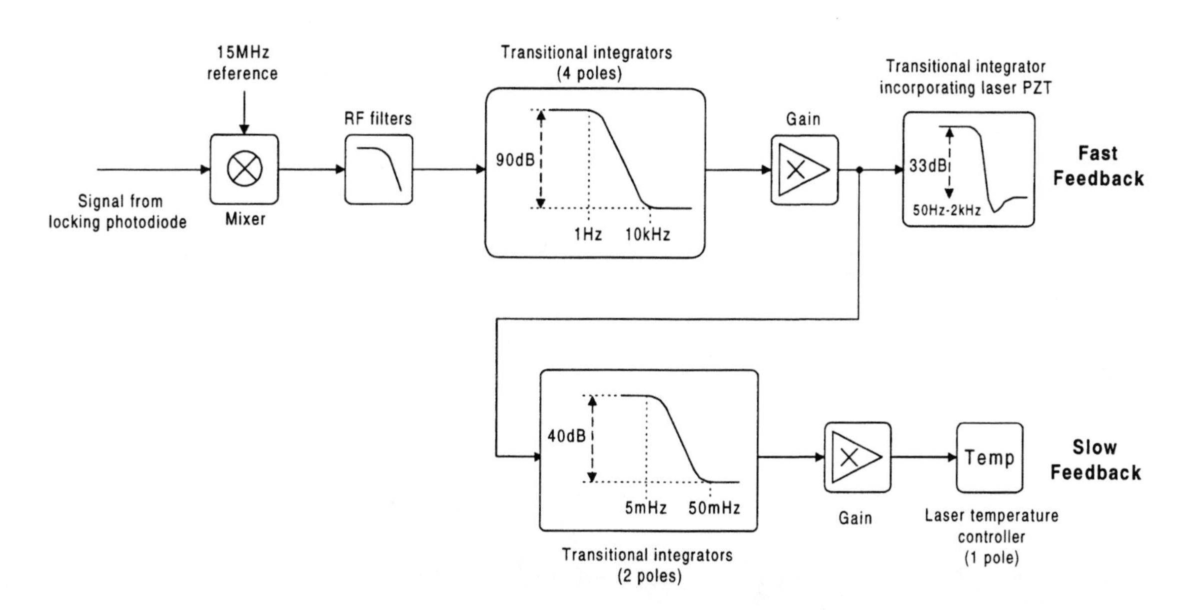

FIGURE 2. Schematic diagram of the phase-locking servo system.

A schematic diagram of the feedback system is shown in Figure 2. The feedback signal acts *via* two elements. Fast signals (above $\sim$ 1 Hz) are fedback to a piezo transducer that changes the frequency of the slave laser by stressing the YAG crystal. Slower signals are used to drive the temperature controller of the YAG crystal. Transitional integrators are used to provide adequate low frequency gain in the servo system while ensuring stability at the unity gain frequency of $\sim$ 20 kHz.

During acquisition of lock the master laser remains at a fixed frequency, while the temperature controller of the slave laser is used to tune the frequency of the laser over many GHz to find the region at which the frequencies of the two lasers are close to being separated by the 15 MHz offset frequency. The output of the mixer gives a periodic signal at the difference frequency between the laser beat note and the offset frequency. As the lasers approach the locking point, the output signal from the mixer moves to lower frequency. Within the bandwidth of the phase-locking servo, the net effect is to produce an asymmetric output from the mixer, and hence a net DC level that drives the slave frequency towards the offset master frequency; phase lock occurs automatically once the mixer inputs reach the same frequency.

The measurement signal is also detected on a resonant photo-detector, and demodulated using the same 15 MHz reference. The mixer output is amplified and measured using both a commercial spectrum analyser and a computer data acquisition system. The phase of the reference signal to the mixer is adjusted to give nominally zero DC output from the mixer, in order to maximise the signal to noise ratio of the measurement system.

RESULTS

The experiment was conducted with master and slave laser powers at the locking photodiode of 17 nW and 5 mW, respectively. These power levels were deduced from measurements of the DC and RF photocurrents produced by the locking photodiode. The calculated shot noise limit to the achievable relative phase noise using these powers was $3.6 \times 10^{-6}\,\mathrm{rad}/\sqrt{\mathrm{Hz}}$.

Figure 3 shows a spectrum of the relative phase noise of the slave laser, over a frequency range from 1 Hz to 10^3 Hz. Three traces are shown, together with a horizontal line at the level of the calculated shot noise limit for the powers used. The uppermost trace shows the unstabilised relative phase fluctuations and is derived from the feedback signal in the phase-locking loop. The lowermost trace is an upper bound to the calibrated error point of the phase-locking loop; this trace is dominated by measurement noise below $\sim$ 100 Hz. The intermediate trace is obtained from the independent measurement system and demonstrates that the phase-locking system achieved near shot-noise-limited performance over a wide frequency range from 800 Hz down to 10 Hz.

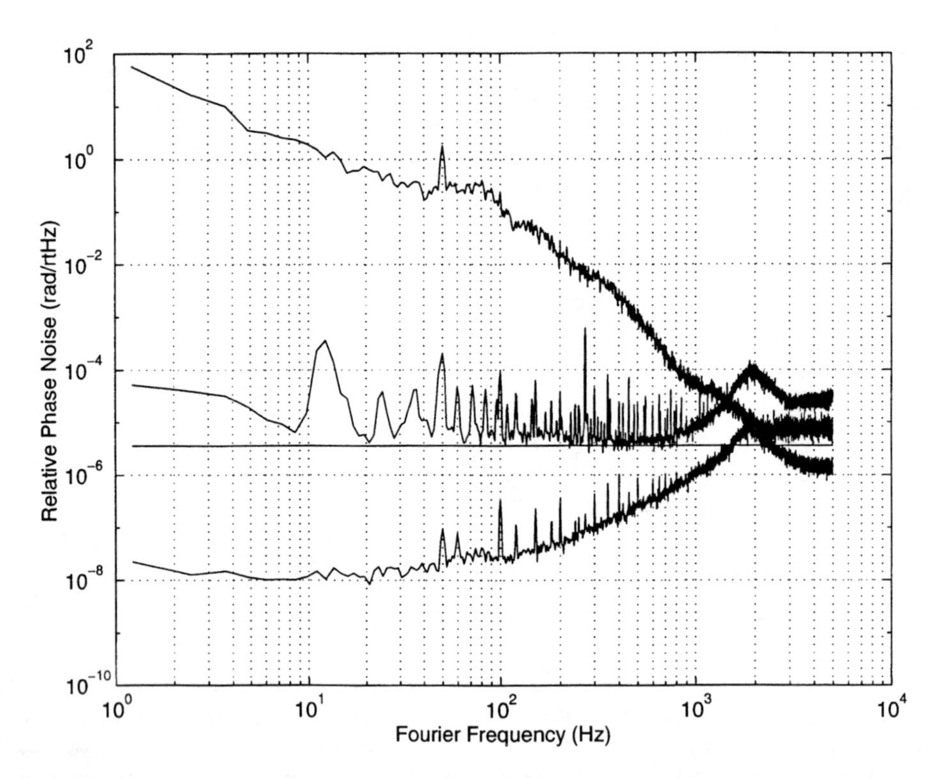

FIGURE 3. Performance of the phase-locking system: the uppermost trace shows the unstabilised noise; the middle trace is the independent measurement; the lowermost trace is an error point measurement from the locking loop; the shot noise limit is indicated by a horizontal line.

Figure 4 shows further detail of the residual phase noise over a frequency range extending down to 10^{-4} Hz. Excess noise is observed - reaching two orders of magnitude above the shot-noise limit at 10^{-4} Hz. This excess noise was not reduced by active temperature stabilisation of the sensitive mixers and post-detection amplifiers in the locking and measurement electronics. Seismic noise and temperature effects on the optical layout are the most probable causes of the excess noise.

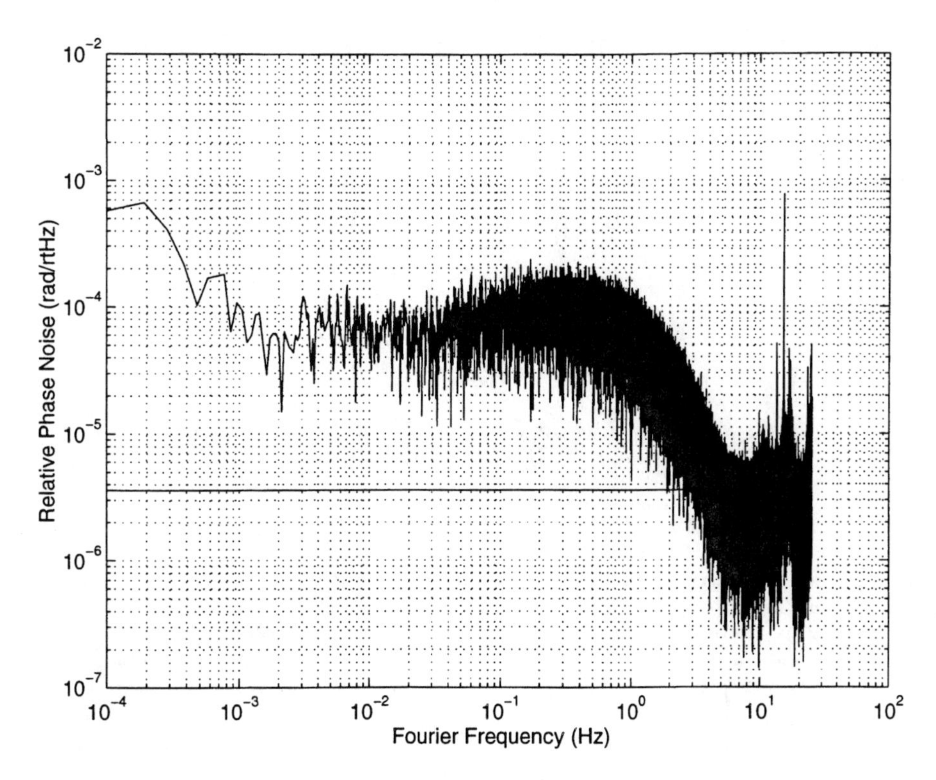

FIGURE 4. Independent measurement of the low frequency relative phase noise; the shot noise limit is indicated by the horizontal line.

DISCUSSION

The results presented are a significant step towards a demonstration of the LISA phase locking goals.

Extrapolation of the results to LISA power levels would involve a weak-light power reduction from that used in these experiments by a factor of ~ 240. This reduction would raise the shot noise limited relative phase noise stability by a factor of ~ 16 and would mean that the demonstrated stability was within a factor of 6 of the LISA goal down to 10^{-4} Hz. Further work in this direction is indicated.

One important caveat remains. Separate experiments [1,2] aimed at demonstrating the frequency stabilisation requirements for LISA have shown levels of unstabilised laser phase noise that are significantly higher than the apparent unstabilised relative phase noise demonstrated in the phase-locking experiments. One possible interpretation of these findings is that there is some degree of injection phase locking due to light leaking back to the lasers. Experiments with improved optical isolation will be needed to investigate this hypothesis.

ACKNOWLEDGEMENTS

The authors wish to thank the Particle Physics and Astronomy Research Council, the Scottish Higher Education Funding Council, the Rutherford Appleton Laboratory, and the University of Glasgow for their continued support during this work.

REFERENCES

1. McNamara, P. W., Ward, H. and Hough, J., *Class. Quantum Grav.*, **14**, 1543-1547 (1997)
2. McNamara, P. W., Ward, H. and Hough, J., submitted to *Advances in Space Research*
3. Kane, T.J. and Byer, R.L. *Optics Letters* Vol. **10**, No. 2 65-67 (1985).

Laser development and laser stabilisation for the space-borne gravitational wave detector LISA

M. Peterseim[1,2], O.S. Brozek[2], K. Danzmann[2], I. Freitag[1], P. Rottengatter[1], A. Tünnermann[1] and H. Welling[1]

(1)Laser Zentrum Hannover e.V.
Hollerithallee 8, D-30419 Hannover, Germany
FAX:+49-511-2788-100, mp@lzh.de

(2) Max-Planck-Institut für Quantenoptik, Außenstelle Hannover
Callinstrasse 38, D-30167 Hannover, Germany

Abstract. For the interferometric readout of the 10^7 km optical path length, the LISA mission will rely on a compact, reliable and highly efficient source of stable radiation. The strain sensitivity of the instrument will be limited by photon shot noise in the mHz frequency regime and therefore an output power of at least 1 W is required. The noise added by power and frequency fluctuations of the laser is negligible only for a relative power stability of $\widetilde{\delta P}/P < 2 \times 10^{-4}/\sqrt{\mathrm{Hz}}$ and for a frequency noise spectral density of $< 30\,\mathrm{Hz}/\sqrt{\mathrm{Hz}}$ within the LISA detection band. The only light source approaching the performance demanded for LISA is a stabilised monolithic diode pumped Nd:YAG ring laser. We report on recent progress in designing and stabilising this laser type for the LISA mission.

I THE LISA LASER SYSTEM

A Monolithic miniature Nd:YAG ring laser

The laser system to be used in the LISA mission is a diode laser-pumped monolithic miniature Nd:YAG ring laser (NPRO) which can generate a diffraction limited beam at $1064\,\mu$m of up to 1.5 W in the configuration described below. The optical beam path in the NPRO crystal is determined by three internal total reflections and one reflection at the negatively curved front surface [1]. The front surface is dielectrically coated, reflecting about 97 % of the 1064 nm laser radiation and highly transmitting the pump radiation at 808 nm (see Fig. 1).

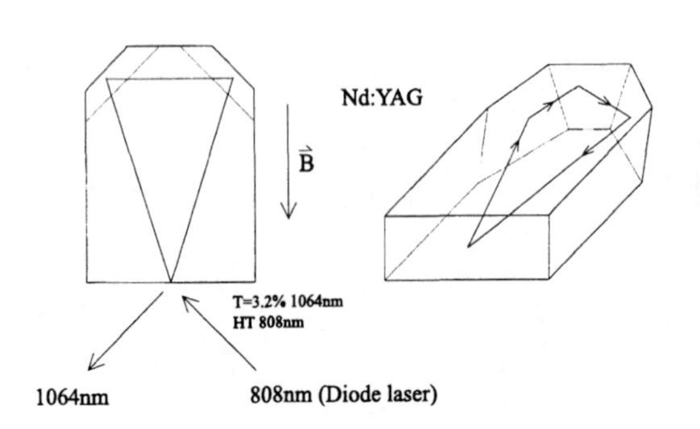

The high frequency stability required for the LISA mission can only be achieved because of the high intrinsic stability of the NPRO. This stability results from the monolithic and compact design of the resonator and from the outstanding thermal properties of the host material YAG (Yttrium Aluminium Garnet $Y_3Al_5O_{12}$). The low coefficient of thermal expansion ($7 \times 10^{-6}\,\mathrm{K}^{-1}$) and the low temperature dependence of the index of refraction ($9.05 \times 10^{-6}\,\mathrm{K}^{-1}$) make the laser rather insensitive to temperature fluctuations.

FIGURE 1. The beam path in the monolithic, non-planar ring resonator is determined by three total internal reflections and one reflection at the dielectrically coated front surface.

CP456, *Laser Interferometer Space Antenna*
edited by William M. Folkner
 1-56396-848-7/98/$15.00

To date there are only few alternatives to Nd:YAG as the active crystal medium. Ytterbium can be used as the active ion, because the efficiency is higher than for Nd^{3+}(Neodymium)-ions, but the pump power requirements are also higher because of the extremely high threshold. This results from Yb^{3+} being a three level-system rather than the four level Nd^{3+}. Yttrium Vanadate (YVO_4) is an alternative candidate for the host material, but its polarisation selectivity interferes with the optical diode required to enforce unidirectional oscillation. Also the thermal properties of YVO_4 are much worse than those of the YAG.
However, the main advantage of Nd:YAG is its availability in radiation hardened quality, which can be used in deep space without significant change of performance.

B LISA laser system design

The laser system consists of two major components: the laser head and the supply unit. Both the laser head and the supply unit are mounted on a carbon-carbon radiator to radiate away the heat that is due to the consumed electrical power not converted into optical power.

The **laser head** consists of a Nd:YAG NPRO pumped by two long life aluminium-free InGaAsP laser diodes. These single stripe devices have maximum cw output power of 2000 mW. The nominal single-mode, cw output power of the NPRO in this configuration is 1500 mW, but this is downrated for LISA to 1000 mW to improve lifetime and reliability properties. The nominal constant power consumption for the 1000 mW of output power of the complete laser system is approxemately 10 W.
The pumplight from each laser diode is transferred into the crystal by imaging the diodes' emitting area of $1\,\mu m \times 200\,\mu m$ at unity magnification onto the entrance surface of the crystal, using two identical lenses with plano-convex surfaces to minimise spherical aberration (best form lens shape). A polarising beamsplitter is inserted between the lenses to combine the pump light from the two diodes, which are orthogonal in polarisation. The optical elements are made of fused silica, which is proven to be resistant to radiation levels as encountered in deep space.

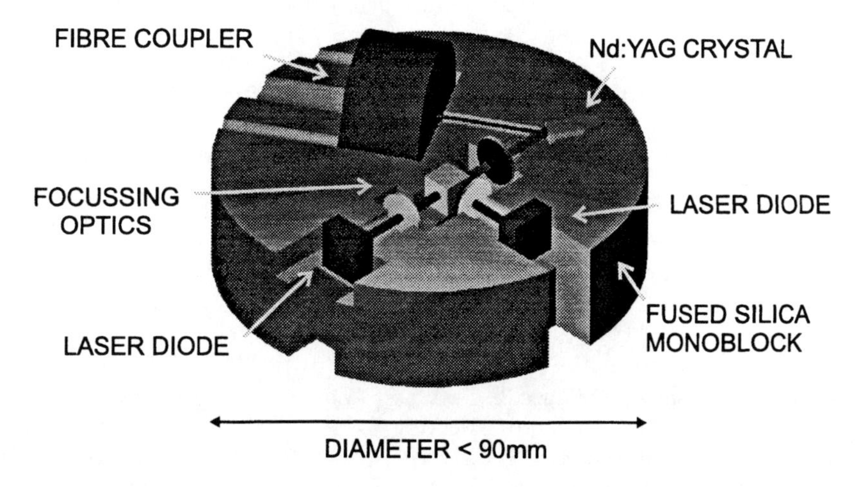

FIGURE 2. LISA laser system fused silica monoblock design. The laser head consists of a Nd:YAG NPRO pumped by two laser diodes. The nominal single-mode, cw output power of the NPRO in this configuration is 1500 mW.

All of the above mentioned components are glued to a solid spacer to ensure mechanical stability (cf. figure 2). That spacer is manufactured from a single block of fused silica. Glued to the backs of the diodes and to the bottom of the crystal are three heat sinks (not shown in figure 2), which serve as the mechanical and thermal interface between the laser head and the radiator plate. There is a heater integrated in each heat sink to control the operating temperature of the diodes and the crystal.

The **supply unit** mainly contains two current sources for the laser diodes, three temperature controllers, two for the diodes and one for the crystal, the mixer and feedback servo for the frequency stabilisation and the feedback circuit for power stabilisation. The supply unit power interface to the spacecraft power subsystem will be the only power interface between the laser system and the spacecraft.

II FREQUENCY STABILISATION FOR LISA

A Review of requirements

The presence of laser frequency noise can lead to an error in the measurement of each arm length. If the difference in two arm lengths is Δx and the relative frequency stability of the laser is $\widetilde{\delta f}/f$, the apparent displacement noise is given by:

$$\widetilde{\delta x} = \Delta x \frac{\widetilde{\delta f}}{f} . \tag{1}$$

For the 5×10^6 km arms of LISA, a maximum value of Δx of the order of 10^5 km is likely. For a relative arm length measurement of $2 \times 10^{-12}\,\mathrm{m}/\sqrt{\mathrm{Hz}}$, which is desired to achieve the envisaged overall sensitivity, a laser stability of $6 \times 10^{-6}\,\mathrm{Hz}/\sqrt{\mathrm{Hz}}$ is required.

The current LISA baseline foresees as the primary method of stabilisation to lock the frequency of the laser system onto a resonance of a Fabry-Perot cavity mounted on the optical bench. With the temperature fluctuations inside each craft limited in the region of 10^{-3} Hz to approximately $10^{-6}\,\mathrm{K}/\sqrt{\mathrm{Hz}}$ a cavity formed of ULE allows a stability level of approximately $30\,\mathrm{Hz}/\sqrt{\mathrm{Hz}}$. This level of laser frequency noise is clearly much worse than the required $6 \times 10^{-6}\,\mathrm{Hz}/\sqrt{\mathrm{Hz}}$ and a further correction scheme is required. Such a correction is provided by comparing the mean phase of the light returning in two adjacent arms with the phase of the transmitted light. The phase difference, measured over the time of flight in the two arms, allows an estimate of laser frequency noise to be made. A detailed analysis can be found in [4].

Extensive work has been performed in the field of frequency stabilising non-planar Nd:YAG ring lasers using optical resonators. Unfortunately, most of the work was focussed on the acoustic frequency regime, ranging from a few Hertz to several kilohertz, where the residual frequency fluctuations of the laser have been reduced to $10^{-2}\mathrm{Hz}/\sqrt{\mathrm{Hz}}$ [2] [5].

In the frequency regime below 1 Hz the temperature fluctuations of the reference resonator usually set a very stringent limit to the achievable stability. Using an optical resonator at cryogenic temperature is one possibility to overcome this problem and has been successfully applied to reduce the frequency fluctuations below $1\,\mathrm{Hz}/\sqrt{\mathrm{Hz}}$ at Fourier frequencies down to the millihertz regime [6]. Another approach to reduce the thermal fluctuations of the reference resonator, that has been applied in this work, is to place the resonator inside a very stable thermal shielding and operate it at room temperature.

B Experimental setup

The basic experimental setup is shown in figure 3. Two laser systems (model *Innolight GmbH Mephisto 800*) are stabilised to two identical reference resonators making use of a rf-reflection locking scheme known as Pound-Drever-Hall scheme: Light from the laser is directed through an electro-optic modulator, which phase modulates the light at 12 MHz or 29 MHz respectively. About 10 mW is split off and mode-matched into the resonator, which consists of two coated ULE substrates that have been optically contacted to a hollow, cylindrical ULE spacer of 210 mm length and a diameter of 80 mm. The substrate coatings both have a transmission of 700 ppm and negligible losses, leading to a finesse of roughly 10,000, which is comparable to what will be chosen for the LISA reference resonator. The light reflected off that cavity is detected on a photodiode and demodulated, producing the bipolar error signal, that is amplified and fed back to the laser frequency actuators. Fast correction signals are sent to a piezo-electric transducer mounted on top of the Nd:YAG crystal and slow signals are fed back to the crystal temperature control.

The most crucial part of the experiment is the ultra-stable housing of the two reference cavities. The purpose of this housing is to simulate the thermal conditions that will be encountered on the LISA optical bench. That means to reduce the thermal fluctuations to a level of $10^{-6}\mathrm{K}/\sqrt{\mathrm{Hz}}$ at 1 mHz, which is about the expected temperature stability on the bench, as recent calculations have shown. To reach that goal, the cavity plus shielding is placed inside a cylindrical vacuum chamber of 500 mm diameter and 600 mm length. Inside the

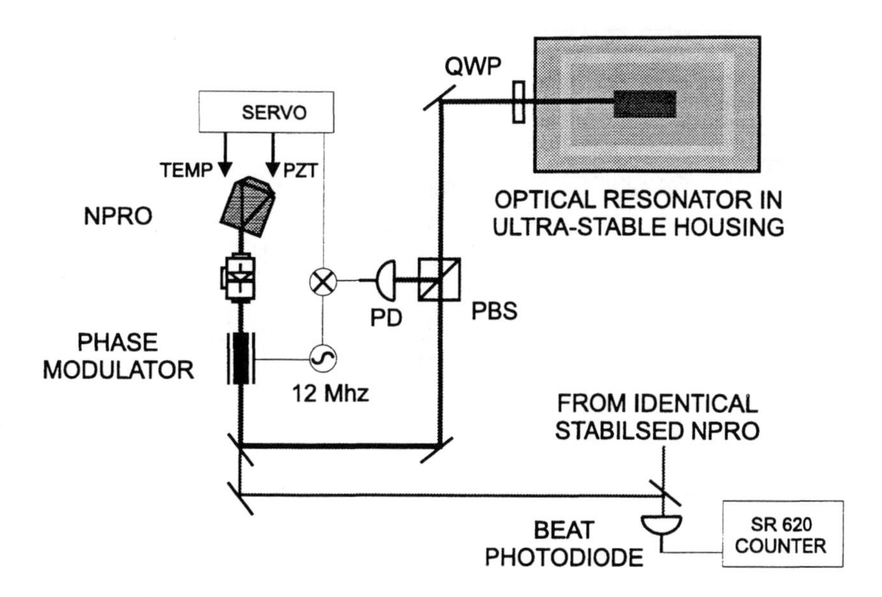

FIGURE 3. Scheme of frequency noise reduction setup. Light from the monolithic non-planar Nd:YAG ring laser is phase modulated at 12 MHz by a resonant modulator. About 10 mW are split off and mode-matched into a high finesse ULE optical resonator. The resonator is placed inside a vacuum chamber and 4 layers of cold-coated stainless steel for optimum thermal isolation to outside temperature fluctuations. The light reflected off the cavity is detected on a photodiode and used as a bipolar error signal for the frequency locking of the laser system.

tank the cavity is surrounded by four concentric cylinders of gold-coated stainless steel of very high surface quality that have an excellent reflectivity for thermal radiation at room temperature. The pressure inside the chamber is below 10^{-6} mbar, so the heat transfer by convection is negligible compared to radiation and conduction heat transfer. The contribution of the latter is reduced by mounting the individual steel cylinders on very thin ceramic spacers made of the poor heat conductor Macor.

In summary the three inner cylinders are the equivalent of three thermal low-passes. These consist of the heat capacity of the steel and the heat resistance of the low thermal coupling between the cylinders. Regarding the specific material constants, all three low-passes should have corner frequencies below 10^{-5} Hz and therefore, neglecting any heat leaks, outside temperature fluctuations should be reduced by at least a factor of 10^{-6} at a Fourier frequency of 1 mHz. Assuming an outside temperature stability of $1\,\mathrm{K}/\sqrt{\mathrm{Hz}}$ at 1 mHz (corresponding approximately to 1 $\mathrm{K_{rms}}$ over 1000 s) and a thermal expansion coefficient of ULE of 10^{-8}, this experimental setup should allow a reduction of the frequency fluctuations of the NPRO to $30\,\mathrm{Hz}/\sqrt{\mathrm{Hz}}$ at 1 mHz.

C Results

The performance of the stabilisation has been evaluated by measuring the beat frequency of two identical laser systems, each stabilised to its own resonator. The two resonators have been thermally decoupled by placing them in individual vacuum chambers and thermal isolation systems. Figure 4 shows the time series of the beat frequency, that has been measured with the *Stanford Research Systems* frequency counter model *SR620*. Although the mean frequency of the free running system shows a significant drift of yet unclear origin, the frequency has been considerably reduced. Frequency drift and frequency fluctuations of the stabilised system are below 10 kHz maximum deviation from mean.

The requirements on the LISA laser system are defined in the Fourier domain. Therefore the linear spectral density of the frequency fluctuations is of major interest. Figure 5 shows the linear spectral density for the free running and the stabilised laser system. The required maximum level for the frequency noise is shown as well. Although the frequency fluctuations have been significantly reduced, the residual noise still is a factor of 10

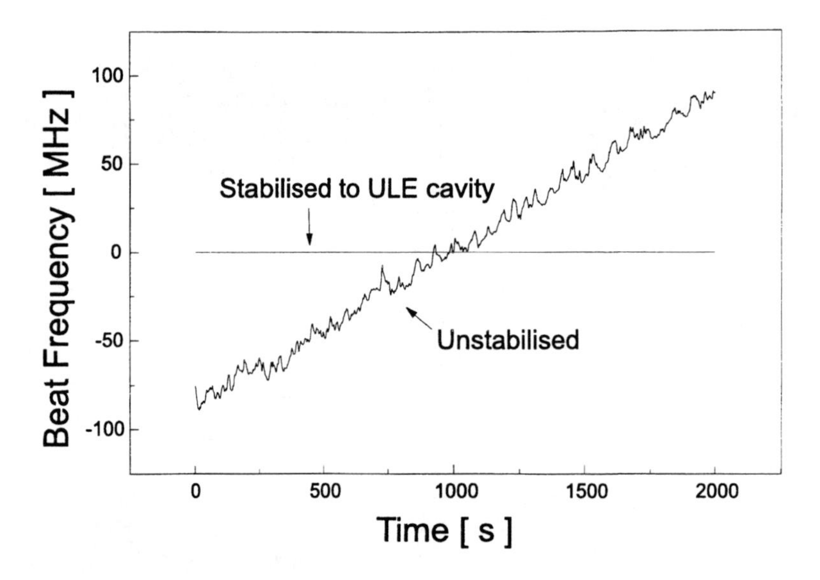

FIGURE 4. Time series of the beat frequency measured with two identical monolithic Nd:YAG ring lasers. One graph shows the case where both lasers are unstabilised. The other graph shows the case where both lasers are stabilised to individual cavities located in their individual thermal shielding.

above the required noise level. In future steps to improve the experiment we will overcome this problem and reduce the residual noise below the LISA specifications. These steps comprise an enhancement of the feedback gain, as the achieved stability is very close to the gain limit. The next step foresees the implementation of an active temperature control system of the vacuum chambers.

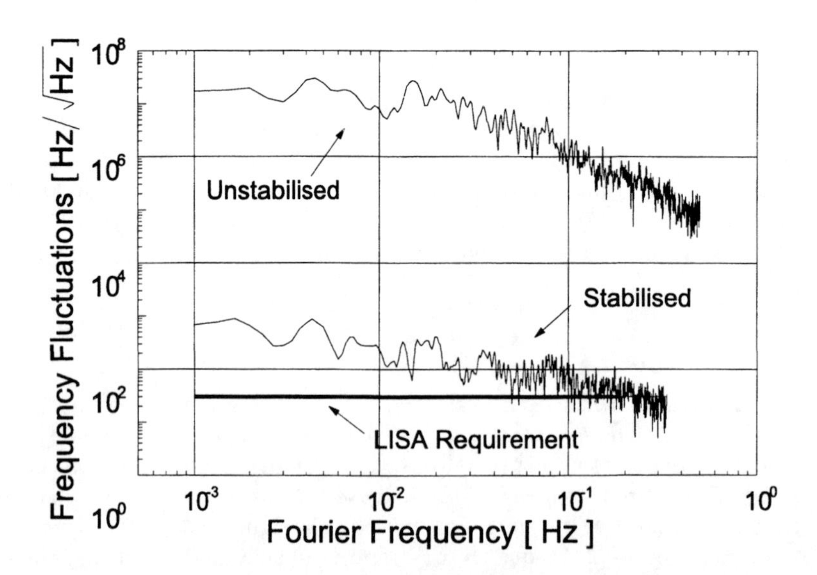

FIGURE 5. The graphic shows the linear spectral densities of frequency fluctuations of the free-running and the stabilised monolithic Nd:YAG ring laser. The measurement was done stabilising two lasers to individual cavities and detecting the beat frequency between the lasers. The residual frequency fluctuations of the stabilised system are below the required noise level for Fourier frequencies above 200 mHz.

III POWER STABILISATION FOR LISA

A Review of requirements

The tolerable limit to laser power noise is set by the radiation pressure effects of the beam to the proof mass in the inertial sensor. The LISA sensitivity goal requires that spurious accelerations of the proof mass are below a level of $10^{-16}\,\mathrm{m\,s^{-2}}/\sqrt{\mathrm{Hz}}$. For a proof mass of 1.3 kg and a reflected light power of 100 μW, the proof mass will undergo a steady acceleration of $5 \times 10^{-13}\,\mathrm{m\,s^{-2}}$. To keep the fluctuating acceleration $< 10^{-16}\,\mathrm{m\,s^{-2}}/\sqrt{\mathrm{Hz}}$ the power stability of the reflected light, and hence of the laser, must be better than $\widetilde{\delta P}/P = 2 \times 10^{-4}/\sqrt{\mathrm{Hz}}$.

The fundamental limit of the power noise is set by the quantum properties of light. In principle diode-pumped solid-state lasers offer the potential to reach this quantum noise limit (QNL). However, in the LISA frequency band the noise of the free running system is mainly due to noise transfer from pump laser diodes, which is orders of magnitude larger than the QNL.

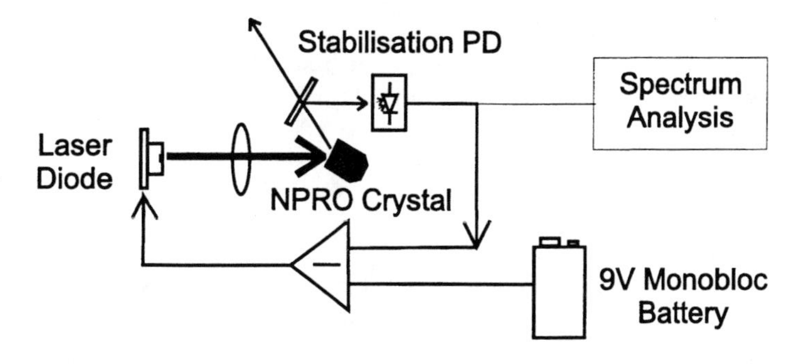

FIGURE 6. Setup of the power stabilisation experiment. A fraction of the laser light (1 mW) is detected with a photodiode. The measured photovoltage across the diode is subtracted from a very stable voltage taken from a monobloc battery. The difference voltage is appropriately amplified and fed back to the pump laser diodes.

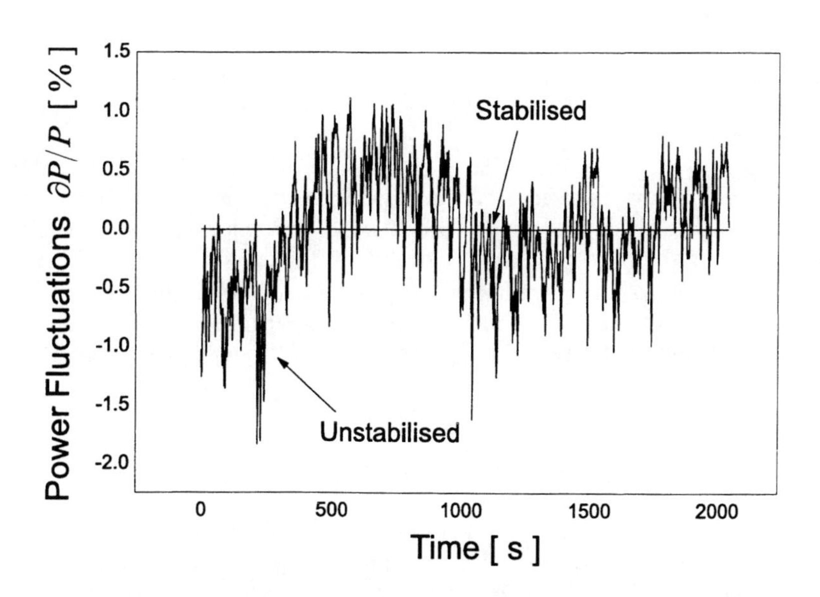

FIGURE 7. Time series of the power relative fluctuations for the free-running NPRO and for the stabilised system. The typical long term power fluctuations of a few percent have been reduced by three orders of magnitude

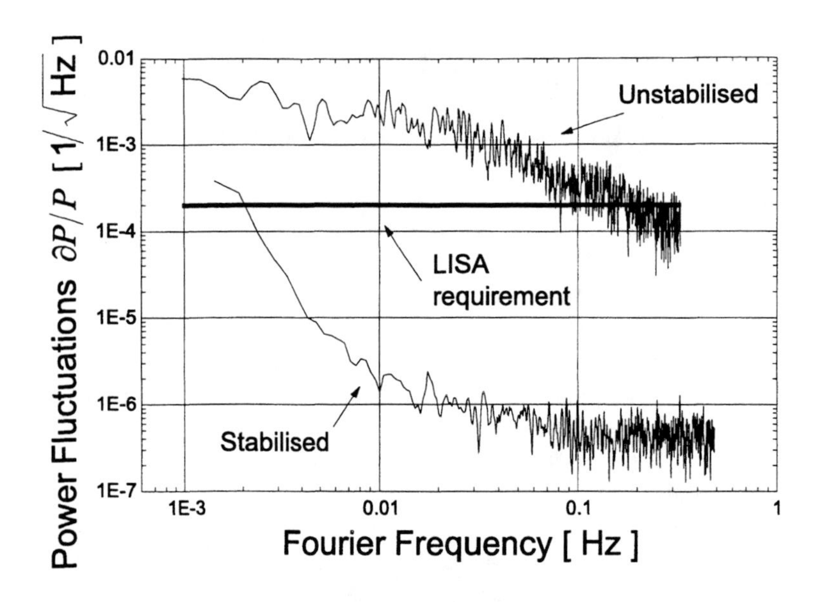

FIGURE 8. Linear spectral density of power fluctuations for the free-running NPRO and the stabilised system. The power fluctuations have been reduced to below $10^{-6}/\sqrt{\text{Hz}}$ down to a Fourier frequency of 20 mHz. The steep increase below 20 mHz is due to temperature fluctuations of the voltage reference.

Substantial power noise reduction has been demonstrated for Nd:YAG ring lasers by application of electronic feedback loops [3]. The noise was reduced to less than 10 dB above the quantum noise limit down to a frequency of 10 KHz, corresponding to a relative power noise of less than $10^{-7}/\sqrt{\text{Hz}}$. To reach the LISA specifications the existing noise reduction scheme was slightly modified and extended to the low frequency regime.

B Experimental setup and results

Figure 6 schematically shows the setup of the power stabilisation experiment. A fraction of the laser light (1 mW) is detected with a photodiode. The measured photovoltage across the diode is subtracted from a very stable voltage reference. In this case the reference is a monobloc battery. The difference voltage is appropriately amplified and fed back to the pump laser diodes. The stability is measured as the relative power fluctuations detected with the stabilisation photodiode.
Figure 7 shows the time series of the power relative fluctuations for the free-running NPRO and for the stabilised system. The typical long term power fluctuations of a few percent have been reduced by three orders of magnitude. In the Fourier domain the linear spectral density of power fluctuation have been measured to be $5 \times 10^{-5}/\sqrt{\text{Hz}}$ at 1 Hz increasing to lower frequencies like $1/f$ (cf. figure 8). The power fluctuations have been reduced to below $10^{-6}/\sqrt{\text{Hz}}$ down to a Fourier frequency of 20 mHz. The residual power fluctuations show a steep increase below 20 mHz. The reason for that increase are temperature fluctuations of the voltage reference. The usage of a temperature stabilised reference in the future should reduce the residual low frequency noise well below the envisaged LISA noise level.

IV CONCLUSIONS

Starting with the well known design of the non-planar Nd:YAG ring-laser, we have developed a compact monobloc laser head design. This design seems well suited for an application as the laser system of the space-borne gravitational wave detector LISA in terms of laser output power, mechanical stability and radiation resistance.

A laboratory prototype of the LISA laser system has been stabilised in frequency and power. The frequency fluctuations have been reduced below the required noise level of $30\,\mathrm{Hz}/\sqrt{\mathrm{Hz}}$ at Fourier frequencies above 200 mHz and are about a factor of 10 above the requirement a 1 mHz. A variety of possible improvements make it likely that the noise will be significantly reduced below the requirements in the near future.

The power fluctuations of the stabilised system have been reduced more than two orders of magnitude below the LISA requirements above Fourier frequencies of 20 mHz. Below that, the achieved stability is yet limited by temperature fluctuations of the voltage reference.

V ACKNOWLEDGEMENTS

This work has been supported by a grant from the *Wernher von Braun-Stiftung*, Schopfheimer Straße 17, D-14165 Berlin. We are also grateful to John L Hall and Matthew Taubman from the Joint Institute for Laboratory Astrophysics, Universitity of Colorado, for valuable discussions.

REFERENCES

1. T.J. Kane, A.C. Nilsson, R.L. Byer:
   Frequency stability and offset locking of a laser-diode-pumped Nd:YAG monolithic nonplanar ring oscillator,
   OPTICS LETTERS / Vol. 12 / 175(1987)
2. N. Uehara, K. Ueda:
   Ultrahigh-frequency stabilization of a diode-pumped Nd:YAG laser with a high-power-acceptance photodetector,
   OPTICS LETTERS / Vol. 19, No. 10 / May 15, 1994
3. C. Harb, H. Bachor, R. Schilling, P. Rottengatter I. Freitag, H. Welling:
   Suppression of the Intensity Noise in a Diode-Pumped Nd:YAG Nonplanar Ring Laser,
   OPTICS LETTERS / Vol. 19, No. 10 / May 15, 1994
4. G. Giampieri, R.W. Hellings, M. Tinto, and J.E. Faller:
   Algorithms for Unequal-Arm Michelson Interferometers,
   Opt. Commun., in press (1995)
5. F. Bondu, P. Fritschel, C. Man, A. Brillet:
   Ultrahigh-spectral-purity laser for the VIRGO experiment,
   OPTICS LETTERS / Vol. 21,No. 8 / April 15, 1996
6. S. Seel, R. Storz, G. Ruoso, J. Mlynek, and S. Schiller:
   Cryogenic optical resonators: a new tool for laser frequency stabilization at the 1 Hz level,
   Phys. Rev. Lett. 78, 4721 (1997)

Optical Engineering Requirements for the LISA Wavefront Error Budget

Martin Caldwell*, Paul McNamara[◊] and Anna Glennmar[#]

* *Optical Systems Group, Space Science Dept., Rutherford Appleton Lab., UK; e-mail: m.caldwell@rl.ac.uk*
[◊] *Dept. of Physics and Astronomy, Glasgow University, UK*
[#] *Dept. of Space and Climate Physics, University College London, UK*

Abstract: Part of LISA's interferometer length error budget is assigned to the wave-front error (WFE) of the optical beam. That is, the aberration of the ideal beam due to the non-ideal effects of the real optical system. These produce interferometer length errors when they are combined with the non-ideal pointing effects (accuracy and stability) of the spacecraft. The aberration effects include fabrication errors such as in mirror surface form and alignment, and added effects in-flight, such as thermo-mechanical deformation. The LISA telescope is the most critical sub-system in this regard, and previously the error budget has been converted into a WFE for the primary mirror, for the case of de-focus aberration alone. Due to in-flight focus correction this effect reduces to that of astigmatic form error, with the result being an estimated requirement for WFE $< \lambda/30$. There is ultimately the need for a more comprehensive budget to include all aberration effects, especially since the above WFE requirement is very challenging from an optical engineering viewpoint. It is also appropriate to construct this budget at the same time as the spacecraft design is made, so that critical effects can be allowed for and, if necessary, traded-off in the design. Recent work has realized (a) a ray-trace model of the system to allow the full geometry and its deformation to be analysed (b) an assessment of the size of the expected deformations, from the spacecraft structural and thermal designs. This model and its WFE predictions will be presented in this paper.

INTRODUCTION

Previously the LISA wavefront error (WFE) requirement has been determined for the case of defocus errors, using analytic and beam-mode analysis methods (1, 2). The purpose of this work is to extend the analysis to include WFE contributions from other opto-mechanical effects, so that a comprehensive WFE budget may be compiled for LISA. This is needed for the purpose of sub-system specification; from the system-level opto-mechanical tolerance analysis, the manufacturing and assembly requirements on each sub-system can be derived.

To include all of the important effects (aberrations, beam clipping by edges, diffractive propagation) in a single model, a modern ray-trace software package is used (3). This also allows integration with the other CAD models, e.g. mechanical, thermal and structural.

This paper describes a mathematics model of the LISA beam interferometry, needed to relate the WFE to the LISA path length error requirement. The ray-trace model of WFE is then described, as well as its verification by comparison with previous work. Finally the error effects themselves are considered and a preliminary budget is compiled.

INTERFEROMETER MODEL

The LISA interferometer operation is also described elsewhere in this issue (4). The part of the overall LISA interferometer length error allocated to optics aberration effects is $\delta z = 1$ pm/$\sqrt{\text{Hz}}$. This relates only to *steady-state* geometric non-ideality in the optical wavefronts, coupled into the temporal error spectrum via the pointing jitter (units nrads/$\sqrt{\text{Hz}}$). It therefore does not include the various non-steady state effects in the optics, e.g. temporal stability of optical path lengths.

The important characteristic for the aberration effect is the wavefront at the receiving spacecraft, shown in Fig. 1a. In a perfect case the transmitted beam is a pure gaussian mode, for which the received wavefront is essentially spherical, with waist position at the send craft. Since the receive craft is in the far-field, the received wavefront has its centre-of-curvature situated at the send craft. The beam waist ω (at 1/e amplitude) is approx. equal to the radius of the telescope aperture (0.15 m), and the corresponding far-field divergence angle is $\theta_{div}=\lambda/\pi\omega$ (=2.26 μrads at λ=1.064 μm).

CP456, *Laser Interferometer Space Antenna*
edited by William M. Folkner
 1-56396-848-7/98/$15.00

Figure 1b shows the LISA pointing behaviour, where the angles θ are measured with respect to the beam direction of the send craft. θ_o is the intended pointing angle, θ_{dc} is the d.c. error in pointing (±15 nrads) and $\delta\theta$ is the jitter (6 nrads/√Hz) (5). The nominal receiver aperture of 0.3m represents an angle of only 0.06 nrads. The figure shows that interferometer length error δz is only generated if the received wavefront deviates from a referencesphere centred on the send craft. Otherwise (as in the 'perfect' case above), δz=0 regardless of pointing performance.

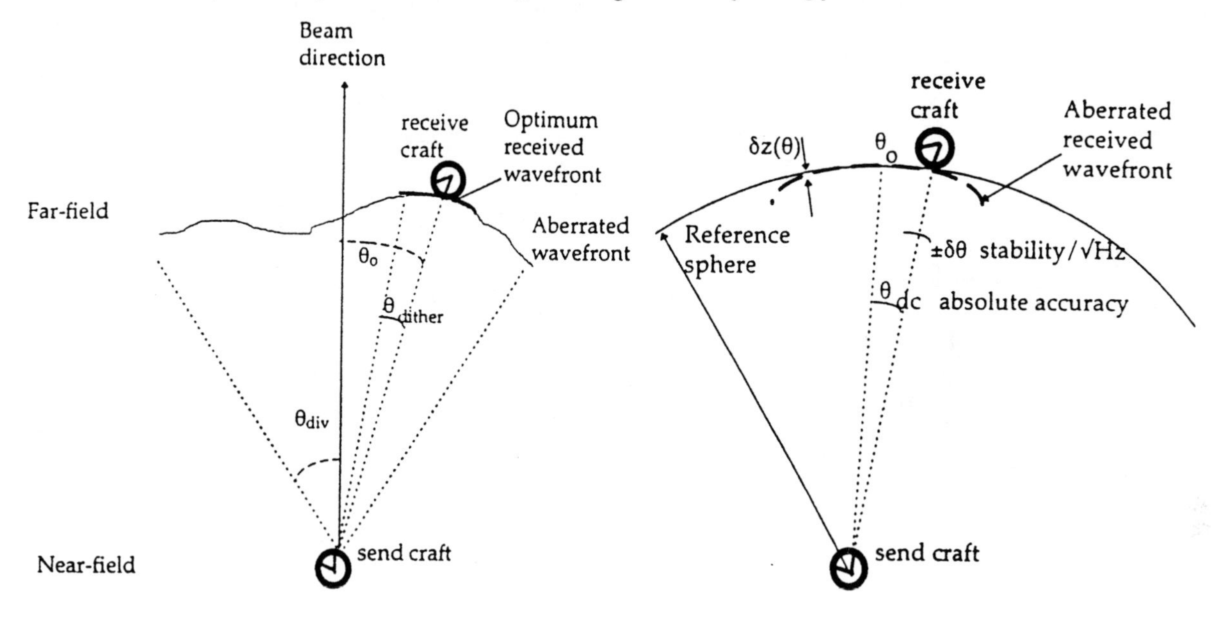

FIGURE 1. (a) Full beam pattern, showing tilt-correction. (b) Scanned portion, with pointing deviations.

In the real case of an aberrated wavefront there are 2 corrections, made by spacecraft operations:

1. Tilt correction: pointing adjustment of the send craft such that the portion of the beam hitting the receive craft always has its direction coming from the send craft (Fig. 1a).
2. Focus adjustment: Movement of the transmitter optical fibre on the send craft, such that the wavefront COC is always restored to its optimum position, i.e. at the send craft.

To make these corrections, the wavefront must be measured over some finite angular range, and this is done by modulating the send craft pointing to scan the beam across the receive craft over an angular range $\pm\theta_{dither}$. This parameter is not yet chosen, but we require $\theta_{dither} >> \delta\theta$ to avoid excessive noise, and there is usually an optimum value of $\theta_{dither} < \theta_{div}$, which depends on the type of aberrations (shape of wavefront) present. Here we use $\theta_{dither} = \theta_{div}/4$.

After processing the scan data to determine the required pointing θ_o and focus correction, the send craft can be commanded to make the pointing and focus adjustments to the transmitted beam. It is assumed that these correction operations can be done quickly enough that the system remains well corrected at all times. E.g. if the aberration is produced by changes in environment, then the timescales of significant changes in this environment are long compared to the time taken for the correction operation to be made.

In our model δz is evaluated from the received wavefront numerical data $z(\theta)$ as

$$\delta z(\theta) = \frac{\partial z}{\partial \theta}\delta\theta \qquad (1)$$

The nominal pointing position θ_o is by definition a stationary point on the wavefront, with $\partial z/\partial\theta=0$, so $\delta z=0$ at this point even for an aberrated wave. For a worst-case analysis, however, we must consider the pointing at its maximum deviation from nominal, i.e. at $\theta=\theta_o+\theta_{dc}$.

It is convenient to expand $\partial z/\partial\theta$ as a Taylor series about θ_o , and the equation for δz then becomes

$$\delta z = \left[\left.\frac{\partial z}{\partial \theta}\right|_{\theta_o} + \left.\frac{\partial^2 z}{\partial \theta^2}\right|_{\theta_o} (\theta_{dc}) + \left.\frac{\partial^3 z}{\partial \theta^3}\right|_{\theta_o} \frac{(\theta_{dc})^2}{2!} + \ldots \right] . \delta\theta \qquad (2)$$

For modeling of the larger optics of the telescope, the effects of non-gaussian beam shape (due to clipping) and surface deformations are important, and to spatially sample these the beam must be modeled using many modes (rays). This telescope model is shown in Fig. 3a, in side view, for the outward-propagation path of the send craft. The telescope design is a Richey-Chretian with 0.3m aperture, 3.48 m focal length, and F/1 relative aperture at the primary. With a collimating lens present at the centre of the primary mirror, the system couples a 3 mm diameter beam from the optics bench to the 0.3 m aperture. The lens is not included in the ray-trace shown.

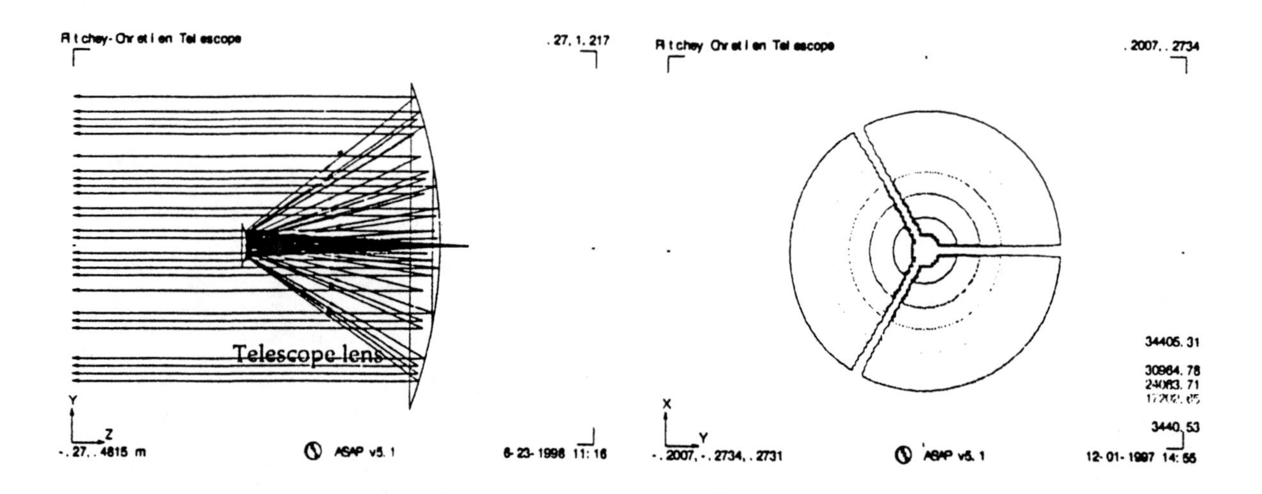

FIGURE 3. Telescope (a) ray-trace (b) Beam pattern at output (near-field), in contour plot.

The beam complex amplitude pattern is computed from the ray-set. This is shown in Fig. 3b and Fig. 4. at the telescope output in the 'nominal' case, i.e. before deformations are applied.

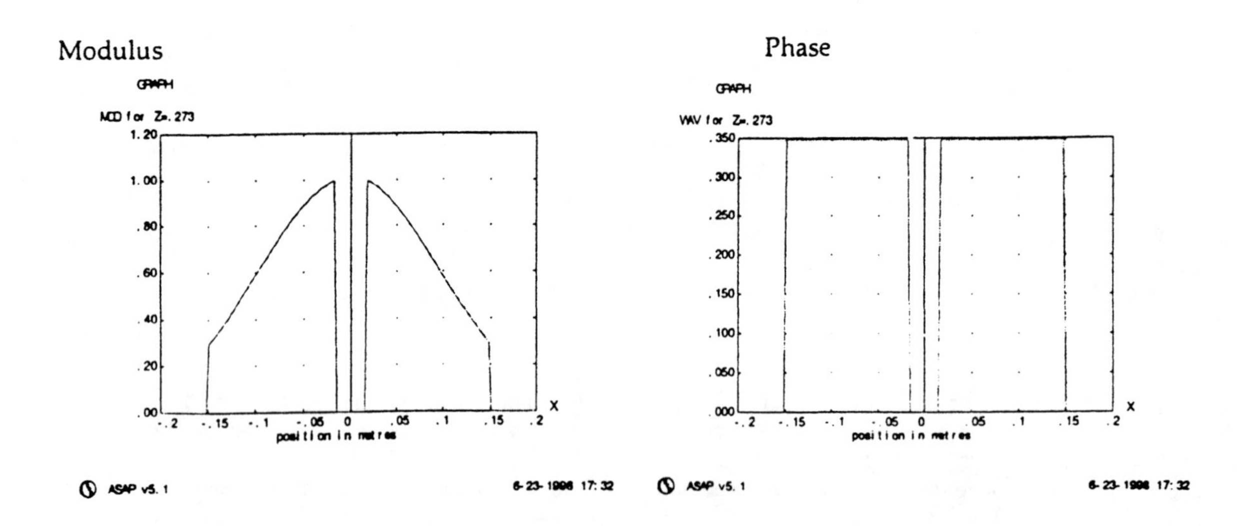

FIGURE 4. Beam profiles at send telescope (near-field), (a) Modulus, (b) Phase (no. of waves), X-direction. It can be seen that the phase distribution is constant over the part of the beam where energy is present, as expected when the beam has its waist at the telescope output and is aberration-free.

δz is calculated from the numerical $z(\theta)$ data by first finding θ_o, and then evaluating the Taylor expansion terms at this point. The 1st derivative is always zero as explained above (tilt correction). To apply the focus correction 'automatically', the second Taylor term is neglected. This is an approximation which is valid for small defocus error, whereas an exact calculation would require re-computing the wavefront with the required focus correction applied to the geometric model.

For detailed aberration analysis the model can be applied to the wavefront data in two-dimensions. For aberration considered here however, a one-dimension slice through the data is sufficient, e.g. Fig. 4 and 5 later.

RAY-TRACE MODEL

The complete LISA optical design is described elsewhere in this issue (4), and here only the relevant parts are shown in detail. The model used (3) is one in which any coherent beam pattern can be 'decomposed' into a set of modes, the propagation of which can be described by the tracing of a small set of rays (typically 4 or 8 per mode). Beam properties such as flux (e.g. for stray light analysis) and polarisation can be included, and the system components can be given corresponding surface and bulk optical properties. This allows the full functionality of the system to be modeled, and by interfacing the model to other CAD tools, such as mechanical, thermal and structural models, its performance under deformations can be analysed in a systematic and comprehensive way.

Figure 2 shows a model of the LISA beamsplitting arrangement, in oblique view. Here the laser beam is modeled as a pure gaussian mode, for which just one input ray is shown, with mixed linear polarisation (horizontal and vertical). The horizontal component passes straight through the beamsplitter whereas the vertical component is reflected, passing along the telescope transmit/receive path. On its return is first transmitted by the beamsplitter and then reflected at the test mass before being recombined with the horizontal beam at the beamsplitter. The various polarisation states used for this beam-routing are also depicted in the trace.

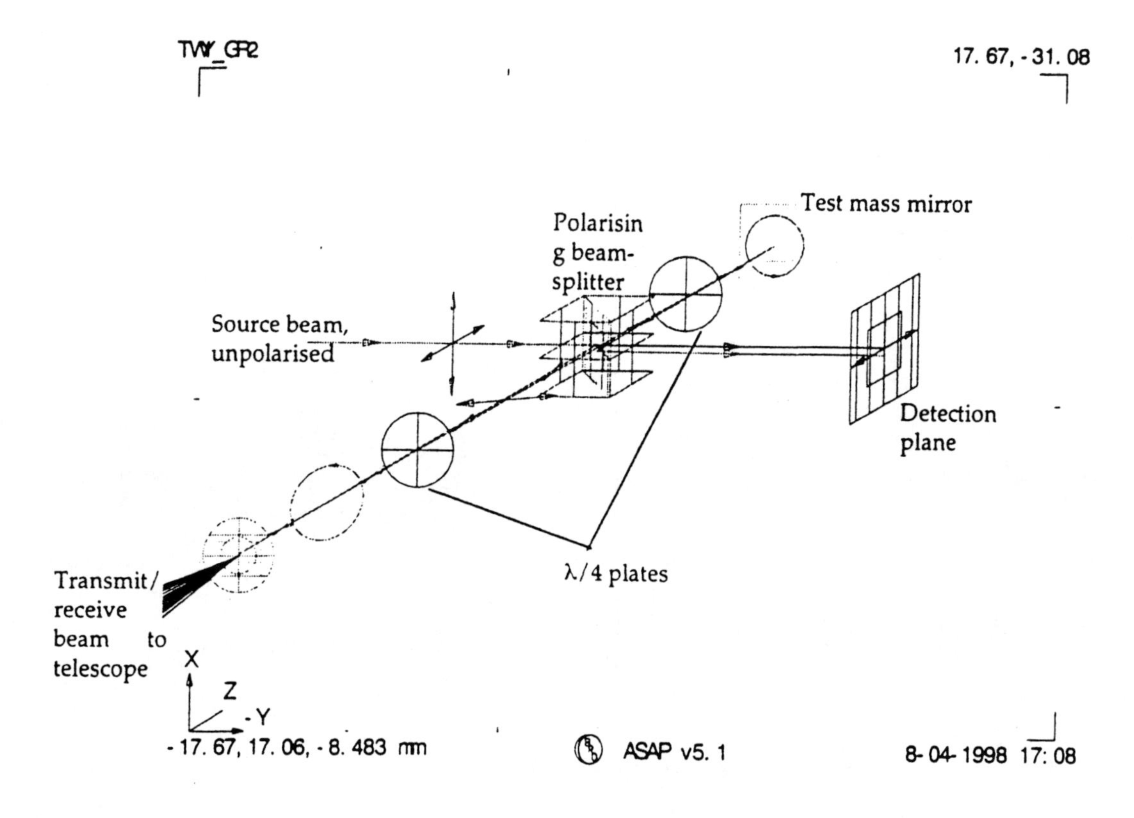

FIGURE 2. Ray-trace model of LISA beamsplitting, indicating ray-splitting and polarisation states, with tilt misalignment added for clarity.

In order to compute the beam pattern at the receive craft (far-field), the ray-set is passed through an ideal lens. The beam pattern at the focal plane is found, and its position co-ordinate is converted to off-axis angle θ by dividing by the lens focal length. This is equivalent to taking the Fourier transform of the near-field pattern. An example is shown in Fig. 5, where the beam pattern (without central obscuration or spider), is shown in near and far fields for the case where a small defocus is applied to the beam (by changing the telescope mirror separation).

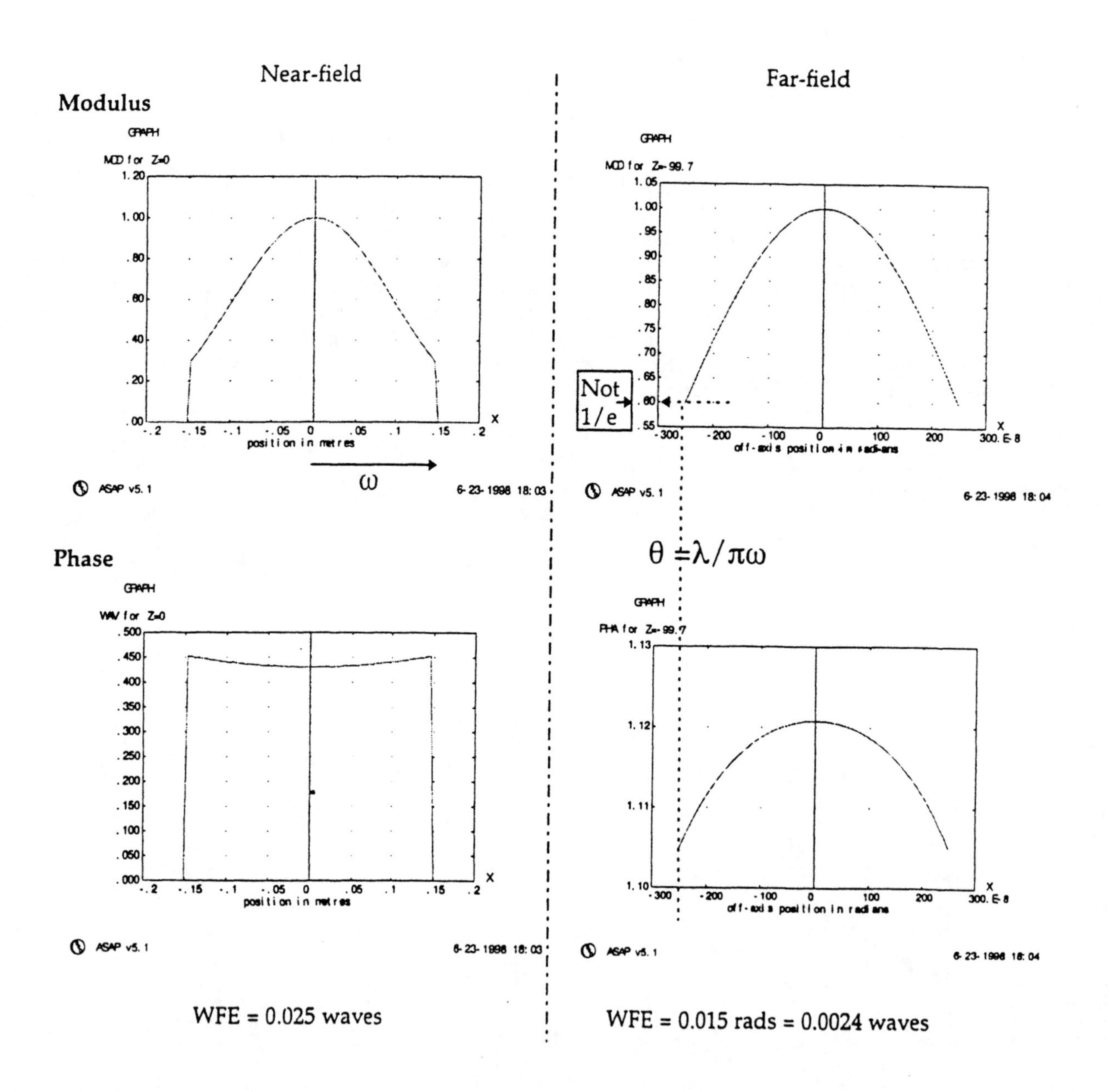

FIGURE 5. Beam-shape: Clipping at 1/e amplitude, with defocus

Verification

Since the model is a numerical one it is checked as far as possible against analytic and previous analysis results. The near-field WFE is calculated as the phase difference between the beam centre and the 1/e point ω (beam edge). We consider first the case without clipping, for which the far-field WFE can be measured in the same way, and the result from the model is that the WFE due to small defocus is approximately equal in the near- and far-fieldcases, in the limit of large analysis aperture (negligible clipping). This is in agreement with the analytic result obtainable from the gaussian beam equations (in which case by 'small' defocus we mean such that the beam waist size isn't significantly changed).

A second verification test is the case of defocus with clipping as shown in Fig. 5. Here in the far-fieldpattern the relative amplitude level at θ (the 1/e point of the unclipped case) is no longer 1/e=37%; it is increased to 60% , indicating a significant widening of the far-fieldbeam. This is as expected from diffraction theory; as the clipping decreases the beam size in near-field, it increases its angular spread in far-field (the source looks more like a 'pinhole'). The changing beam shape raises a question for the definition of far-field WFE; at what beam size (aperture) is the WFE to be measured? We take as the reference point the original 1/e angle θ, and as shown in Fig. 5. the WFE at this point is approx. 1/10 of its value in the near-field. This reduction in WFE from near to far-field, is evidently a property of the clipping (depending on degree of clipping, and the resulting far-fieldbeam shape relative to the chosen reference aperture). The reduction has been noted in previous analysis (2), where a factor 1/4 was found. The differencebetween that result and ours may be due to the far-fieldreferencepoints used being different, and to a change in the design telescope size (previously 0.38 m aperture, now 0.3 m). Reduction in WFE due to clipping can be considered as an effect of the increased spatial coherence as the source becomes more 'pinhole-like', and it works in favour of reducing LISA's sensitivity to aberrations.

ERROR BUDGET

In compiling a budget of all contributors to aberrations, the approach is to consider firstly the larger contributors and those from the parts of the design already fixed. These are listed in Table 1, which is not an exhaustive budget since much of the detailed sub-system design is not yet made. The contributors are classed into two types, those in the telescope (larger beam sizes) and those in the optics bench (beam sizes a few mm). The detailed ray-trace model has only been applied to the former effects, with the results for the latter (optics bench) effects being inferred by extrapolation. The effectreferredto in the table as 'beam shape' is that of clipping of the gaussian beam by the telescope aperture, central obscuration and spider.

TABLE 1 Budget for aberration-induced path-length error δz . The total allowed path length error is 1 pm/√Hz (5).

Item		**Conditions**	**Error contributed** pm/√Hz
Telescope	Beam-shape	Clipped at edge, centre and spider.	0.0015
	Defocus	Zerodur structure, with re-focusing by fibre	< 0.01
	Tilt	As-built (1 arcsec=5μrad) In-flight (ΔT= 0.5 °K, ⇒ 3 μrad) Point-ahead	0.004 0.002 0.004
	Mirror form	spectral density model, λ/25 total, times 2 surfaces	0.2
Optics bench	Tilt	As-built and in-flight, 2 μrad	0.002
	Collimator form	WFE λ/35 in transmission, from (8)	0.05
	Flat components	WFE λ/300 transmission per surface x √ (number of surfaces), from (8)	0.02
	Bulk inhomogeneity	unknown	unknown

Telescope defocus

The recent spacecraftstructural design and thermal analysis (6) gave predicted operating temperatures around the instrument. Most problematic for the telescope is a difference ΔT along its length, i.e. between the primary and secondary, due to the radiative cooling of the latter to the nearby space aperture, plus the limited thermal conductivity of the spider legs. According to the CTE of the CFRP legs the ΔT causes a length change ΔL (from ground to flight) of approx. 5 µm in the current design.

In LISA such a length change could be in principle be tolerated, since it can be corrected using the re-focus mechanism of the optical fibre (for a 13 mm focal length fibre collimating lens the adjustment range required being approx. 1µm and can be achieved using a piezo-electric actuator). However this has several drawbacks: it degrades the mode-matching efficiencyof the beams at the interferometric detector, correction is not perfect as it introduces some higher order aberrations, and the mechanism will have a finite accuracy. For these reasons it is considered best to minimise ΔL in the telescope design.

This can be done e.g. using the low CTE of zerodur, and such a telescope , with similar size and design to LISA, has been built and space qualified with $\Delta L<2$ µm. This is for a similar laser beam transceiver function on a mission for optical communications in space (7). For the LISA application the corresponding residual error after focus correction is small, as indicated in Table 1.

Component tilts

The several effects classed as 'tilt' errors in Table 1 are:

- Tilt misalignment between telescope mirrors, due to a) as-built alignment accuracy b) in flight deformation due to lateral ΔT across structure.
- Transmit beam tilted with respect to telescope axis due to: Point ahead requirement. Deformations in optical bench.

In each case the size of tilt involved is given in the table, in terms of µradians in the transmitted beam. The data are derived from:

As-built telescope: Diffraction limit in testing with visible light.

In-flight telescope: Lateral ΔT from (6), structure modeled as simple parallelogram.

Point-ahead: Due to photon time-of-flight, the transmit beam must be pointed-ahead by $\alpha \approx 3$µrads. Each telescope has to be oriented to the receive beam direction (for wavefront tilt-correction), and the telescope field-of-view angle for the transmit beam is then 2α.

Alignment of optics bench and components. As-built alignment and stability of optical contacts has been demonstrated for laser components similar to those for LISA, to the µ-radian level (8).

The error sensitivity to these effects is calculated using the ray-trace model, and here the far-field beam profile used is that in the plane of the tilt. The table shows that the LISA error due to these effects is small, and this is mainly because of the tilt-correction of the system. Although here each tilt error is listed separately, in the actual system the tilts are combined, with the net error likely to be less than the sum of the individual terms, except in the worst-case combinations.

Component form error

This item refers to deviations in the shape of each optical surface from its ideal form (flat, spherical or aspheric), usually specified at low spatial frequency such as is measurable using interferometer component testing equipment. Although optics manufacturers often use only a single figure-of-merit for form error (either root-mean-square deviation over the surface (σ_{RMS}) or peak-valley (σ_{PV})), for applications involving diffraction it is often appropriate to use the more detailed specification of power spectral density (PSD), (9). This is the spatial spectrum of surface deviations.

In LISA for example the effect of the deviations on instrument error depends greatly on their spatial size relative to the aperture, as this determines the far-field diffraction behaviour. In Fig. 6a the spectral shape for a known surface is shown (10), and this shape will vary depending on the surface manufacturing and finishing processes used. The figure indicates the normal situation in which the deviations (or ripples) decrease in amplitude with increasing spatial frequency. At very high frequency the deviations effectively become micro-roughness, with the scatter becoming incoherent, and dealt with separately as part of the stray-light error budget. The area under the spectral curve is equivalent to the RMS form error σ_{RMS}, and in the example shown it is set equal to $\lambda/25$, a value known to be near the limit achievable in mirrors of the size and form needed in the LISA telescope.

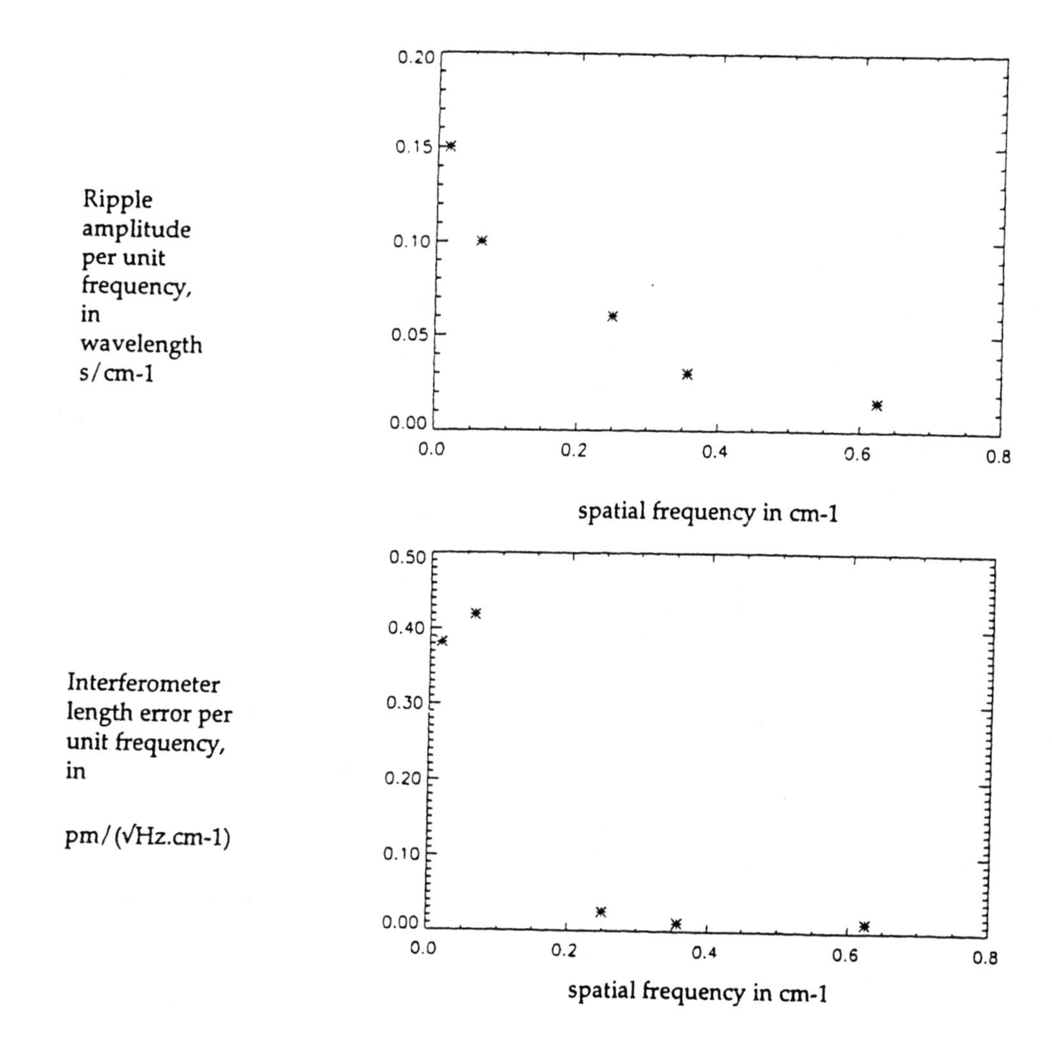

FIGURE 6. (a) Spectral surface form error used, using shape adapted from (10), and area normalised for $\sigma_{RMS}=\lambda/25$); (b) corresponding spectrum of interferometer error δz.

The ray-trace model for calculation of δz is executed separately for a primary mirror rippled at each frequency as per Fig. 6a, and a spectrum of errors is thereby generated, shown in Fig. 6b. The net error δz is then calculated as the total area under this curve. The ripple deformation functions applied to the mirror are symmetric in radius and in rotation about the axis. To account for the actual random nature of the ripple and its non-axial symmetry (e.g. astigmatic form error), the normal assumption of focus correction is not applied to this case (the system cannot correct asymmetric defocus errors such as astigmatism).

It should be noted that the net δz computed here applies only to one specific surface model and PSD shape. For use in specification of mirrors this could be used to specify an 'envelope' PSD which should not be exceeded at any frequency. However, a better alternative to avoid over-specification would be to obtain PSD characteristics for actual candidate manufacturer's processes and feed those into the model for a case-by-case assessment.

For the other optical components the net δz error is estimated by assuming that the spectral shape is the same for all components and that $\delta z \propto \sigma_{RMS}$. For the transmissive components of the optics bench the σ_{RMS} specified is component wavefront error rather than the surface form, and in comparing this with the mirror case, the effect of reflection has to be taken into account. In table 1 the major contributor for the bench is the laser collimator lens, but this has acceptably low error. The WFE for flat transmissive components is approx. an order of magnitude better than the lens (8), but in this case there are many more components present, approximately 20 surfaces in the path between the collimator and the telescope. Assuming the resulting errors add randomly the resultant is taken as the root-sum-squared of the component errors.

A final item for transmissive optics is the WFE due to bulk material inhomogeneities. This may be particularly important for the phase modulator devices but no data is available at this time.

CONCLUSION

From Table 1 it appears that the overall budget of 1 pm/√Hz is likely to be met with some margin. For the optical engineering such margin is welcome at this stage since many of the effectsand sub-system performancesare yet to be verified. In particular the above budget is dependent on the LISA pointing performances(accuracy 15 nrads, stability 6 nrads/√Hz), and all errors would increase with poorer performance, according to equation (2).

ACKNOWLEDGMENTS

The authors are grateful for the help of D. Robertson and W. Winkler in defining the LISA requirements, and S.ÊPeskett, .M Whalley and P. Gray for inputs on the thermal, structural and optical aspects.

REFERENCES

1. Robertson, D. I. et al. "Optics for LISA" *Class. Quantum Gravity* vol.14,p.1575 (1997)
2. Winkler, W. "A truncated gaussian beam in the far-field" *Class. Quantum Gravity* vol.14,p.1579 (1997)
3. *Advanced Systems Analysis Program , ASAP©* , Breault Research Org. , Tucson,AZ,USA.
4. Robertson, D. I. "Interferometry for LISA - principals" This journal.
5. *LISA pre-phase A report*. Pub. MPQ-208. Feb.96.
6. *LISA pre-phase A report*. 2nd Edition. July.98. Chapter 5 Payload design.
7. Juranek,H.J. et al "Use of glass ceramic as a structural material for a high precision space telescope" SPIE Vol.2210. (1994).
8. Leptre, F. "High stability optical components for SILEX project." SPIE Vol.2210, p.83 (1994).
9. Stover, J.C. *Optical Scattering*. Publ.McGraw-Hill. New York (1990).
10. Breckinridge, J.B. et al."Space telescope low-scattered-light camera: a model" Opt.Eng. Vol.23,No.6,pp.816-820 (1984).

Interferometry for ELITE

David I. Robertson* and William M. Folkner†

*Max-Planck-Institut für Quantenoptik, Callinstraße-38, D-30167, Germany
†Jet Propulsion Laboratory, California Institute of Technology, Pasadena, CA 91109

Abstract. The purpose of the interferometry in the ELITE mission is to measure the differential displacement between the two inertial sensors mounted in the spacecraft. It will thus provide an independent verification of the performance of the inertial sensors. The interferometric system should function under all the likely operating modes of the inertial sensors. In one likely mode all the suspension forces in one of the sensors would be switched off and the sensor allowed to drift freely over perhaps 10s of microns.

Here we present a system to achieve this measurement and give some of the constraints on the laser performance and spacecraft environment that must be satisfied in order to provide a useful measurement.

INTRODUCTION

LISA is a proposed space mission to measure gravitational waves with frequencies from $10^{-4}\,Hz$ to $1\,Hz$. It requires significant advances in the performance of a number of technologies such as: inertial sensors, low frequency laser interferometry and drag-free satellite operations using field emission ion thrusters. The ELITE [1,2] mission is proposed as a low cost flight to test these key technologies. Its aim is to demonstrate their operation at a level close to the final LISA requirements. The LISA mission is outlined by Danzmann in these proceedings [3] and in more detail in [4].

The payload of the ELITE microsatellite will consist of two inertial sensors, with laser interferometry to measure their separation. The spacecraft will be maintained in a drag-free orbit by using the signals from the sensors to control proportional ion thrusters.

The two inertial sensors are mounted in a ULE plate. This plate also serves as a high stability optical bench with all optical components mounted on the bench.

The laser used to illuminate the interferometer is a Nd:YAG laser similar in design to that proposed for the LISA mission and is described in [5]. The laser is mounted away from the optical bench and the light is brought to the bench by a single mode optical fibre.

INTERFEROMETER LAYOUT

The interferometry for the ELITE mission should measure the relative displacement of the two inertial sensors to a level of a few $\mathrm{pm}/\sqrt{\mathrm{Hz}}$ at frequencies above 1 mHz and thus confirm that the sensors are indeed inertial to this level of accuracy. This will also verify the techniques required to do low frequency laser interferometry between inertial masses for the LISA mission.

A schematic of the optical layout is shown in Fig. 1. Light from the laser is introduced onto the optical bench by the optical fibre. Some of this light is phase modulated by an electro-optic modulator (*eom*) and then introduced to an reference cavity. The laser is frequency stabilised to this cavity using reflection locking. The remaining light is split into two paths. In the measurement path the light is frequency shifted by about 80 MHz in an acousto-optic modulator (*aom*) and is then reflected from the two inertial masses before being recombined with light from the reference path at a beamsplitter. In the reference path the light is frequency shifted by a second *aom* by 80.01 MHz and is then recombined at the beamsplitter. Thus at the photodiode we have a beat signal at 10 kHz. The phase of this signal varies with changes in the length of the measurement

CP456, *Laser Interferometer Space Antenna*
edited by William M. Folkner
 1-56396-848-7/98/$15.00

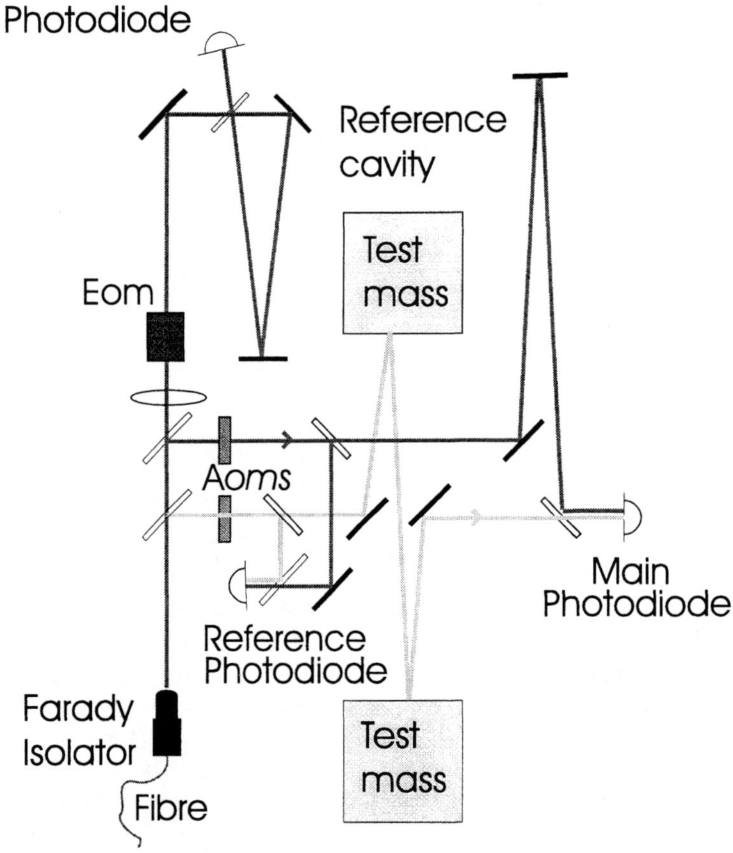

FIGURE 1. A schematic diagram of the ELITE optical layout. The measurement path is shown in gray and the reference path in black.

path. To achieve a measurement accuracy of $1\,\text{pm}/\sqrt{\text{Hz}}$ we need to measure the phase of this 10 kHz signal with an accuracy of $6 \times 10^{-5}\,\text{rad}/\sqrt{\text{Hz}}$.

INTERFEROMETER REQUIREMENTS

A number of different noise sources could couple in to the measurement of the relative displacement of the two inertial sensors δx. Here we consider the most important noise sources and derive the constraints they imposed on the spacecraft and its systems.

Laser Frequency Noise

If there is a path length difference of Δx between the measurement path and the reference path then laser frequency noise will couple directly to the measurement of δx.

$$\delta\nu \leq 9 \times 10^4 \left(\frac{3\,\text{mm}}{\Delta x}\right)\left(\frac{\delta x}{1\,\text{pm}/\sqrt{\text{Hz}}}\right)\text{Hz}/\sqrt{\text{Hz}} \tag{1}$$

This, in turn imposes a constraint on the stability of the reference cavity (see equation 3).

Laser Power Noise

The laser power P reflected from the inertial sensors (each of mass m) produces a force on the masses. Fluctuations δP in the laser power will produce fluctuating accelerations on the inertial sensors which could limit the sensitivity of the measurement. For 1 mW of light in the measurement arm we get a required laser power stability at a Fourier frequency of 10^{-3} Hz of:

$$\frac{\delta P}{P} \leq 1.2 \times 10^{-4} \left(\frac{1\,\mathrm{mW}}{P}\right)\left(\frac{m}{0.4\,\mathrm{kg}}\right)\left(\frac{\delta x}{1\,\mathrm{pm}/\sqrt{\mathrm{Hz}}}\right)/\sqrt{\mathrm{Hz}} \quad (2)$$

This requirement can be satisfied by existing Nd:YAG lasers [5].

Thermal Stability

Temperature changes on the optical bench can produce spurious signals by a number of different routes, including direct effects on the inertial sensors. Here we consider only influences on the interferometry.

Reference Cavity

The length stability of the reference cavity could limit the frequency stability of the laser. Using the frequency stability requirement from equation 1 and a thermal expansion coefficient of α we get:

$$\delta T \leq 1 \times 10^{-2} \left(\frac{3 \times 10^{-8}/\mathrm{K}}{\alpha}\right)\left(\frac{\delta x}{1\,\mathrm{pm}/\sqrt{\mathrm{Hz}}}\right)\mathrm{K}/\sqrt{\mathrm{Hz}} \quad (3)$$

Path Length Changes

The measurement accuracy could be degraded by thermally driven changes of the distances on the optical bench that differentially affect the two arms of the interferometer. For an optical path of length L in each arm and assuming that the difference in temperature fluctuations between the two beam paths is a factor of β below the absolute fluctuations, the relevant formula is:

$$\delta T \leq 7 \times 10^{-4} \left(\frac{3 \times 10^{-8}/\mathrm{K}}{\alpha}\right)\left(\frac{0.5\mathrm{m}}{L}\right)\left(\frac{\beta}{10}\right)\left(\frac{\delta x}{1\,\mathrm{pm}/\sqrt{\mathrm{Hz}}}\right)\mathrm{K}/\sqrt{\mathrm{Hz}} \quad (4)$$

Optical Thickness Changes

A number of optical elements are used in transmission. The optical thickness of these elements will be temperature dependent, both through refractive index changes and physical thermal expansion. Taking a combined coefficient of α and an optical thickness of d then we get:

$$\delta T \leq 3 \times 10^{-4} \left(\frac{1 \times 10^{-5}/\mathrm{K}}{\alpha}\right)\left(\frac{5\mathrm{mm}}{d}\right)\left(\frac{\beta}{10}\right)\left(\frac{\delta x}{1\,\mathrm{pm}/\sqrt{\mathrm{Hz}}}\right)\mathrm{K}/\sqrt{\mathrm{Hz}} \quad (5)$$

The relevant paths in equations 4, 5 are from the beamsplitter in front of the reference photodiode to the beamsplitter in front of the main photodiode down the reference arm and down the measurement arm.

CONCLUSIONS

We have presented an interferometry scheme to measure the differential displacements between the inertial sensors in the ELITE spacecraft. The system has been designed to minimise the effect of thermally driven changes in the optical path length, however these still impose a temperature stability requirement of $\sim 3 \times 10^{-4}\mathrm{K}/\sqrt{\mathrm{Hz}}$ at a frequency of 10^{-3} Hz. Further modelling of the spacecraft is required to see if this stability can be achieved.

ACKNOWLEDGMENTS

We would like to thank our colleagues in the ELITE science team, particularly those in Hannover, Glasgow and Garching, for much helpful advice. The work described in this paper was in part carried out by the Jet Propulsion Laboratory, California Institute of Technology, under a contract with the National Aeronautics and Space Administration

REFERENCES

1. Jafry Y., *ELITE: European LISA Technology Demonstration Satellite*, these proceedings.
2. Danzmann K.,*ELITE Proposal*, RAL, Didcot, 1998.
3. Danzmann K.,*LISA and ground-based detectors: An overview*, these proceedings.
4. Bender P., *et al.*, *LISA Pre-Phase A Report, (Second edition)*, Garching: MPQ, **MPQ 233** 1998.
5. Peterseim M., *et al.*, *Laser system for ELITE*, these proceedings.

Experimental demonstration of some aspects of LISA interferometry

J. A. Giaime, R. T. Stebbins, P. L. Bender, J. E. Faller, J. L. Hall

JILA, University of Colorado and National Institute of Standards and Technology, Campus Box 440, Boulder CO 80309

Abstract. Plans are described to experimentally demonstrate at JILA a possible phase measurement technique for use in LISA interferometry. This demonstration will include use of a compact system for fringe generation and photodetection with realistic beam powers in a Mach-Zehnder interferometer, and development of a stable digital fringe phase measurement system.

INTRODUCTION

As is described in detail elsewhere [1,2], LISA will seek to measure gravitational-wave strain signals by making precision interferometric measurements of changes of the length differences of its three arms. In each spacecraft there are two optical assemblies, each pointed towards one of the other two spacecraft. A locally stabilized laser [3] provides a beam that is sent towards a distant receiving assembly on the other spacecraft, where another laser is phase-locked to it (frequency-shifted) and its beam is sent back. About 0.1 nW of the returning beam is captured and interfered with the original beam, producing an RF fringe signal on a photodetector. These fringes are then timed with respect to a local ultra-stable oscillator (USO), and the data are reduced and sent to Earth as described in Reference [4].

A crucial requirement for the entire optical and electronic receiver path is low phase delay noise in the LISA band from 10^{-4} Hz to 1 Hz. The propagation of a signal may be a function (perhaps non-linear and time-varying) of the signal's amplitude, its spectral components (*i. e*, dispersion), a component's temperature or temperature gradient, or even time derivatives of these things. The designers of LISA are paying particular attention to the thermal engineering issues in the passive optics and modulation components; the design issues under study include thermal insulation and shielding as well as stabilization of RF and optical power absorption. For the fringe phase measurement electronics, similar sources of error have to be considered.

With these needs in mind, the JILA group is beginning a table-top experimental study using a simplified interferometer with realistic beam powers and a candidate for the fringe counting system.

I COMPACT INTERFEROMETER

We plan to use a small and simple symmetric Mach-Zehnder (MZ) interferometer to simulate LISA-like fringes with some natural noise sources. This is sketched in Figure 1. A 1.06 μm laser is the light source for the interferometer. Each MZ arm includes a double-passed acoustooptic modulator (AOM) tuned near the RF frequency, f_0. Also shown are a single-sideband mixer and bandpass filter that cause one AOM to be driven Δf away from the other. If the MZ arms are held at mid-fringe, equal and opposite intensity signals at Δf will be seen on the two outputs, shown entering a matched differential photodiode receiver. One beam can be attenuated in power by decreasing the RF power to its AOM and by decreasing the reflectivity of the mirror used in the double passing. A dashed line in Fig. 1 encloses the components to be mounted on a low-CTE block ($\approx 10^{-8}$/K) block within a hermetically sealed and temperature controlled chamber, in order to reduce temperature and air-pressure induced path length changes. Operation in a vacuum chamber probably will be needed later. The arms can be adjusted for equal path lengths and the photodiode can be balanced to reduce

CP456, *Laser Interferometer Space Antenna*
edited by William M. Folkner
 1-56396-848-7/98/$15.00

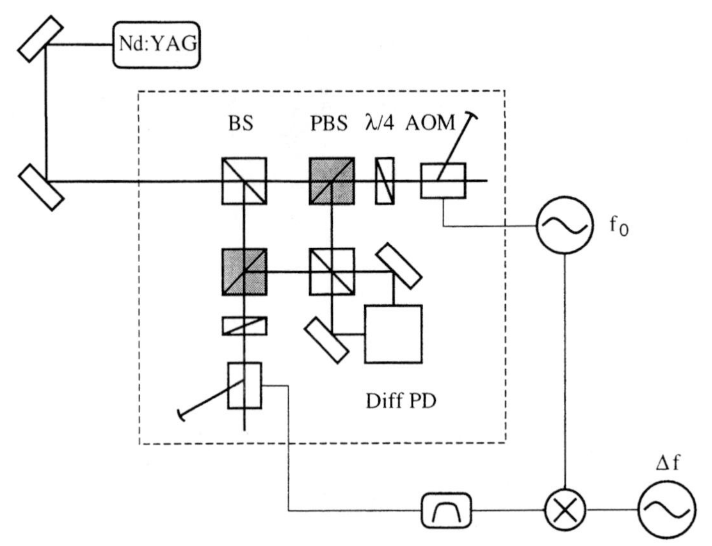

FIGURE 1. Interferometer experimental layout.

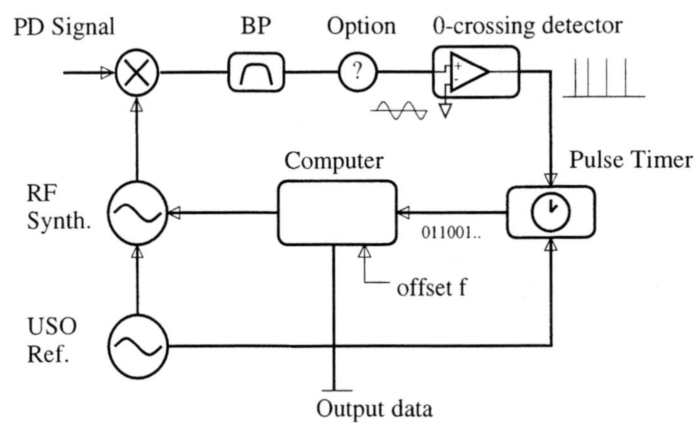

FIGURE 2. Fringe timing system block diagram.

noise from AM and FM in the laser. The goal for the interferometer is to avoid introducing errors in the production of fringes due to widely different beam powers or to temperature-induced optical length changes.

II FRINGE PHASE MEASUREMENT ELECTRONICS

The other component of JILA's experimental plans is the design, construction, and testing of a system to precisely time the interference fringes formed between the local stabilized laser and the beam returning from the distant spacecraft. We plan to test at least one scheme to do this, shown in Figure 2. An alternate scheme based on very rapid sampling of the PD signal and digital processing of the resulting amplitude has been suggested by W. M. Folkner and colleagues at JPL, and probably will be tested there.

The LISA spacecraft are to be placed in orbits whose relative velocities approximately set the detector fringe rate at up to 15 MHz. Using 1 MHz $\Rightarrow 1.6 \times 10^{-7}$ s/rad as an example, and the required 6×10^{-5} rad/$\sqrt{\text{Hz}}$ phase measurement precision, this means that direct fringe timing would need to have a jitter noise and systematic error level of less than 10 ps/$\sqrt{\text{Hz}}$. Instead we plan to mix down the photo-current signal with a local synthesizer that derives its output from a USO reference as commanded by a digital controller. The output of this mixer, at a chosen offset frequency of perhaps 5 kHz, will need to be phase-timed to a precision of 2 ns/$\sqrt{\text{Hz}}$. This can be done by bandpass filtering, converting the zero-crossings to pulses, and then timing the pulses with respect to the USO phase. These timing data are then used by the controller computer to adjust the synthesizer output to approximately track the fringe signal minus the offset frequency. Thus the measured data are the times of waveform zero-crossing, which are corrected by the computer to represent the phase of the received fringes relative to an ideal sinusoid, define locally by the USO.

There are a number of expected sources of the phase noise that will be seen in each of LISA's received fringe signals, including laser phase noise and shot noise. The former is expected to dominate. This can be simulated in the laboratory interferometer by putting phase noise onto the offset frequency Δf. The synthesizer loop will to some extent track this noise. However, the synthesizer's output phase does not need to stay within π rad of the fringe signal, as would be the case in an analog phase-locked loop, so the gain of this loop can be set as desired to minimize system noise. There are several technical noise sources that are to be studied in the context of the synthesizer loop design. Any element in the timing chain with a phase delay that changes with the frequency or phase of its signal can convert a phase error between the synthesizer and fringe signals to noise in the output. This potential problem would argue for high gain in the synthesizer loop, so that all of the components operate at near nominal conditions at all times. Another difficulty might arise if the act of changing frequency rapidly in the synthesizer creates noise, arguing for less loop gain. These issues are to be studied. The element marked with a "?" can be replaced with a tracking filter (a second order phase-locked loop, perhaps) if excess noise prevents clean zero-crossing measurements.

This work is supported by NASA under the Ultraviolet, Visible and Gravitational Astrophysics Research and Analysis Program, NAG5-6880.

REFERENCES

1. *LISA Pre-Phase A report,* 2nd Edition, Publication MPQ-233, Max-Planck-Institut für Quantenoptik, Garching, Germany, July 1998.
2. D. Robertson, *Interferometry for LISA—principles,* this volume.
3. P. MacNamara, *Phase locking for LISA,* this volume.
4. R. T. Stebbins, P. L. Bender, W. M. Folkner and the LISA Science Team, *LISA data acquisition,* Class. Quantum Grav. **13** No. 11A (1996) A285-A289.

Design Issues for LISA Inertial Sensors

Stefano Vitale[*] and Clive Speake[+]

[*]*Department of Physics, University of Trento, Povo, Trento, I-38050, Italy and* [+]*School of Physics and Astronomy, University of Birmingham, Edgbaston, Birmingham, B15 2TT, United Kingdom*

Abstract. In this paper we discuss a few design issues of the inertial sensor for LISA. These issues include the role of the stiffness and the losses that are introduced by the readout and by other parasitic sources. A possible plan for testing those effects on ground is also discussed.

INTRODUCTION

The very core of the LISA mission (1) is a fiducial test mass that has to be kept in a purely gravitational "free fall" within $S_a^{1/2} \approx 3 \cdot 10^{-15} \left[1 + \left(f/3 \cdot 10^{-3}\right)^2\right] m \cdot s^{-2} Hz^{-1/2}$ in the measurement bandwidth 10^{-4} Hz < MBW < 0.1 Hz. This is achieved, in the mission design, by measuring the displacement of the spacecraft in respect to the mass and by feeding back this information to a set of properly designed thrusters.

It is an easy exercise to calculate, in the case of a single axis control loop of very large gain, that the residual acceleration of the test mass is

$$a \rightarrow g + \frac{f_{int}}{m} + \omega_{int}^2 x_n \tag{1}$$

while the displacement of the spacecraft relative to the test mass is

$$x_{s/c} \rightarrow x_n \tag{2}$$

Here g and m are the gravitational acceleration of the test mass and its mass value respectively while f_{int} is the stray force that intervene between the mass and the spacecraft, like that due, for instance, to thermal noise, to pressure fluctuation etc,. Finally x_n is the sensor noise that, by driving the thrusters, produces a random force via the residual coupling of the test mass to the spacecraft. This coupling is summarized by ω_{int}, the natural frequency of the oscillation of the test mass relative to the spacecraft. Notice that ω_{int}^2 needs not to be positive or real: actually for a standard capacitive readout, ω_{int}^2 has a negative real part and acquires an imaginary part because of losses

It is may be worth remembering that the LISA fiducial mass with its read-out is not aimed at measuring the acceleration of the spacecraft. Would one need to do that, he should make ω_{int} larger than any signal frequency ω in order to have constant noise in the measurement bandwidth. On the contrary, to minimize *the test mass acceleration one* has to minimize the product $\omega_{int}^2 x_n$, i.e. he has to reduce the stiffness of the coupling between the mass and the spacecraft even more than reducing x_n. If for instance x_n is chosen in order to comply with the need that the spacecraft motion relative to the test mass is $S_{x,s/c}^{1/2} \approx 5 \cdot 10^{-9} m \cdot Hz^{-1/2}$ then $\omega_{int} < 5\ 10^{-4}$ Hz.

CP456, *Laser Interferometer Space Antenna*
edited by William M. Folkner
 1-56396-848-7/98/$15.00

DISPLACEMENT SENSITIVITY AND STIFFNESS

As it is clear from the above discussion, LISA inertial sensor has to be designed in order to minimize both stray forces and stiffness. In this configuration the residual stiffness intervening between the test mass and the spacecraft is provided by the motion read-out itself. In fact in the simplest case where a capacitive ac bridge is the motion readout, each electrode facing the test, mass forms with it a two ports electromechanical device whose input output transfer relations are:

$$V(\omega)=\frac{1}{C_0}Q(\omega)+Q_0\frac{\partial 1/C}{\partial x}\left[\frac{x(\omega-\omega_0)+x(\omega+\omega_0)}{2}\right]$$

$$f(\omega)=-Q_0\frac{\partial 1/C}{\partial x}\left[\frac{Q(\omega-\omega_0)+Q(\omega+\omega_0)}{2}\right]-\frac{1}{2}\frac{Q_0^2}{2}\frac{\partial^2 1/C}{\partial x^2}x(\omega) \tag{4}$$

Here f, x, V and Q are the force, displacement, voltage and charge small signals respectively. $\omega_0 Q_0$ is the peak amplitude of the ac current bias of the capacitor that oscillates at the pump angular frequency ω_0. We also use in eqs. (4) the unperturbed values of the capacitance C and of its derivative in respect to displacement $\partial \frac{1}{C}/\partial x$ and we have barred components at frequencies higher than ω_0 like for instance those close to $2\omega_0$.

Eqs. (4) show that indeed the readout provides a stiffness. If for instance the two port device is inserted in a circuit that keeps the voltage constant, such that then $V(\omega)=0$ one gets, for $\omega<<\omega_0$:

$$f(\omega)=\frac{CQ_0^2}{2}x(\omega)\left[\left(\frac{\partial 1/C}{\partial x}\right)^2-\frac{1}{2}\frac{\partial^2 1/C}{\partial x^2}\right] \tag{5}$$

In the same approximation the charge-to-displacement sensitivity at low frequency can be derived from:

$$Q(\omega_0+\delta\omega)=-\frac{CQ_0}{2}\frac{\partial 1/C}{\partial x}\delta x(\delta\omega) \tag{6}$$

As the displacement noise is, within this simplified discussion, inversely proportional to the sensitivity, from eqs. (5) and (6) one easily gets that

$$x_n\omega_{int}^2\propto Q_0\frac{\partial 1/C}{\partial x}\left[1-\frac{1}{2}\frac{\partial^2 \frac{1}{C}/\partial x^2}{\left(\partial\frac{1}{C}/\partial x\right)^2}\right] \tag{7}$$

This contribution then can apparently be suppressed at will by reducing the coupling factor $V_0\,\partial 1/C/\partial x$. In addition special configurations can be found where the stiffness can be reduced, to some extent, independently of the coupling. This can be done by acting on the geometrical factor within the square parenthesis in eq. (7), a factor which is 1 for an ideal, infinite, plane capacitor where the plates move normally to their surface and is zero for an ideal semi-infinite plane capacitor where the plates slide sideways.

However this process of mass "uncoupling" is limited by the fact that, beyond some point, other weak coupling mechanism between the test mass and the spacecraft intervene. Among those a serious candidate are the stray voltages due to inhomogeneous contact potential on the surfaces both of the test masses and of the surrounding electrodes. The effect of this inhomogeneity has been estimated (2) to be, at least in order of magnitude, that of providing a stiffness in all direction of order

$$k = \gamma \frac{\varepsilon_0 A \delta v^2}{d^3} \quad (8)$$

where A is the area of the facing electrode, δv is the rms fluctuation of the contact potential and d is the distance between the electrode and the test mass. γ is the real unknown of the phenomenon, a dimensionless factor that depends on the detailed spatial distribution of these "charge patches"

In addition to that, a stiffness is also introduced by any dc bias of the test mass/electrodes system. For instance, if the electrodes are also used as actuators, a formula similar to eq. (5) holds with V_o now meaning twice the dc voltage difference between the electrode and the test mass.

Moreover any net charge deposited by cosmic rays onto the test mass also provides a stiffness. This additional stiffness depends on the details of the arrangement of the electrodes around the test mass.

Also the gravitational gradient across the test mass provides an additional stiffness. As a reference the resonant frequency for a cubic, 1.3 kg, 4cmx4cm test mass due to the gravitational gradient produced by a 2 g mass lying close to its surface is $\omega_g^2 = 1.110^{-8} (\text{rad/s})^2$. This clearly shows that the sensor has to be gravitationally balanced to this limit in order not to spoil the performance.

THERMAL NOISE AND DIELECTRIC LOSSES

A rather more complicated estimate is that connected to the thermal noise. Thermal noise gives to the mass an acceleration with spectral density:

$$S_{th}^{1/2} = \frac{4k_B T\beta(\omega)}{m^2} = \frac{4k_B Tk''(\omega)}{m^2 \omega} \quad (9)$$

where $\beta(\omega)$, and k"(ω) are defined from the damping force as $\vec{F}_{damping}(\omega) = -\beta(\omega)\vec{v}(\omega) = -ik''(\omega)x(\omega)$.

A few sources of damping are known and can be evaluated in straightforward manner. Among those, residual gas gives a negligible effect in the low pressure régime foreseen for LISA.

A second source of dissipation is potentially linked to dielectric losses occurring within the various capacitors formed by sensing electrodes and the test mass (3). To be specific, any surface layer due to adsorbed gas or to deposited oxide or even to spatial inhomogeneities of the work function, will act as an effective lossy polarizable medium coating the surface of the otherwise loss-free capacitor. In addition any lossy capacitor in parallel to it will also be a source of dissipation. To calculate the effect of this kind of losses, let refer to the scheme in Fig.1:

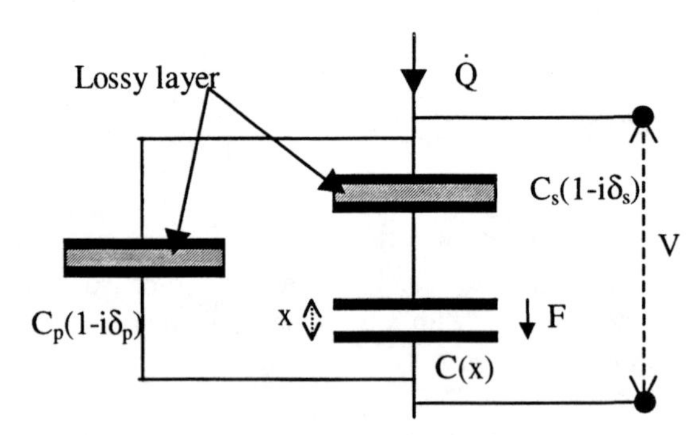

Figure 1. Schematic for calculation of dielectric losses

and let for simplicity consider the case where all voltages and currents are low frequency. The case of the parametric ac bridge gives the same results when the rms values of the ac bias are substituted for the dc ones. Losses in the two capacitors C_s and C_p are summarized by assigning a loss angle to their capacitances: $C_s \to C_s(1 - i\delta_s)$ and $C_p \to C_p(1 - i\delta_p)$. It is a straightforward calculation to show that the circuit obeys, again as a linearized two port device, to the following equations:

$$V(\omega)=\frac{Q(\omega)}{C}\left\{1+\frac{C}{C_s}-\frac{C_p}{C}+i\left[\frac{C}{C_s}\delta_s+\frac{C_p}{C}\delta_p\right]\right\}+x(\omega)Q_o\frac{\partial 1/C}{\partial x}\left\{1-\frac{C_p}{C}+i\frac{C_p}{C}\delta_p\right\}$$

$$F(\omega)=-Q(\omega)Q_o\frac{\partial 1/C}{\partial x}\left\{1-\frac{C_p}{C}+i\frac{C_p}{C}\delta_p\right\}+ \quad (10)$$

$$+x(\omega)Q_o^2\left\{\frac{1}{C}\left(\frac{\partial 1/C}{\partial x}\right)^2\left(1-\frac{C_p}{C}\right)-\frac{1}{2}\frac{\partial^2 1/C}{\partial x^2}-i\frac{1}{C}\left(\frac{\partial 1/C}{\partial x}\right)^2\frac{C_p}{C}\delta_p\right\}$$

Where we have assumed $\frac{1}{C(x)}=\frac{1}{C}+x\frac{\partial 1/C}{\partial x}+\frac{1}{2}x^2\frac{\partial^2 1/C}{\partial x^2}$.... Q_o is the dc value of the charge residing across the device. As we have assumed that C_p/C and C/C_s are both <<1, then $V_o=Q_o/C$ is the dc value of the voltage that biases the sensor.

To get the main features of the phenomenon one can work out the two limiting cases where the device is operated either at constant voltage, $V(\omega)=0$, or at constant charge, $Q(\omega)=0$. In the first case the force to displacement relation becomes:

$$F(\omega)=x(\omega)CQ_o^2\left\{\left(1-\frac{C}{C_s}\right)\left(\frac{\partial 1/C}{\partial x}\right)^2-\frac{1}{2}\frac{\partial^2 1/C}{\partial x^2}-i\left(\frac{\partial 1/C}{\partial x}\right)^2\frac{C}{C_s}\delta_s\right\} \quad (11)$$

while in the second becomes:

$$F(\omega)=x(\omega)CQ_o^2\left\{\frac{C_p}{C}\left(\frac{\partial 1/C}{\partial x}\right)^2-\frac{1}{2C}\frac{\partial^2 1/C}{\partial x^2}-i\left(\frac{\partial 1/C}{\partial x}\right)^2\frac{C_p}{C}\delta_p\right\} \quad (12)$$

In both cases the stiffness provided by the electrodes acquires an imaginary part and becomes then noisy. Notice that this lossy part is always proportional to the square of the input/output coupling factor:

$$CQ_o^2\left(\frac{\partial 1/C}{\partial x}\right)^2\approx\frac{1}{C}\left(\frac{\partial Q}{\partial x}\right)^2 \quad (13)$$

as the noise basically results from conversion of electrical thermal noise into a mechanical one.

To this imaginary spring constant it is associated, for instance in the constant voltage bias case, a thermal noise with spectrum:

$$S_{th}=\frac{4k_BT}{m^2\omega}CQ_o^2\left(\frac{\partial 1/C}{\partial x}\right)^2\frac{C}{C_s}\delta_s\approx\frac{4k_BT}{m^2\omega}\frac{C}{C_s}\delta_s\frac{1}{C}\left(\frac{\partial Q}{\partial x}\right)^2 \quad (14)$$

and thus proportional to the square of voltage-to-displacement gain. This is one of the main results of this paper: in order to reduce the effect of the losses one should reduce the gain of the electromechanical transducer more than simply its stiffness. The two quantities only coincide for the geometry where the capacitor plate moves normally to its own surface. In addition losses get diluted by the factor C/C_s a strong argument in favor of lowering coupling by using large capacitor gaps.

The constant charge bias case is probably of some relevance to understand the spoiling of mechanical Q observed in dc charge biased, resonant motion transducer used in gravitational wave detectors. There, at large bias values, a decrease of the Q factor proportional to the square of the charge is indeed observed (4).

The frequency dependence of δ_s and δ_p, as far those reflect surface properties of the motion sensor have to be the subject of dedicated testing. However some prediction can be made and indeed recently a test with an electrostatically suspended torsion pendulum (5) seem to confirm those predictions. In many field of condensed matter a general behavior of the generalized susceptibility at low frequency has been found (6). The susceptibility is found to be:

$$\chi(\omega)=\chi_o+\frac{T^*}{T_o}\chi_{1/f}\left[-\ln\left(\left|\frac{\omega}{\omega_o}\right|\right)-i\frac{\pi}{2}\mathrm{Sign}[\omega]\right] \tag{15}$$

where χ_o, $\chi_{1/f}$,ω_o and T_o are constants and T^* is an effective temperature that, in common cases around room temperature, coincides with the thermodynamic one. By generalized susceptibility we mean for instance a magnetic or dielectric susceptibility but also the stiffness of a mechanical spring $-k(\omega)=\chi(\omega)$. Eq. (15) predicts indeed a frequency independent imaginary susceptibility and then, at least approximately by assuming $\chi_o >> \frac{T^*}{T_o}\chi_{1/f}$, a frequency independent loss angle $\delta(T)$ that only may depend on temperature. If this dependence is indeed assumed for δ_s in eq. (14), one gets a thermal noise with spectrum:

$$S_{th}=\frac{4k_BT}{m^2|\omega|}CQ_o^2\left(\frac{\partial 1/C}{\partial x}\right)^2\frac{C}{C_s}\delta_s(T)\approx\frac{4k_BT}{m^2|\omega|}\frac{C}{C_s}\delta_s(T)\frac{1}{C}\left(\frac{\partial Q}{\partial x}\right)^2 \tag{16}$$

i.e. a 1/f thermal noise. Notice that this kind of noise is a thermal equilibrium one that does not need any "bias" to the sample. It has not to be confused with the flicker noise observed in resistors and other devices under proper bias condition. Such a noise has been reported for magnetic (7) and dielectric (8) materials and in torsion fibers (9).

The ubiquitous nature of the phenomenon is understood by considering that this is generally attributed to the existence in the system under study, of a distribution of independent simple relaxators with relaxation time constants spread on many decades of the time axes. When each relaxator is associated with a two level system (a flipping magnetic or electric dipole, a dislocation switching between two pinning sites etc.) hopping over a barrier by thermal activation, this model also recovers the frequently observed temperature dependence where T^* is simply $T^*=T$.

If the residual loss angle in the sensing electrodes would turn to be frequency independent, it is easy then to calculate that, with $1/C(\partial V/\partial x)^2$ of order $\approx 10^{-8}$ N/m, LISA would meet its mission goal if $\delta_s \le 10^{-6}$. For this low level of dielectric loss, experimental information on the subject is still partly lacking. Recently a loss angle of $\approx 10^{-5}$ has been estimated (10) for capacitors with ≈mm gaps. As the loss angle originates from a similar loss angle in the capacitance of the various electrode pairs that constitute the read-out, in principle one can set up purely electrical measurements to assess the value for a given electrode configuration. Phase measurement with accuracy of order 10^{-6} rad at mHz frequency are challenging but do not look beyond the reach of present technology.

GROUND TESTING

The demonstrated performances of accelerometers are limited by the "drag-free" ability of the vibration isolation of the test bench to be worse than $\approx 10^{-10}\,\mathrm{ms^{-2}Hz^{-1/2}}$ at some 0.1 Hz and to rapidly deteriorate at lower frequencies.

Also, unfortunately, accelerometer designed for space are often tested by levitating the test mass by means of an intense electric field ($10^6 \div 10^7$ V/m) . The horizontal degrees of freedom are then tested for functionality or performance. The electric field though provides a very stiff coupling between the test mass and the platform. The coupling makes the 1g field to leak into the horizontal degrees of freedom via the platform tilting limiting the minimum detected output noise to some $\approx 10^{-10}\,\mathrm{ms^{-2}}\sqrt{\mathrm{Hz}}$. The large electric field also provide a large stray stiffness that does not allow to test if the sensing device provides a low enough stiffness as that requested by LISA.

For a sensor like that in LISA a torsion balance seems the favored choice for ground testing. The main advantages of the balance is that the suspension is highly uncoupled from the horizontal degree of freedom.

The differential mode, i.e. the torsional mode of the balance can be made very soft (1 mHz is not out of reach) so that a small added stiffness like the one provided by the readout can be detected as a change in the resonant frequency.

In addition torsion balances have proven to be the most sensitive weak force detectors at low frequencies. Thanks to the mechanical amplification due to high Q's, the force noise can be brought above the sensor noise. Force sensitivities, at resonance of $10^{-12}\,m\cdot s^{-2}Hz^{-1/2}$ have been achieved in best instruments. If a ribbon is used as a torsion element the resonant frequency of the pendulum is(11):

$$\omega_{res}^2 \approx \frac{gb^2}{12d^2L} + \omega_{int}^2 \tag{17}$$

Where g is the gravitational acceleration, b is the ribbon width d is the armlength of the pendulum and L is the ribbon length. ω_{int}^2 is the intrinsic (negative) stiffness of the sensor as defined in the preceding section. Eq. (17) is derived in the limit of a thin ribbon and shows that the restoring force is provided by the lifting of the inertial member due to the ribbon twist. With b ≈500 μm, L ≈1 m, d ≈0.2 m the pendulum would have an intrinsic frequency of ≈ 2 mHz falling in the operating range of the inertial sensors we are discussing here.

The first, comparatively simple test is for the stiffness provided by the electrodes when those are biased by the readout voltages or when the test mass is charged. Due to the very low natural frequency of the pendulum, the correction is around ≈ 0.2 % when the bias voltage becomes comparable to the operating one. A 0.2% shift is well within the precision of a torsion pendulum period measurement.

A figure of 0.2% is actually likely to be within the reproducibility of the oscillation period between different experimental runs. If this would prove to be true, it would open the possibility to test for the stiffness provided by the electrodes without any charge or bias, an information of high value. Comparison of resonant frequencies between a run without electrodes and one with the full instrument assembled would provide this information.
Testing the dissipation provided by the electrodes is a more difficult task. At very low bias voltage most of the stiffness is provided by the torsion element so that the Q factor of the pendulum is given by:

$$\frac{1}{Q} \approx \frac{1}{Q_o} + \frac{|\omega_{int}|^2}{\omega_{res}^2}\delta \tag{18}$$

where δ is the loss angle of the readout and Q_o is the intrinsic Q factor of the pendulum.
In order to put a significant upper limit on δ, a figure that should be of order $\delta \approx 10^{-6}$, one need to have a pendulum with comparable value of 1/Q and make the stiffness provided by the sensing electrodes say a 10% of the overall one. This is because Q factors are usually measured with low overall accuracy.

Q factors in excess of 10^6 have been reported for silica fibers. At a voltage level of 1 V, with the same parameters as those used above the correction to the resonant stiffness due to the electrodes becomes ≈ 25% of the total stiffness allowing for the measurement of δ.

ACKNOWLEGMENTS

It is a pleasure to acknowledge many useful discussions with P. Touboul, M. Rodriguez and E. Willemenot

REFERENCES

1 Bender, P. et al., LISA PPA2 Report, Max Plank Institute for Quantum Optics, MPQ233-1998.
2 Speake, C.C., Class. and Quantum Grav.**13**, A291 (1996).
3 Vitale, S., Zendri J.P., and Maraner, A., "Gradiometer design review" Contractor report to Alenia Spazio, ESA GOCE pre-phaseA study (1995). Maraner, A., Vitale, S., and Zendri, J.P., Class. and Quantum Grav.**13**, A129 (1996)
4 Zendri , J.P., PhD Thesis, University of Trento (1992).
5 Willemenot, E., PhD Thesis, Université de Paris-Sud, (1997)
6 For a review see for instance : Vitale S., Cerdonio M., Prodi, G. A., Cavalleri, A., Falferi, P., Maraner, A., in "Quantum tunneling of Magnetization-QTM '94" L. Gunther and B. Barbara Eds. NATO ASI Series **E301**, 157 (1995)
7 Prodi, G. A., Vitale, S., Cerdonio, M., Falferi, P., J. Appl. Phys. **66**, 5984 (1989); Ocio, M., Bouchiat, H. , Monod, P., J. Phys. Lett. **46** L-674, (1985); Wellstood, F.C. et al, Phys. Rev. Lett. **70**, 89 (1993).
8 Israeloff, N.E., and Wang, W., Rev. Sci. Instrum. **68**,1543 (1997)
9 Quinn , T.J., Speake, C.C. and Brown, L.M., Phil. Mag. **A65**, 261 (1992); Gonzales G.I., and Saulson P. , Phys. Lett. **A201**, 12 (1995)).
10 Speake, C.C, Davis, R.S., Quinn, T.J. and Richmann, S.J., BIPM preprint
11 Quinn, T.J., Speake, C.C., and Davis, R. S. Metrologia, **34**, 245 (1997)

The Design and Testing of the Gravity Probe B Suspension and Charge Control Systems

Saps Buchman, William Bencze, Robert Brumley, Bruce Clarke, and G.M. Keiser

W.W. Hansen Experimental Physics Laboratory, Stanford University, Stanford, California 94305-4085

Abstract. The Relativity Mission Gravity Probe B (GP-B), is designed to verify two rotational effects predicted by gravitational theory. The GP-B gyroscopes (which also double as drag free sensors) are suspended electrostatically, their position is determined by capacitative sensing, and their charge is controlled using electrons generated by ultraviolet photoemission. The main suspension system is digitally controlled, with an analog backup system. Its functional range is 10 m/s^2 to 10^{-7} m/s^2. The suspension system design is optimized to be compatible with gyroscope Newtonian drift rates of less than 0.1 marcsec/year (3×10^{-12} deg/hr), as well as being compatible with the functioning of an ultra low noise dc SQUID magnetometer. Testing of the suspension and charge management systems is performed on the ground using flight gyroscopes, as well as a gyroscope simulator designed to verify performance over the entire functional range. We describe the design and performance of the suspension, charge management, and gyroscope simulator systems.

I. INTRODUCTION

The Relativity Mission, also known as Gravity Probe B (GP-B), is first in a series of challenging basic science space space-experiments. Taking advantage of the space environment to reduce gravitational disturbances, these missions are designed to measure relativistic effects with an accuracy of four to seven orders of magnitude better than those achievable on the ground. State of the art technology must be developed, to meet the challenge of attaining the experimental accuracy and to compensate for the demanding space environment.

GP-B (1) is a test of the rotational effects of gravity designed to measure the geodetic and frame dragging relativistic rotational precessions. The local frame of reference in a polar orbit of 650 km, determined by high precision gyroscopes, is compared with the fixed frame of reference of distant stars, determined by a telescope. Figure 1 is a schematic representation of the GP-B experimental concept. In General Relativity the magnitudes of the geodetic and frame dragging precessions are 6.6 arcsec/yr and 0.042 arcsec/yr, to be measured in the Relativity Mission with an accuracy of 0.3 marcsec/yr or better. STEP and LISA, two major space tests of gravitational theories presently in progress, will utilize technologies similar to those developed for GP-B.

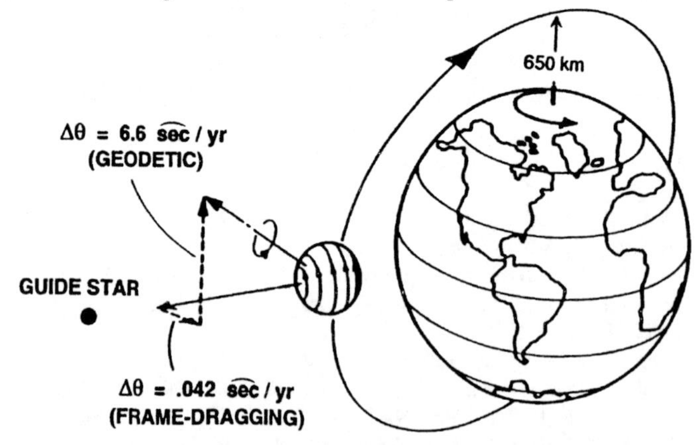

Figure 1. GP-B experimental concept

CP456, *Laser Interferometer Space Antenna*
edited by William M. Folkner

 1-56396-848-7/98/$15.00

Chapter II describes the gyroscopes, their Newtonian disturbances, and the requirements they impose on the suspension system. Chapter III contains the requirements, implementation, and performance of the Gyroscope Suspension System (GSS), while chapter IV details the testing approach of the GSS. We describe the charge measurement and control method, using UV photoemission, in chapter V.

II. GYROSCOPES / DRAG-FREE SENSORS

The most demanding goal of the Relativity Mission is the measurement of the parameterized post-Newtonian parameter _ to one part in 10^5; with _ determined from the measurement of the geodetic effect in Earth orbit. The knowledge of _ to one part in 10^5 will extend the search for a possible scalar interaction in gravity by two orders of magnitude, and allow a test of the critically damped version of the Damour-Nordtvedt (2) "attractor mechanism". This goal implies a measurement of the geodetic precession to 0.67×10^{-5}, and combined gyroscope drift and read-out noise of less than 50 μarcsec/yr (1.7×10^{-12} deg/hr). GP-B system testing shows that the instrument meets all requirements, and is expected to achieve the stated goal. Results from more than 100,000 hours of gyroscope operation indicate that residual Newtonian drift is less than 0.14 marcsec/yr for a supported gyroscope in 10^{-9} m/s^2, and less than 0.02 marcsec/yr for a fully inertial orbit. Low temperature bake-out is used in conjunction with a sintered titanium cryopump to achieve a vacuum level measured to be less than 3×10^{-12} Pa (2×10^{-14} torr).

The gyroscope readout system is based on measuring their London moment using dc-SQUID magnetometers and has demonstrated a noise performance of 5×10^{-29} J/Hz at 5 mHz (the spacecraft roll frequency). This is equivalent to an angular resolution of 1 marcsec for an integration period of four hours, thus fully meeting GP-B requirements. Through the use of normal and superconducting shields, the dc magnetic field of the science instrument is reduced to less than 10^{-7} G, with an attenuation of the ac field in excess of 10^{13}.

Four high precision cryogenic gyroscopes are used in GP-B to determine the inertial reference frame in the vicinity of Earth. Three of the gyroscopes are electrostatically suspended, while the fourth is used as the drag free sensor for the experiment. (Note that we use the term gyroscope interchangeably for gyroscopes and drag free sensor, unless clarification is required.) Residual torque is reduced to a minimum by compensating for the drag of the satellite (3) and by carefully controlling the sphericity of the gyroscope and its housing. Figure 2 shows a schematic view of an exploded gyroscope.

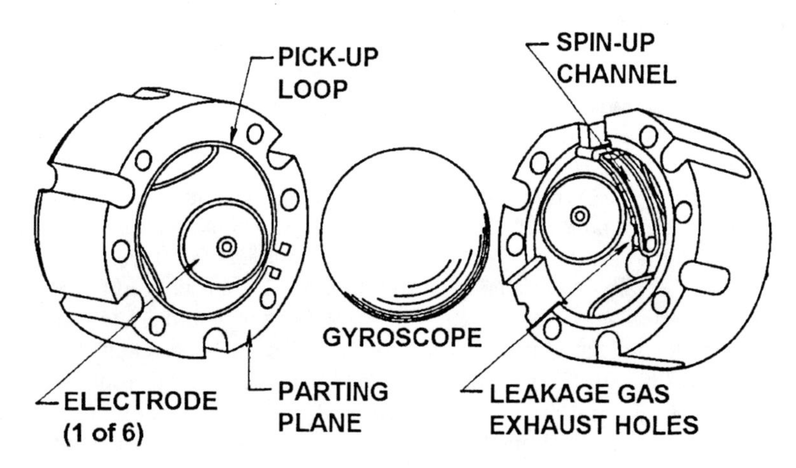

Figure 2. Schematic view of gyroscope

Table 1 summarizes the disturbance precessions for supported and unsupported gyroscope (4). The unsupported gyroscope, also used as the drag free sensor, does not need electrostatic support, thus eliminating the largest disturbance precession. Note however that the largest disturbances for both the supported and unsupported gyroscopes are due to electrostatic suspension and rotor-charge induced torques, emphasizing the need for optimization of both the electrostatic suspension and the charge control systems. Ground testing of the gyroscopes has shown performance consistent with the Relativity Mission requirements, indicating that the performance of the unsupported gyroscope will allow the measurement of _ to one part in 10^5.

Table 1. Gyroscope disturbance precessions

	GYROSCOPE SUPPORT	
DISTURBANCE TYPE	**Supported (marcsec/yr)**	**Unsupported (marcsec/yr)**
Mass Unbalance (12nm)	< 0.007	< 0.001
Electrostatic Suspension	< 0.140	< 0.010
Residual He Gas(10^{-11} torr)		
Differential Damping	< 0.006	< 0.006
Brownian Motion	< 0.001	< 0.001
Rotor Charge (10pC)	< 0.010	< 0.010
Gravity Gradient	< 0.001	< 0.001
Cosmic Radiation	< 0.001	< 0.001
Magnetic	< 0.001	< 0.001
Photon Gas	< 0.001	< 0.001
ROOT SUM SQUARE	**< 0.140**	**< 0.016**

III. GYROSCOPE SUSPENSION SYSTEM (GSS)

As shown in the previous section, the gyroscope suspension system (GSS) is the primary source of disturbance torques on the gyroscopes during the Science Mission phase of the GP-B experiment. The key challenge for the GSS design is to hold these disturbance torques to Science Mission compatible levels, while always maintaining positive suspension of the gyroscope under the effects of on-orbit disturbances and within the constraints imposed by the experimental apparatus.

To meet the overall objective set for the GP-B mission, the forces that the GSS needs to exert the gyroscope must span a huge dynamic range – eight orders of magnitude – and the system must compensate for a diverse set of disturbances during the various phases of the mission. It is not practical to span this large force and disturbance space with a simple, fixed control scheme. Thus, a multi-level scheme has been developed to address the pertinent requirements in each of three control modes: 1) Science Mission, 2) Spin-up and alignment, and 3) Ground-Test. Figure 3 presents a visual picture of the GSS controller set, the specific forces and electrode voltages required in the various modes, and the regions of influence of the primary disturbances operating during the mission.

The suspension system operates by generating electrostatic forces on the rotor via the application of a coordinated set of quasi-static voltages to the six electrodes on the gyroscope housing; see Figure 2. Since the three electrode axes are orthogonal, a judicious choice of the electrode voltage set allows the force on one axis to be decoupled from the other two, thus simplifying the design of the controller dynamics into an identical set of three one-DOF controllers rather than a single three-DOF controller. Even in this case, the resulting voltage-to-force map is still highly non-linear in both rotor position and in the magnitude of the applied voltage. An active non-linearity inversion scheme is employed to linearize the dynamics of the gyroscope plant to that of a simple inertial mass. This further simplifies the design of the controller dynamics, especially in the analog backup modes. The position of the rotor is measured with three capacitance bridges, one per axis, using a 40 mV-pp (400 mV-pp in the Spin-up and Ground Test modes), 34 kHz sinusoidal sense signal superimposed onto the drive electrodes. The high precision, low-noise design of the bridge results in an operational noise floor of 0.1 nm/√Hz, a resolution that allows the control system to meet the rotor centering requirements for the mission. The magnitude and frequency of the sense signal was chosen to be compatible with the ultra low noise operation of the SQUID magnetometers used to measure the orientation of the rotor's spin axis.

Figures 4 and 5 present a detailed block diagram of the GSS for a single gyroscope; each gyroscope is suspended by its own suspension electronics set. The overall design is partitioned into two physical enclosures. Figure 4 shows the makeup of the Forward Suspension Unit (FSU) enclosure. It primarily houses the precision analog electronics suite needed by the suspension system but also contains a bank of 16-bit A/D and D/A converters to translate the analog drive and position sense signals to their digital equivalents for use in the aft-mounted computer. Extensive internal shielding is employed to minimize the propagation of high-frequency digital noise to the sensitive analog electronics set. Thermal stability of the analog electronics is of a prime concern, especially in a space environment. This stability is achieved through a passive, multi-layer insulation scheme that will hold the electronics to within

0.05 K at the critical roll period of the spacecraft (1-3 min). The FSU also includes a multiplexed A/D converter for telemetry monitoring of the main control signals in the forward enclosure.

A radiation-hardened configuration register holds the state of all the configurable elements of the FSU enclosure. The 16-bit data word is stored in a set of radiation-impervious latching relays, which is changed only after a successful majority-voted succession of update commands to this register sent from the computer. It is powered independently of the computer; thus aft power failures or computer faults will not affect the configuration of FSU.

The aft block diagram presented in Figure 5 shows the bulk of the digital circuitry of the GSS system. At the heart of the Aft Control Unit (ACU) is a RAD6000-based CPU which acts as a DSP for control calculations, as well as a system monitoring and communication channel to the main spacecraft computer. This enclosure includes spacecraft timing synchronization circuitry to lock the computer clock and all derived clocks to the master spacecraft timing generator. This circuitry also generates the A/D sample and D/A convert triggers for the converter bank in the FSU and also performs a telemetry data gathering function via another multiplexed A/D converter in the aft enclosure. The GSS processor communicates with the FSU via the aft portion of the GSS Forward/Aft Bus, or GFAB.

Power is generated for the GSS by a very low noise DC-to-DC converter system mounted in its own enclosure next to the aft box. It generates a warm-redundant set of analog and digital supply voltages for the FSU, a non-redundant set of voltages for the ACU, and survival heater power to keep both the forward and aft electronics warm enough so that they will reliably start on orbit. The aft-created set of FSU supply voltages are further filtered and regulated forward to ensure that conducted EMI is minimized.

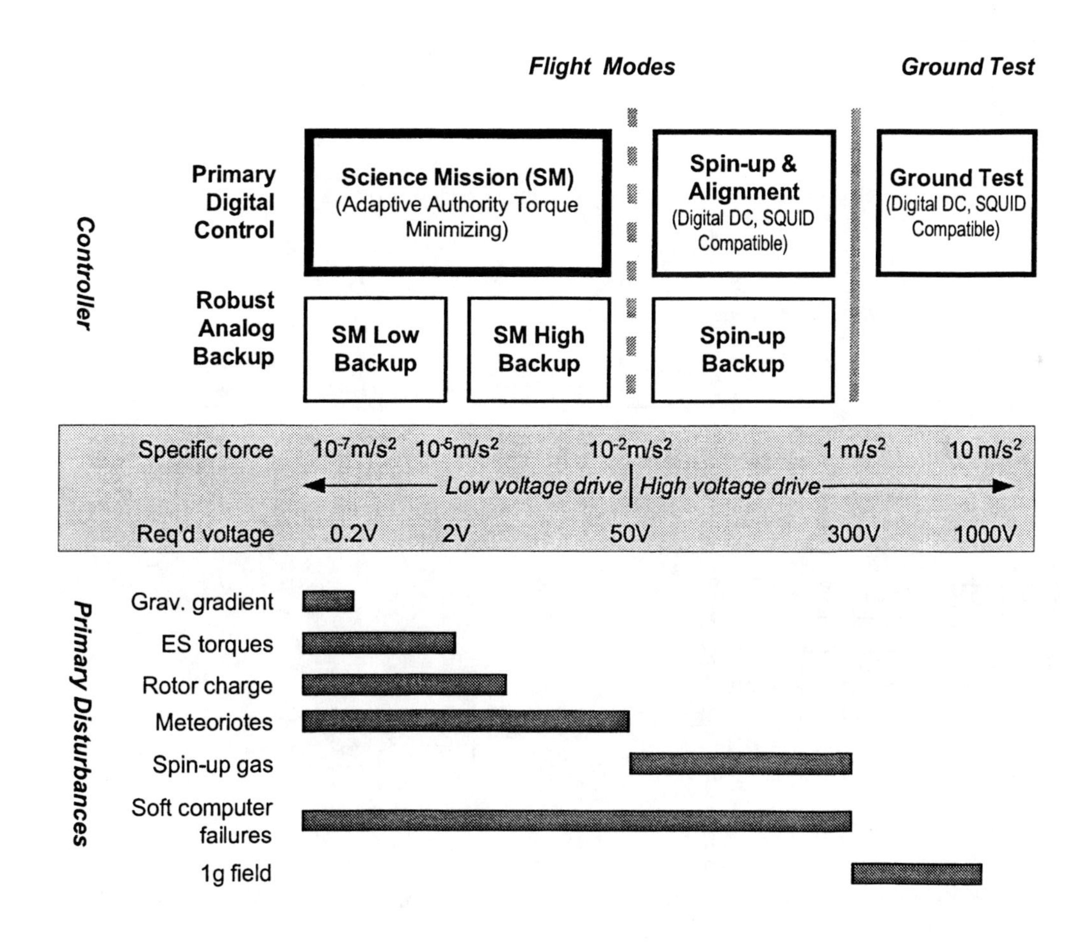

Figure 3. GSS controller set, required forces/voltages by mode, and primary disturbance set

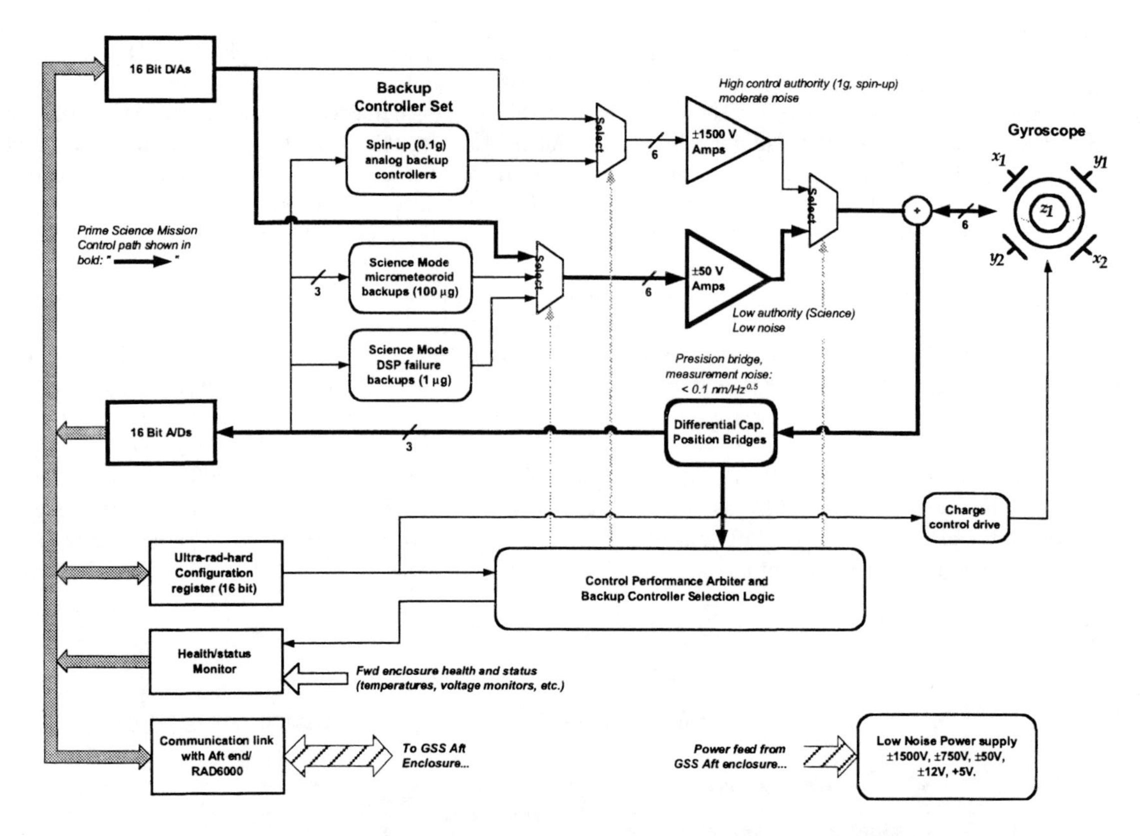

Figure 4: Block diagram of the forward GSS electronics enclosure (FSU)

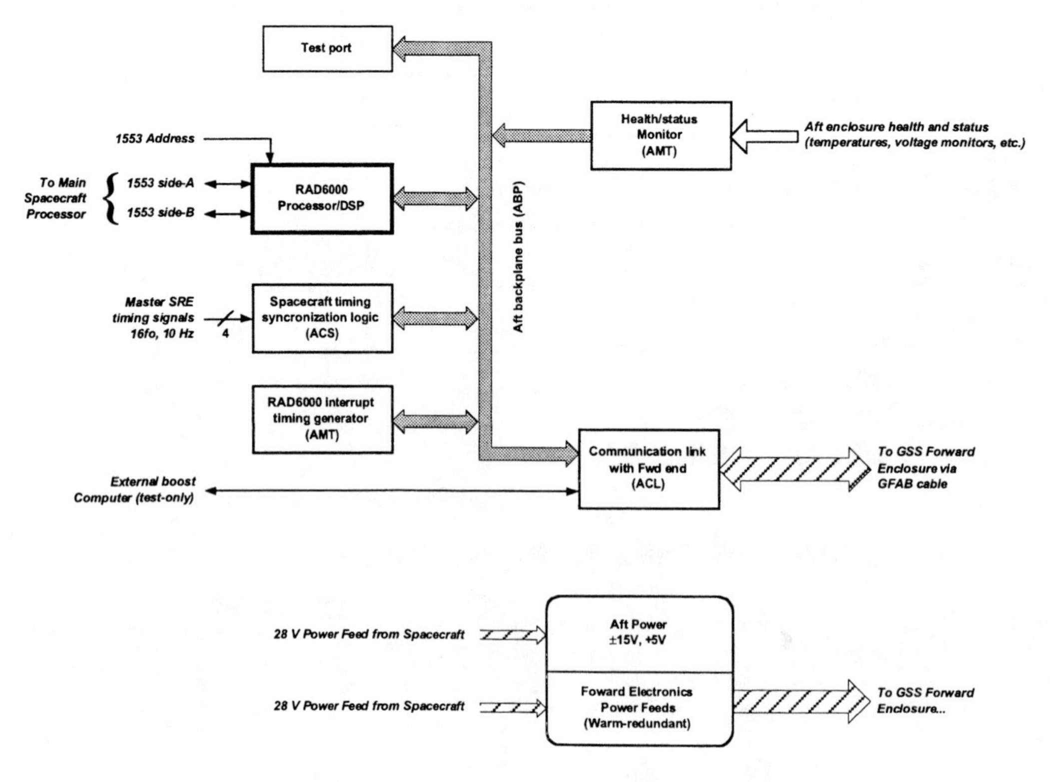

Figure 5: Block diagram of the aft GSS electronics enclosure (ACU)

The primary controller for the GSS system – the controller which shall be operating 99%+ of the mission – is the Science Mission digital controller running the GSS computer. The primary function of this controller is to minimize the electrostatic suspension-generated torques acting on the gyro, but it must also effectively handle the disturbance set active in this mode. This set consists primarily of slowly time-varying gravity gradient accelerations and the effects of rotor charging as well as the impulsive disturbances of micrometeoroid impacts. It adjusts to these disturbances by actively adjusting the bandwidth of the control loop so as to provide the minimum needed integrated force on the gyroscope to keep it within its centering requirements. At each control sample, it computes an electrode voltage set which provides the required force vector on the rotor while minimizing the maximum magnitude of the control voltages. These two strategies minimize the operational torques on the gyroscope.

This controller adapts to the disturbance environment and becomes increasingly aggressive as the magnitude of the disturbance increases. It smoothly transforms itself from a soft, linear controller to a maximum authority *bang-bang* controller to handle disturbances from micrometeorites with momenta as large as 1.0 kg-m/s; these occur with a probability of less than once per mission for the GP-B orbit. The prime science drive path is shown with a bold line in Figure 4. The D/A converters directly drive a low voltage (±50V), low noise amplifier to the gyro and the bridge feeds the position measurement to the A/D converter set for use by the control algorithm in the aft-located processor.

The primary Spin-up and Ground Test controllers are also implemented in the GSS processor, but are of a less sophisticated design. The main character of both of these modes is that they must handle a quasi-DC disturbance as part of their nominal operation. The Spin-up controller is required to hold the rotor near the spin-up channel in the presence of a specific force on the order of 1 m/s^2 during the time spin-up gas is flowing. In a similar way, the Ground Test controller must suspend the gyroscope against the 9.81 m/s^2 acceleration of the Earth's gravitational field in the laboratory. These high specific forces require high suspension voltages, and thus, a Spin-up/Ground Test high voltage amplifier is needed to provide the necessary voltages. During spin-up, the amplifier is designed to provide up to ±750 V to the electrodes, while on the ground up to ±1500 V is required for suspension. The aft-mounted power system will provide these voltages when needed, but will be shut off during the bulk of the mission in order to save power.

While it is desired to run the entire mission via the digital control algorithms in the GSS processor, radiation-induced soft failures of the processor which require a full reboot are likely to occur 1 to 3 times per mission per gyroscope. Thus a backup control system is required. In the Science Mission mode, two separate backup controllers are provided. A low control authority controller is provided to take over from the processor during normal operation where only gravity gradient and rotor charge disturbances are in effect. This controller is a simple PD (proportional-derivative) linear controller that will generate specific forces on the rotor up to the order of 10^{-5} m/s^2. While the torque performance of this simple PD controller is worse than that of the baseline digital algorithm, it is designed to meet as nearly as possible the Science Mission specifications. It is predicted that the gyroscope can remain in this low backup mode for 1-2 months before adversely affecting the precision of the measurement.

Though the low backup controller is nearly Science Mission compatible, it is unable to handle the random impulsive disturbances caused by micrometeoroid impacts. For this case, and aggressive high authority backup controller is included. This controller aggressively catches and re-centers the gyroscope in the event of a large micrometeoroid impact. It is a *bang-bang* controller and is constructed to use the full ±50 V range of the low-noise science drive amplifier. While greater control forces can be generated through the use of the high voltage amplifier, the system is designed so that it will never need to use these amplifiers in an emergency situation. Similarly, the primary digital spin-up controller also has a PD backup control system in the event of a soft computer failure during spin-up operations.

The selection of which controller is in operation at any one time is determined by the control system arbiter, a simple, robust analog computer in the forward electronics enclosure. This arbiter continuously monitors computer activity as well as the position of the gyroscope. In the event of a computer failure or a gyro excursion out of a pre-determined safety zone about the center of the cavity, the arbiter autonomously will switch between from prime digital controller to one of its backup systems. It will return control of the gyroscope to the computer only after it confirms that the computer is operational and is functioning properly. It is powered separately from the processor by the FSU warm-redundant power supply, and thus will faithfully execute its function regardless of the condition of the aft electronics set.

Laboratory tests to date indicate that this system will meet the overall gyroscope suspension, calibration, and operational needs of the GP-B experiment.

IV. GSS TESTING

A significant problem facing GP-B, is checking the performance of the suspension system levitating the on-orbit gyroscopes without actually having to perform an additional full fledged space experiment. It is necessary to test the closed-loop response of the electrostatic suspension, in order to determine the subtleties of the integrated system and to verify compliance with all flight requirements. While the gyroscopes are designed for ground levitation capability needed for functionality checking, the four orders of magnitude separating the ground and on-orbit suspension voltages (1000 V versus 0.1 V) make the ground testing unsuitable for complete verification of on-orbit performance. Computer simulations, with their dependence on idealized models, would provide a less than satisfactory test of the systems.

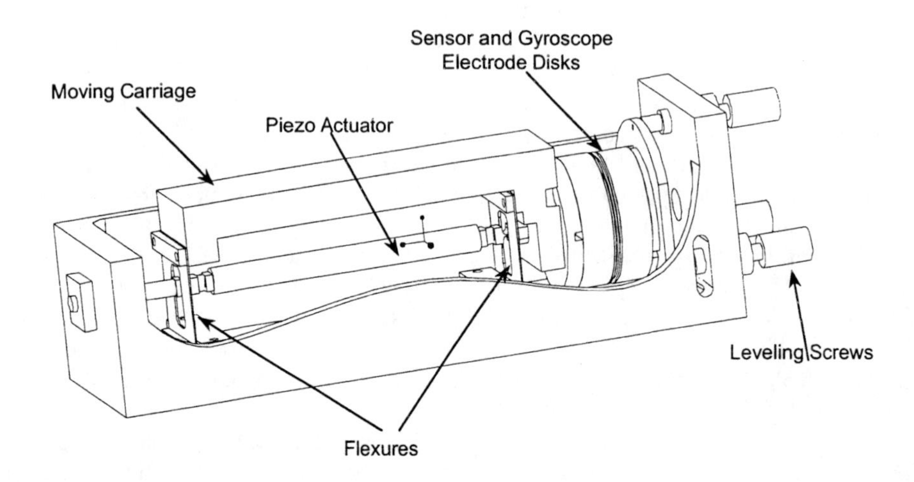

Figure 6. Schematic view of testbed actuator design

The solution implemented combines precision engineering with modern control techniques, to create a device with the dynamics of a gyroscope, which operates in a fully defined and controllable environment. The gyroscope 'testbed' consists of six electrode pairs on quartz disks that simulate the six gyroscope electrodes, by generating the required electrode to gyroscope capacitance. Piezoelectric actuators control the spacing between the quartz disk pairs, using the position information supplied by additional 'measurement' electrodes. Complex shielding between the gyroscope and the measurement electrodes on the quartz disks is needed in order to minimize cross talk between the two feedback systems. Figure 6 shows a schematic view of a complete testbed actuator.

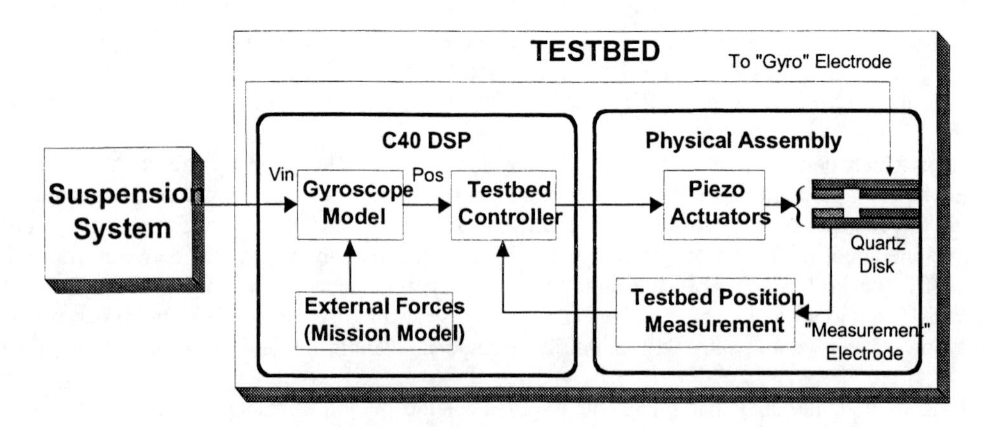

Figure 7. Schematic view of testbed concept

A C40 DSP contains the testbed controller, the gyroscope model, and a science mission model including all on-orbit disturbances, while the suspension system couples directly into the testbed. Figure 7 shows a representation of the entire testbed concept. The testbed allows the integrated testing of all functions of the suspension system, including position measurement and control, charge measurement and control, and spin axis alignment using space vehicle roll rate modulation of the suspension voltages. It also allows the testing of all suspension regimes from 1 g to 10^{-7} g, while incorporating all on orbit disturbances during both the initial setup and gyroscope spin-up stages, as well as the Science Mission data acquisition period.

Figure 8 demonstrates that the position resolution of the testbed is better than 0.1 nm, by showing the result given by the measurement electrodes for a commanded 0.3 nm peak-to-peak sinusoid. Verification of the testbed functionality is performed by comparing the gyroscope positions for all three axes during ground 1 g levitation of a real gyroscope and of the testbed. Figure 9 shows the results of this test.

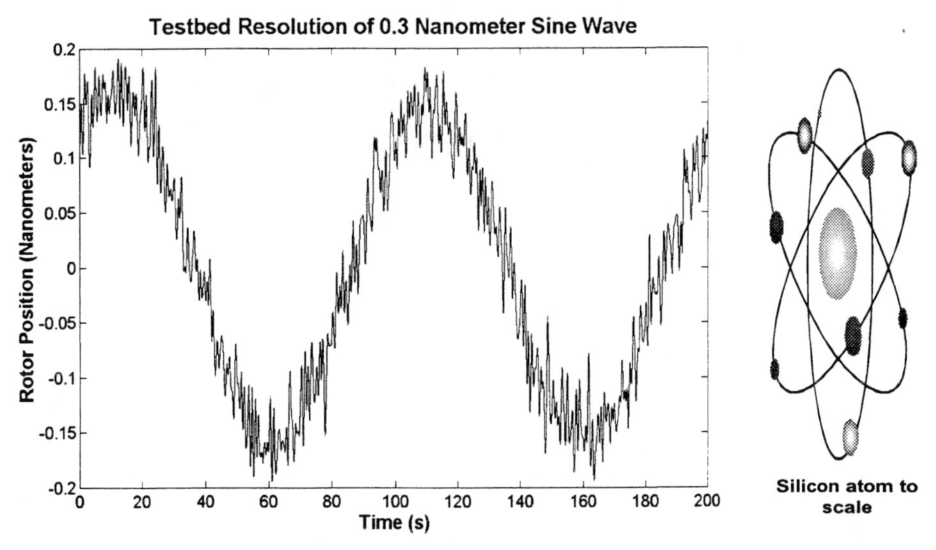

Figure 8. Testbed resolution of 0.3 nm sinusoid

V. CHARGE MEASUREMENT AND CONTROL

In the 650 km polar orbit used by GP-B the main contributions to the radiation environment are due to charged particles trapped in the Earth's magnetic field (5) and to charged particles generated by solar flares (6). Note that mainly proton trapping affects the gyroscopes, causing the cosmic radiation charging to be positive. The shielding provided by the spacecraft stops most primary electrons, while secondary emission of electrons has a yield of less than unity for the GP-B environment. During experiment-initialization stages two additional mechanisms can cause gyroscope charging: the separation of dissimilar metals during gyroscope levitation off the housing, and charge deposition by ionized helium during the spin-up of the gyroscopes. The polarity of the charging due to gyroscope levitation and spin-up are difficult to predict, making it necessary for the charge control technique to have bipolar capability.

In order to reduce heating and charging during solar flares, a radiation shield of 10 $g{\cdot}cm^{-2}$ was added, resulting in a total shielding for the gyroscopes of 20 $g{\cdot}cm^{-2}$ aluminum equivalent. Standard space technology is used to mitigate the effects of cosmic radiation on the electronics, the cryogenic probe, and the satellite. However, the gyroscope rotors are mechanically isolated systems spinning in ultrahigh vacuum, thus making it necessary to use non-contact methods for charge control and to rely on thermal radiation for cooling.

Gyroscope requirements, due to torque and acceleration considerations, limit the rotor charge to 10 pC, or equivalently to a 10 mV potential (for the 1 nF rotor capacitance). The total charge accumulation over the 1.5 year mission is about 600 pC making it necessary to monitor the gyroscope potential and use active charge control.

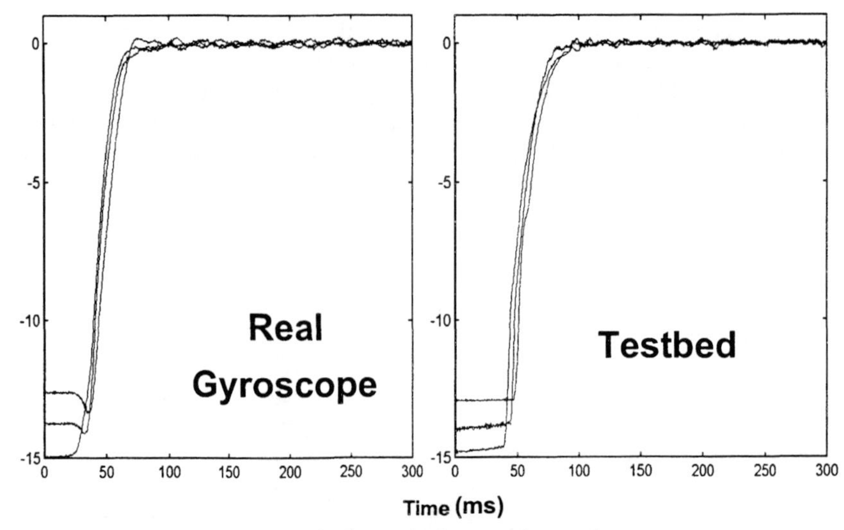

Figure 9. 1 g testbed suspension results

The force modulation technique is insensitive to gyroscope miscentering, is independent of the ambient acceleration, and achieves an accuracy of better than 5 mV for an integration time of 100 s, making it suitable for use for the GP-B mission. GP-B uses two levels of charge measurement excitation. a) A 10 mV level for continuous monitoring that achieves the 5 mV accuracy in 100 s. b) A 100 mV level for measurements during active control using the UV system, which achieves the 5 mV accuracy in about 1 s, thus making the control loop much simpler to implement.

UV photoemission is the method used by GP-B to generate the electrons used for charge control (7). Rotor and biasing electrode are illuminated with UV light, and the electrons generated by photoemission from both these surfaces are added or removed from the rotor using a dedicated biasing electrode. The direction of the charge flow is controlled by biasing of the charge control electrode to ±3 V with respect to the gyroscope surface. The gyroscope surface is a sputtered thin film of niobium, while the charge control electrode surface is electroplated gold. Experimental considerations pose additional constraints on the hardware near the gyroscope: low remnant magnetization, very high standards of cleanliness, a superconducting transition temperature below 1.5 K, and compatibility with the 2 K GP-B experimental temperature. Figure 10 is a schematic representation of the fixture mounted in the gyroscope housing which directs the UV light onto its inner surface and onto the gyroscope, while also fulfilling the function of the charge control bias electrode.

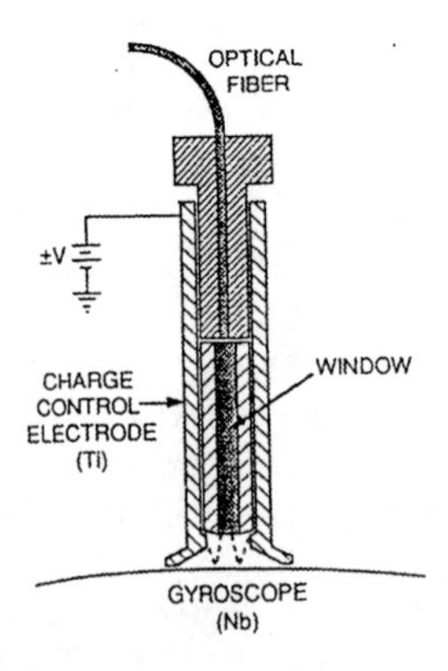

Figure 10. Schematic of UV counter electrode

The UV source is an rf-discharge mercury lamp manufactured by Resonance Ltd. Canada (8). About 10 μW of 254 nm mercury light is coupled from the lamp into each of twelve 300 μm UV fibers. For redundancy GP-B will fly two lamps, with two fibers from each lamp capable to illuminate the two UV fixtures on each gyroscope. Each of the eight fibers going to the four gyroscopes has a UV compatible switch developed by GP-B, thus allowing the choice of one of the two lamps for illumination. The photoemission efficiency is strongly dependent on the exact surface conditions of the transmitting and reflecting system elements, and varies between 50 and 1000 fA/μW for the gyroscope and between 100 and 2000 fA/μW for the UV fixtures.

Ground testing of the charge control system proceeds by two techniques. In the first method the gyroscope is not levitated, and the photoemission current is measured directly through the connection to the ground-plane of the housing on which the gyroscope is resting. Figure 11 gives the photo current as a function of the bias voltage for a flight gyroscope, for room temperature, 300 K, and low temperature, 4 K. The second technique uses the voltages generated by the electrostatic suspension system to measure the gyroscope potential variations under UV illumination. Flight hardware testing confirms that charge management, using measurement by force modulation and electrons generated by ultraviolet photoemission, are the solutions for the GP-B gyroscope-charging problem.

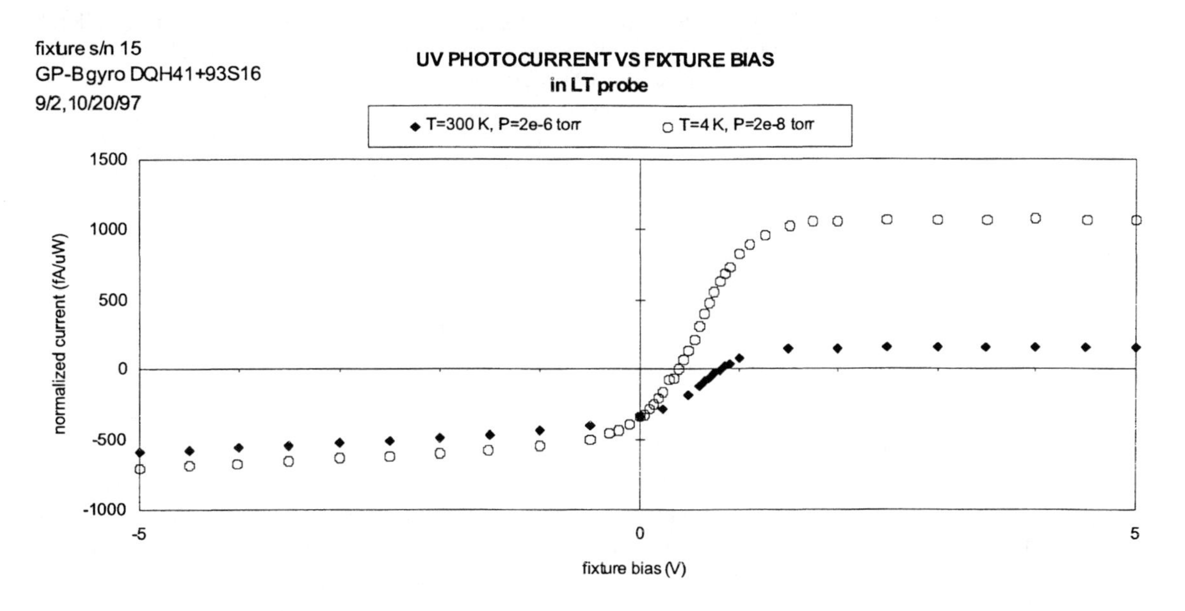

Figure 11. UV photo-current vs fixture bias

ACKNOWLEDGEMENTS

This work was supported by NASA contract NAS8-39225. The authors wish to thank the members of the Stanford University GP-B group for many stimulating discussions.

REFERENCES

1. J. P. Turneaure *et al.*, Adv. Space Res. **9**, 29 (1989)
2. T. Damour, K. Nordtvedt, Phys Rev Lett 70, 2217 (1993)
3. D.B. DeBra, *DISCOS Description* (private communication, Stanford University, 1970)
4. C. W. F. Everitt and S. Buchman, *Particle Astrophysics Atomic Physics and Gravitation*, pp. 467 (Editions Frontieres, Cedex-France, 1994)
5. J. I. Vette, *The NASA/National Space Science Data Center Trapped Radiation Environment Model Program (1964-1991)*, NSSDC/WDC-A-R&S 91-29, (1991)
6. J. Feynman, *et al.*, Journal of Geophysical Research **98**, 13281 (1993)
7. Saps Buchman *et al.*, Rev. Sci. Instrum. **66**, 120 (1995)
8. Resonance Ltd., 143 Ferndale Drive North, Barrie, Ontario L4M 4S4, Canada

The Expected Performance of Gravity Probe B Electrically Suspended Gyroscopes as Differential Accelerometers

G. M. Keiser[†], Saps Buchman[†], William Bencze[†], and Daniel B. DeBra[†‡]

[†]*W. W. Hansen Experimental Physics Laboratories and*
[‡]*Department of Aeronautics and Astronautics,*
Stanford University, Stanford, CA 94303

Abstract. Four cryogenic gyroscopes on the Gravity Probe B satellite will be used to measure the precession of the local inertial reference frame with respect to a distant inertial reference frame. One of these four gyroscopes will serve as the drag-free sensor for the satellite. The other three gyroscopes, which are separated from each other by 8.25 cm, will be electrostatically supported by a digital control system. Although the gyroscopes and the electrostatic suspension system are designed to measure a precession as small as 0.1 mas/yr, any pair of these gyroscopes may also be used as a differential accelerometer. This paper analyzes the expected performance of these gyroscopes as differential accelerometers for accelerations in the frequency band from 2×10^{-3} to 2×10^{-2} Hz. The three contributions to the specific force on any one of the gyroscopes are the residual acceleration of the spacecraft, the specific forces acting between the gyroscope and the satellite, and the noise in the capacitance bridge which senses the position of the gyroscope relative to its housing. The dominant source of noise in this frequency band is found to be the quantization noise in the D/A converter used in digitally controlled electrostatic suspension system for the supported gyroscopes.

I. INTRODUCTION

The Gravity Probe B Relativity Mission (1) is a NASA supported program designed as a high precision experimental test of the General Theory of Relativity. Four gyroscopes will be used to measure two relativistic effects on gyroscopes placed in a circular, polar orbit at 650 km altitude. According to General Relativity, the geodetic or de Sitter (2) effect will cause each gyroscope to precess 6.6 arc seconds per year (as/yr) in the orbital plane, while the frame dragging, or Lense-Thirring (3), (4) effect, will cause the gyroscopes to precess 42 milliarcseconds per year (mas/yr) in a direction perpendicular to the orbital plane. The gyroscopes will be mounted in a quartz block, which is rigidly attached to a telescope. This instrument assembly will be operated at 2.3 K. During those parts of the orbit that the chosen guide star is visible, the satellite's attitude control system will keep the telescope pointed at a guide star to within 20 mas rms. In addition, the satellite will slowly roll about the line of sight to guide star at a roll rate between 0.1 and 1.0 rpm. Analysis of potential sources of error in the experiment, including the uncertainty in the proper motion of the guide star, indicates that the expected measurement accuracy will be approximately 0.2 mas/yr.

The Gravity Probe B gyroscopes are designed to measure rotations of the local inertial frame as small as 0.1 mas/year. However, these gyroscopes are also inertial sensors which may be used to measure the external forces on the spacecraft and the difference in the specific forces on the gyroscopes. One of the gyroscopes will be used as a drag-free sensor, and thrusters, which use the helium boil-off gas from the superfluid liquid helium dewar, keep gyroscope's rotor centered within its cavity. Then, the acceleration of the spacecraft itself will be a sensitive measure of the forces on the unsupported gyroscope, and the control effort of the drag free control system may be used to measure the external forces on the spacecraft (5).

M. Tapley (6), (7) carefully analyzed the use of the gyroscopes as a gravity gradiometer which could be used to measure the higher harmonics of the earth's gravitational potential. He showed that measurements of the differential specific force on two supported gyroscopes in the GP-B satellite at periods between one and ten minutes

CP456, *Laser Interferometer Space Antenna*
edited by William M. Folkner
 1-56396-848-7/98/$15.00

could significantly improve the knowledge of the higher harmonics of the earth's geopotential. The Gravity Probe B gyroscopes and their electrostatic suspension system are designed to minimize the torques on the gyroscopes, reliably suspend the gyroscopes, and operate compatibly with the SQUID readout system. As such, the operating conditions of the gyroscopes are significantly different from those of an accelerometer designed to reduce the specific forces on a proof mass, and they are also different from a system which is optimized to measure the gradient in the gravitational field. Nevertheless, measurements of the differential acceleration on any pair of gyroscopes will provide useful information. This paper updates Tapley's original estimates of the sensitivity of the gyroscopes as gravity gradiometers and also extends Tapley's original analysis to include several additional specific forces acting on the gyroscopes and the effects of the noise in the position readout on the noise of the gradiometer.

The following section describes the configuration of the Gravity Probe B gyroscopes in more detail. Section III is a discussion of the operation of the drag-free and electrostatic control loops. Section IV discusses physical sources of forces and force gradients acting on the gyroscopes, which determine the performance of the drag-free system. Section V summarizes the expected performance of the gyroscopes as differential accelerometers.

II. CONFIGURATION OF THE GRAVITY PROBE B GYROSCOPES

The Gravity Probe B satellite will contain four gyroscopes, three of which are electrostatically suspended and a fourth which will act the drag-free proof mass for the satellite. As shown in Figure 1, each of these four gyroscopes has three mutually perpendicular pairs of electrodes. These six electrodes are used both to sense the position of the gyroscope rotor relative to the housing and, for the supported gyroscopes, to apply a control voltage to keep the rotor centered with respect to the housing. There is also a gas spin up channel, which is used to spin the gyroscopes up to their operating speed of approximately 150 Hz. The d.c. rotor potential is determined by applying a low frequency bias which is 180^0 out of phase on opposite electrodes and by measuring the control effort required by the suspension system to prevent the rotor from moving (8). This d.c. rotor potential may be controlled by shining ultraviolet light, which enters the housing through an optical fiber, on both the rotor and the housing and using a bias electrode to control the current between the rotor and the housing (8). All surfaces on the interior of the housing are coated with a conducting coating which is electrically tied to the reference ground for the electrostatic suspension system. The radius of the rotor is 19 mm, and the gap between the rotor and the electrode is 31 μm.

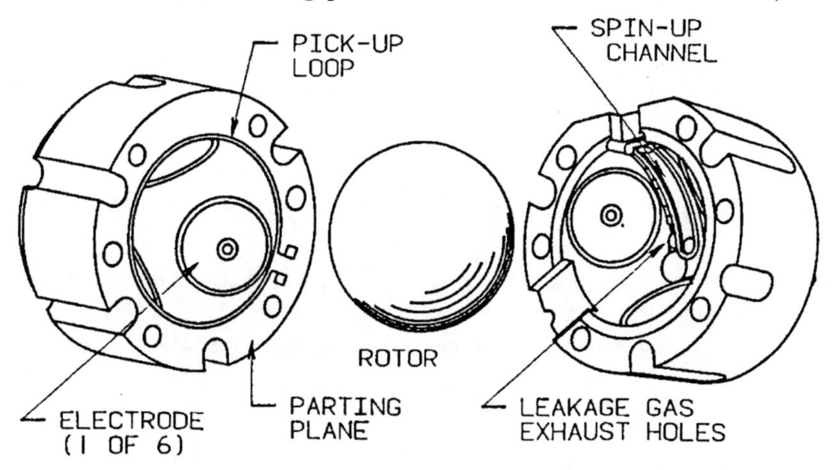

FIGURE 1. The Gravity Probe B Gyroscope

The four gyroscopes are enclosed in a fused quartz block, which is bonded to a telescope, as shown in Figure 2. The quartz block and the telescope operate within a liquid helium dewar at a temperature of 2.3 K. During those periods which the guide star is valid, the telescope remains pointed at a reference star, and the satellite rolls about the line of sight to the reference star at a fixed rate between 0.3 and 1 rpm. The rms pointing accuracy is expected to be less than 20 mas. Star sensors and rate gyroscopes external to the liquid helium dewar maintain the roll rate constant to 1 part in 10^5.

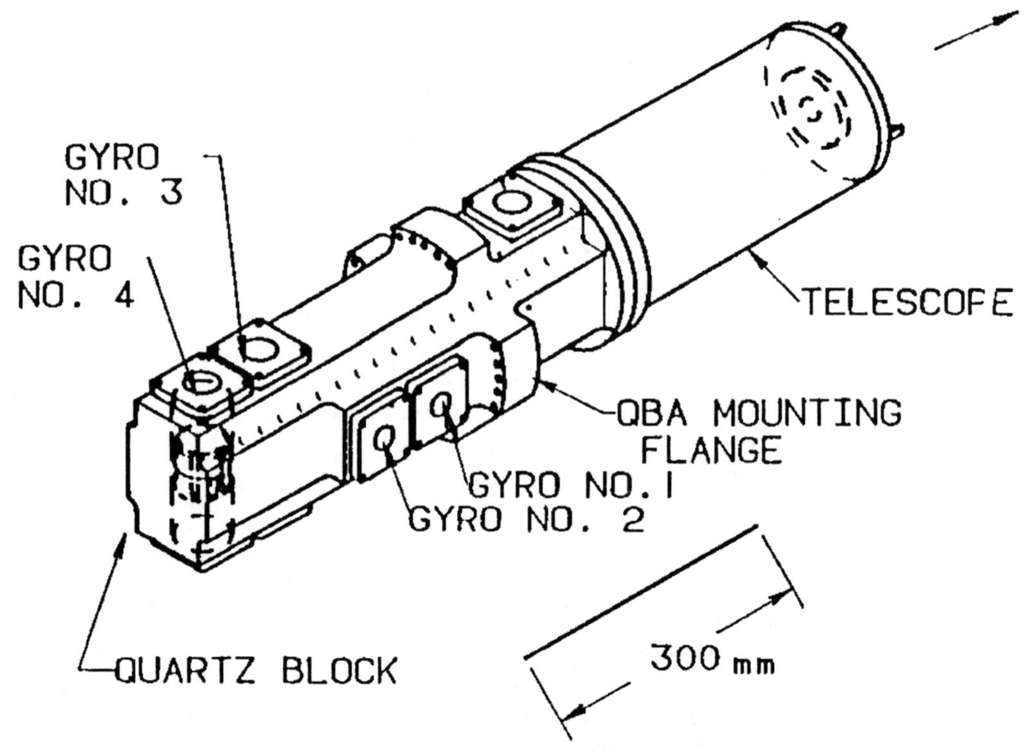

FIGURE 2. The Quartz Block Assembly. The satellite roll axis lies along the direction to the guide star.

For the unsupported gyroscope, the position of the rotor is sensed with a capacitance bridge which operates at 35 kHz and applies 40 mV peak-to-peak to the each pair of electrodes. After demodulation, the position signal is converted to a digital signal with a 16 bit A/D converter operating a 220 Hz. A digital control loop is then used to control the thrusters on the satellite so as to maintain the position of the spacecraft fixed with respect to the gyroscope rotor. For the supported gyroscopes, another digital control loop is used to vary the control voltage applied to the electrodes. The electrostatic force due to this applied voltage maintains the position of the rotor with respect to the housing.

III. EQUATIONS OF MOTION AND OPERATION OF THE DRAG FREE CONTROL SYTEM AND ELECTROSTATIC SUSPENSION SYSTEM

Under most conditions the effect of the forces due to the proof mass on the satellite may be ignored since they are considerably smaller than the other forces acting on the satellite. With this assumption, the equations of motion for the geometric center of the housing, **c**, and the center-of-mass of the rotor, **r**, are

$$M\frac{d^2\mathbf{c}}{dt^2} = \mathbf{F}_{\mathrm{ext}} + \mathbf{F}_{\mathrm{f}}$$
$$m\frac{d^2\mathbf{r}}{dt^2} + \beta\frac{d\mathbf{x}}{dt} + k\mathbf{x} = \mathbf{f} + \mathbf{f}_{\mathrm{f}} \tag{1}$$

where m and M are the masses of the gyroscope and satellite, respectively. The vector **x** is the displacement of the center-of-mass of the rotor from the geometric center of the housing, **r=c+x**. Here the geometric center of the housing is assumed to coincide with the pick-off null for the position readout system. Any d.c. offset does not change the analysis that follows, and a.c. offsets are included as noise in the position readout system. The coefficients β and k are the damping coefficient and the effective spring constant for any velocity dependent or position dependent forces acting on the rotor due to the housing. The external forces, $\mathbf{F}_{\mathrm{ext}}$, include external forces acting on the satellite as well as inertial forces due residual angular accelerations that are not removed by the attitude control system. The feedback forces supplied by the satellite's thrusters are denoted by $\mathbf{F_f}$. Any force acting on the gyroscope rotor including gravitational forces due to gradients in the gravitational field is represented by **f**, with the exception of the feedback control force applied by the electrostatic suspension system, which is denoted $\mathbf{f_f}$. Solutions of these differential equations are discussed below for the case of the unsupported gyroscope ($\mathbf{f_f}$=0) and the supported gyroscope, where the force due to the thrusters, $\mathbf{F_f}$, is determined by the drag-free control system.

Unsupported gyroscope

For the gyroscope which acts as the drag-free sensor, there is no control force supplied by the electrostatic suspension system, $\mathbf{f_f}$=0. In this case, the feedback force on the satellite due to the helium thrusters, $\mathbf{F_f}$, supplies a force on spacecraft so that the center of the gyroscope cavity coincides with the center of the proof mass, **x**=0. The dynamic solution for the equations of motion along any axis may be found by taking the Laplace transform of the equations of motion. For the drag-free control system, the Laplace transform of the feedback force is given by

$$F_f(s) = -H(s)(x(s)+n(s)) \tag{2}$$

where H(s) is the Laplace transform of the compensation network and n(s) is the Laplace transform of the position noise. These equations may be solved to find the relative position of the satellite and the drag-free gyroscope, the feedback force supplied by the drag-free control system, and the acceleration of the spacecraft and the drag-free gyroscope.

Below the bandwidth of the drag-free control system, the accelerations center of the gyroscope housing is given by:

$$s^2c(s) \approx \frac{F_{ext}(s)}{G(s)H(s)M} + \frac{f(s)}{m} - (ms^2 + \beta s + k)\frac{n(s)}{m} \tag{3}$$

where

$$G(s) = \frac{m}{M(ms^2 + \beta s + k)}, \tag{4}$$

and the approximation has been made that the open loop gain of the servo system, G(s)H(s), is much larger than unity.

From equation (3), it can be seen that the residual acceleration of the gyroscope housing has three contributions. The first is the residual acceleration of due to external forces acting on the gyroscope and inertial forces due to angular motion of the attitude control system. The second is the contribution due to specific forces acting on the gyroscope, and the third is due to the noise in the position sensing bridge. It is interesting to note that the acceleration of the unsupported gyroscope may be considerably smaller than the acceleration of the housing. The contributions due to the residual forces and the noise in the position sensing bridge are smaller by a factor of k/ms^2 for the gyroscope than for the housing provided $k<ms^2$.

Supported gyroscope

From equation (1), the equation of motion for a supported gyroscope is

$$m\frac{d^2\mathbf{x}}{dt^2} + \beta\frac{d\mathbf{x}}{dt} + k\mathbf{x} = \mathbf{f} + \mathbf{f_f} - m\frac{d^2\mathbf{c_s}}{dt^2} \tag{5}$$

where f_f is the feedback force on the supported gyroscope and the other symbols have the same meaning as before. In this case, $\mathbf{c_s}$ is the position of the center of geometry of the supported gyroscope. The last term on the right hand side of this equation includes the residual acceleration of the unsupported gyroscope, given by equation (3), as well as inertial accelerations due to the angular motion of the housing of the supported gyroscope. If the feedback force is given by

$$f_f(s) = -H(s)(x(s) - n(s)), \tag{6}$$

then the closed loop expression for the feedback force may be found by substituting this expression for the feedback force into the Laplace transform of equations for motion, equation (5). Then, below the bandwidth of the servo system, the feedback force on the supported gyroscope becomes

$$f_f(s) = -f(s) - (ms^2 + \beta s + k)n(s) + ms^2c_s(s) \tag{7}$$

This result shows that below the bandwidth of the servo system, the control effort signal has contributions from the forces acting on the unsupported gyroscope, the noise in the position sensing bridge, and the residual acceleration of the gyroscope housing.

The differential acceleration of two supported gyroscopes may be measured by comparing the control effort on two gyroscopes. In this case, the residual acceleration of the spacecraft will be a common mode signal which may be removed provided the scale factors of the control effort signal are adequately known. Tapley (7) discusses several methods of calibrating the scale factors and argues that the best method is to use the naturally occurring low degree gravity gradient of the Earth. Alternatively, the differential acceleration between one supported gyroscope and the unsupported gyroscope may be measured by monitoring the control effort signals on one of the supported gyroscopes. In this case the residual acceleration of the spacecraft may contribute to the measurement noise. In addition, small errors in the differential acceleration may occur because of differences in the closed loop gains of the drag-free control system and the electrostatic control system of the supported gyroscope. However, with this method only one control effort signal needs to be monitored.

IV. FORCES AND FORCE GRADIENTS

Electrostatic

The force on a gyroscope in the x-direction due to a potential difference between the rotor and the electrode (9) is

$$F_x = +\frac{1}{2}\frac{\partial C}{\partial x}V^2 \tag{8}$$

where C is the capacitance between the rotor and the electrode, and V is the potential difference. For two opposite electrodes along the same axis, the net electrostatic force is

$$F_{net} = \frac{1}{2}\left(\frac{\partial C_+}{\partial x}V_+^2 + \frac{\partial C_-}{\partial x}V_-^2\right) \tag{9}$$

Here, the subscripts + and – denote two opposite electrodes on the same axis. The variation in the net electrostatic force due to changes in the position, x, and the voltages, V_+ and V_-, is then

$$\begin{aligned}\delta F_{net} &= k_x\delta x + \frac{\partial F_{net}}{\partial V_+}V_+\frac{\delta V_+}{V_+} + \frac{\partial F_{net}}{\partial V_-}V_-\frac{\delta V_-}{V_-} \\ &= k_x\delta x + F_{net}\left(\frac{\delta V_+}{V_+} + \frac{\delta V_-}{V_-}\right) + \frac{1}{2}mh\left(\frac{\delta V_+}{V_+} - \frac{\delta V_-}{V_-}\right)\end{aligned} \tag{10}$$

where the negative spring constant is given by

$$k_x = \frac{\partial F_{net}}{\partial x} = \frac{1}{2}\left(\frac{\partial^2 C_+}{\partial x^2}V_+^2 + \frac{\partial^2 C_-}{\partial x^2}V_-^2\right) \tag{11}$$

and the preload acceleration, h, is given by

$$h = \frac{1}{m}\left(\frac{\partial F_{net}}{\partial V_+}V_+ - \frac{\partial F_{net}}{\partial V_-}V_-\right) \tag{12}$$

In equations (10) and (11), the assumption has been made that the electrodes are driven by a voltage source there is no substantial change in the rotor potential with a displacement of the rotor. The noise in the net applied force depends on whether the variations in the voltages, V_+ and V_- are correlated, anticorrelated, or uncorrelated. Below the bandwidth of the servo system, the net force is approximately equal to the external force.

Control Voltages

The system is normally operated so that the voltage on the positive electrode is the sum of a preload voltage, V_p, and a feedback voltage, V_f, while the voltage on the negative electrode is the the preload voltage minus the feedback voltage. The preload acceleration may then be defined as the acceleration required to drive on of the electrode voltages to zero. Under normal operating conditions, the preload acceleration is larger than the specific acceleration on the gyroscopes, so that the preload voltage is larger than the feedback voltage. Then, the negative spring constant is approximately equal to

$$k_x = 2\frac{C_0}{d_0^2}V_p^2 \tag{13}$$

Using a value of 80 pF for the capacitance and 30 microns for the gap, the negative spring constant for a preload voltage of 0.1 volts is 1.8×10^{-3} kg/sec^2. This negative spring constant corresponds to a natural frequency of 0.027 Hz. Below this frequency, the transfer function of the gyroscope is frequency independent thereby reducing the open loop gain. The specific force noise due to the position bridge of 0.1 nm/√Hz and this negative spring constant and is equal to 2.9×10^{-12} m/sec^2√Hz.

Since the feedback voltages applied to the two opposite electrodes are out of phase with one another and are a small fraction of the preload voltage, the noise due to the D/A converters on the two electrodes will be anticorrelated. In this case, the noise force on the rotor due to the D/A converter is given by the last term in equation (10)

$$\delta F_{net} = \frac{1}{2}mh\left(\frac{\delta V_+}{V_+} - \frac{\delta V_-}{V_-}\right) \approx mh\frac{\delta V_q}{V_p} \tag{14}$$

where the quantization noise (10) is given by

$$\delta V_q = \frac{q}{\sqrt{12 f_c}} \tag{15}$$

where q is the quantization step size and f_c is the conversion rate. For a 16 bit D/A converter operating over a range of 100 volts with a conversion rate of 220 Hz, the quantization noise is 30 μV/√Hz. If the preload accelertion is 10^{-6} m/sec^2 with a nominal operating voltage of 0.11 volts, then the specific force acting on the gyroscope is 2.6×10^{-10} m/sec^2√Hz. A considerable reduction in this noise could be achieved either by reducing the range over which the D/A converter operates or by reducing the preload acceleration. The preload acceleration is limited by the expected maximum acceleration on the supported gyroscopes, and the wide range of the D/A converter was chosen to increase the reliability of the suspension system in the event of mirco-meteoroid impacts

The thermal effects on the control voltages are likely to be correlated between the opposite electrodes since the amplifiers are expected to be in the same thermal environment. Then, from equation (10) the contribution to the change in the net force produced by the control voltages will be proportional to the net force acting on the supported gyroscopes. The electrostatic suspension system is designed to maintain the control voltages constant to better than 1 part in 10^5 at roll frequency, which is expected to be the frequency of the dominant disturbance in the frequency band from 2×10^{-3} to 2×10^{-2} Hz. Although thermal variations may produce a coherent signal at roll as large as 10^{-11} m/sec^2, the white noise is expected to be less than 10^{-13} m/sec^2/√Hz when averaged over periods as long as 10^4 seconds.

Position Sensing and Charge Measurement Voltage

A 40 mV peak-to-peak signal at 35 kHz is applies to the same electrodes which are used to apply the control voltage. This signal produces a negative spring constant of 3.5×10^{-5} kg/sec^2 on the both the supported and the unsupported gyroscopes. For the supported gyroscopes, this negative spring constant is small compared to the negative spring constant produced by the control voltages so its contribution to the noise may be ignored. However, for the unsupported gyroscope, this position sensing voltage makes a significant contribution to the negative spring constant. The natural frequency for this negative spring constant is 3.8 mHz, which is close to the lower end of the bandwidth of interest. Then above this natural frequency, from equations (3) and (7), the contribution to the specific force due to the noise in the position sensing circuit increases as the square of the frequency. At the midpoint of the bandwidth of interest, 7×10^{-3} Hz, the specific force noise due to the noise in the position sensing bridge is 1.9×10^{-13} m/sec^2√Hz with a bridge noise of 10^{-10} m/√Hz.

Since the position sensing voltages applied to opposite electrodes are generated from the same frequency oscillator, fluctuations in the rms voltage are likely to generate little net specific force noise. The dominant frequency component of the thermal variations is expected to be at the roll frequency, and the voltages are specified to be stable to less than 20 μV at this frequency. However, there will only be a net force on the rotor if does not lie at the force center of the housing. Assuming a constant miscentering of 4% of the gap between the rotor and the electrode, from equation (9) the d.c. specific force on the rotor is 6.8×10^{-10} m/sec^2. The variation in this miscentering at roll frequency is expected to be less than 0.3 nm, do the variation in the force at this frequency is 5×10^{-13} m/sec^2. Averaged over periods as long as 10^4 seconds, the white noise close to this frequency is expected to be negligible.

The input current noise of the amplifier which is used to measure the output of the capacitive position sensing bridge is approximately 1 pA/√Hz. This current noise, when it is applied to the 15K impedance of the capacitance bridge at the bridge operating frequency of 35 kHz, produces a voltage noise with a spectral density of 1.5×10^{-8} volts/√Hz near 35 kHz. Combined with the 20 mV, 35 kHz bridge excitation voltage, this amplifier current noise produces a specific force noise of 1.2×10^{-14} m/(sec^2√Hz) on the 63 g rotor.

Voltages having an amplitude of 10 mV are applied to opposite electrodes to measure the net rotor potential using the force modulation method (8). These voltages are applied at 0.05 Hz for the supported gyroscopes and 4 Hz for the unsupported gyroscope. These voltages contribute to the negative spring constant but make no additional contribution to the specific force noise due to the noise in the position bridge since the natural frequency is below the frequency band of interest.

Charged Rotor and Asymmetric Ground Plane

If the rotor becomes charged and there is an asymmetry in the ground plane of the electrodes, then the force on the rotor is

$$F_Q = \frac{1}{2}\frac{\Delta C}{d_0}\left(\frac{Q}{C}\right)^2 \tag{16}$$

Here ΔC is the asymmetry capacitance between the rotor and the ground plane, d_0 is the electrode-to-rotor gap, Q is the total charge on the rotor, and C is the total capacitance between the rotor and the housing. For an asymmetry in the capacitance of 50 pF (5% of the total capacitance), a gap of 30 microns, and a rotor potential due to the charge of 15 mV, the specific force on a 63 g rotor is 3×10^{-9} m/sec^2. This charge is expected to vary slowly with time, with typical rates of 1 mV/day. The random noise associated with this slow variation in the charge is not expected to make a significant addition contribution to the noise in the frequency band of 2×10^{-3} to 2×10^{-2} Hz.

The force between the plates of a parallel plate capacitor which has a constant charge is independent of the gap between the electrodes. In this case, the force gradient or effective spring constant is zero. However, when the asymmetry is a small fraction of the total capacitance between the electrode and the rotor, the effective spring constant becomes

$$k_x = -\frac{\partial F}{\partial x} \approx \frac{\Delta C}{d_0^{\,2}}\left(\frac{Q}{C}\right)^2 \tag{17}$$

The negative spring constant associated with the charged rotor where the rotor potential is 15 mV is 1.2×10^{-5} kg/sec^2, which gives a natural frequency of 2.2×10^{-5} Hz.

Patch or Contact Potential Effects

For variations in the rotor potential due to patch fields that are larger than the gap, the electric field is given approximately by

$$E = \frac{V_a(\theta,\phi) - V_b(\theta,\phi)}{d(\theta,\phi)} \tag{18}$$

and the total energy stored in the electric field

$$\begin{aligned} W &= \frac{\varepsilon_0}{2}\int_{volume} E^2 dV = \frac{\varepsilon_0 r_0^{\,2}}{2}\int_{solid\ angle} d\Omega \frac{(V_a(\theta,\phi) - V_b(\theta,\phi))^2}{d(\theta,\phi)} \\ &\approx \frac{\varepsilon_0 r_0^{\,2}}{2d_0}\int_{solid\ angle} d\Omega\left(1 - \frac{\Delta d(\theta,\phi)}{d_0} + \left(\frac{\Delta d(\theta,\phi)}{d_0}\right)^2\right)(V_a(\theta,\phi) - V_b(\theta,\phi))^2 \end{aligned} \tag{19}$$

where $\Delta d(\theta,\phi)$ is the variation in the rotor-to-housing gap and d_0 is the nominal gap. The energy stored in the electrostatic field as a function of the displacement in a direction z is given by setting

$$\Delta d(\theta,\phi) = z\cos\theta \tag{20}$$

Then, the force on the rotor in the z-direction is equal to

$$F_z = +\frac{\partial W}{\partial z}\bigg|_{V=const.} = -\frac{\varepsilon_0 r_0^2}{2d_0^2} \int_{solid\ angle} d\Omega \cos\theta\big(V_a(\theta,\phi) - V_b(\theta,\phi)\big)^2 \tag{21}$$

and the spring constant is

$$k_z = -\frac{\partial F_z}{\partial z} = -\frac{\varepsilon_0 r^2}{d_0^3} \int_{solid\ angle} d\Omega \cos^2\theta\big(V_a(\theta,\phi) - V_b(\theta,\phi)\big)^2 \tag{22}$$

The magnitudes of these quantities are strongly dependent on the size of the patches. A large number of randomly oriented patches, which are small compared to the diameter of the gyroscope rotor, will not contribute significantly to a cosine distribution of the square of the potential difference. Then, the patch effect force and force gradients should be reduced by the square root of the number of patches on the surface of the rotor. In addition, Speake (11) has shown that for parallel surfaces where the patch field is much smaller that the gap, the patch effect fields are exponentially attenuated by the ratio of the gap to patch effect field size. The grain size of the coating on the surface of the housing has been measured to be approximately 6 mircons (12) compared to a gap between the gyroscope rotor and housing of 31μm. Typical values of the patch effect potentials are 0.1 volts (11) (other ref.). Taking these two factors into account, the magnitude of the specific force acting on the gyroscope is approximately 7×10^{-14} m/sec^2, and the force gradient is 1×10^{-10} kg/sec^2, which corresponds to a natural frequency of 3×10^{-6} Hz with a 63 g gyroscope rotor.

The spin and polhode motion of the gyroscope and the roll of the satellite will modulate the force and force gradient on the gyroscope rotor. Since the force and force gradient depend on the square of the difference in rotor potential, patch effect forces and force gradients on the rotor could exist if the housing potential is nonuniform even though the potential on the surface of the rotor was perfectly uniform. In this case, the patch effect forces and force gradients would be constant in a housing-fixed reference frame. A nonuniform potential on the spinning rotor's surface could still produce an average d.c. force. However, this force would be modulated at the polhode period because the orientation of the spin axis in the rotor changes at the polhode period. From the expressions for the force and force gradients, it can be seen that the force and force gradients also depend on the cross correlation between the rotor and housing potentials. Since the rotor spin axis is always aligned with the satellite roll axis to within 100 arc seconds, there should be very little modulation of this cross-correlation at the satellite roll frequency. Since the spin-averaged rotor potential is modulated at the satellite roll frequency, the cross correlation in the rotor to housing potential should also be modulated at this frequency.

Gravitational

The gradient in the gravitational field of the spherical earth will produce an acceleration of each of the three supported gyroscopes. For a gyroscope separated by 25 cm from the unsupported gyroscope, value of this acceleration at a 650 km altitude orbit averaged over the orbital period is 1.4×10^{-7} m/sec^2. In addition, there is a component of this acceleration which has a magnitude of 4.3×10^{-7} m/sec^2 and rotates in the orbital plane at twice the orbital rate. The gravitational acceleration due to the J2 term in the expansion of the earth's gravitational potential will cause a differential acceleration of the gyroscopes which is approximately 1000 times smaller(13).These accelerations are extremely well known and may be used to calibrate the readout any pair of the gyroscopes which are used as differential accelerometers.

The gravitational force of the satellite on the unsupported gyroscope will accelerate the satellite in direction body-fixed direction. Because the satellite rolls about the line of sight to the guide star, the roll-averaged acceleration will lie along the direction of the satellite's roll axis. Although this acceleration will slightly modify the orbit, it will have a negligible effect on the gyroscope drift rate or the performance of the gyroscopes as differential accelerometers. A careful analysis of the mass distribution on the satellite (14) has estimated this acceleration to be as large as 5×10^{-8} m/sec^2. The differential specific force on the gyroscopes due to the gravitational attraction of the satellite has also been estimated to be less than 1.5×10^{-8} m/sec^2 and varies from gyroscope to gyroscope. The gravitational force gradient is estimated to be 5.7×10^{-9} kg/sec^2. This force gradient is small compared to the force gradients due to the electrostatic forces and is expected to make no significant change in the effective spring constant.

The differential gravitational acceleration due to mass motion of the satellite will produce noise in the acceleration measurements. The three independent measurements will allow an estimation of the magnitude of this effect. The variation in the gravitational acceleration due to motion of the superfluid liquid helium will be reduced because of the centrifugal forces on the liquid helium and because of baffles installed within the liquid helium dewar. The differential gravitational acceleration due to the thermal motion of the solar panels has been estimated to be less than 5.7×10^{-13} m/sec^2. In addition, only a small fraction of this gravitational acceleration noise will lie in the frequency band of interest.

Inertial

Transforming from an inertial reference frame to a rotating reference frame, the acceleration of the center of the gyroscope housing is given by

$$\left.\frac{d^2\mathbf{c}_s}{dt^2}\right|_{inertial\ frame} = \left.\frac{d^2\mathbf{c}_s}{dt^2}\right|_{rotating\ frame} + \frac{d\omega}{dt}\times\mathbf{d} + 2\omega\times\mathbf{v} + \omega\times\omega\times\mathbf{d} \tag{23}$$

where $\mathbf{d}$ is the displacement of the supported gyroscope from the unsupported gyroscope. The first term on the right hand side of this equation is the acceleration in the rotating frame; the second term is the tangential rotational acceleration (6); the third term is the Coriolis acceleration; and the fourth term is the centrifugal acceleration. These inertial forces acting on the unsupported gyroscopes have been thoroughly investigated by M. Tapley (6) and are only briefly summarized here.

Centrifugal Acceleration

The magnitude of centrifugal acceleration acting on the gyroscopes due to the rotation of the satellite is given by

$$a_{cf} = \omega^2 d_p \tag{24}$$

where d_p is a small displacement of the center of mass of the gyroscope from the satellite's axis of rotation and ω is the angular rotation rate. The drag free control system will cause the satellite to rotate about the center of geometry of the unsupported gyroscope. For a displacement of the center of mass of any of the supported gyroscopes from the center of geometry of the supported gyroscope as large as 0.01 cm (4 milli-inches) and an angular rotation rate of the satellite of 1 rpm, the centrifugal acceleration is 10^{-6} m/sec^2. The force necessary to counteract this acceleration is expected to be one of the dominant forces acting on the supported gyroscopes. The satellite roll rate is expected to be stable to 1 part in 10^5 at roll frequency, which will produce a variation in the centrifugal acceleration of 2×10^{-11} m/sec^2 at roll frequency. The random noise close to roll frequency is not expected to be a significant contributor to the overall random noise.

The specific force gradient due to the satellite's roll rate is then simply equal to the square of the angular rotation rate. For a rotation rate of 1 rpm, the specific force gradient is 0.01 /sec^2, which corresponds to an effective negative spring constant of 6.3×10^{-4} kg/sec^2 and a natural frequency of 0.017 Hz. For a displacement noise of 0.1 nm/$\sqrt{\text{Hz}}$, the additive readout noise is 10^{-12} m/sec$^2\sqrt{\text{Hz}}$. This negative spring constant only applies to those directions perpendicular to the satellite roll axis for the supported gyroscopes.

Tangential Rotational Acceleration due to Pointing Control System

The next term on the right hand side of equation (23) produces noise due to the pointing control system in a direction perpendicular to the satellite roll axis. The rms noise of the pointing control system is expected to be on the order of 20 mas. However, the tangential rotational acceleration (6) due to this term in the equation of motion, produces noise which is proportional to the angular acceleration. If the noise in the pointing control system is white, then the angular acceleration increases as the square of the frequency up to the bandwidth of the pointing control system. A simulation of the pointing control system (15) shows a coherent pointing error at roll frequency of approximately 2 mas and a random noise close to this frequency of approximately 10 mas/$\sqrt{\text{Hz}}$, which is dominated by the noise in the telescope readout. For a gyroscope separated from the drag free gyroscope by 8.25 cm, the tangential rotational noise is 2.5×10^{-12} m/sec^2 at a roll frequency of 1 rpm with a random noise component of 1.25×10^{-11} m/sec$^2\sqrt{\text{Hz}}$ close to roll frequency.

TABLE 1. Dominant Sources of Specific Force due to Noise in Position Sensing Bridge

	Spring Constant kg/sec^2		**Natural Frequency Hz**		**Specific Force Noise m/(sec$^2\sqrt{Hz}$)**	
Source	Supported	Unsupported	Supported	Unsupported	Supported	Unsupported
Control Voltages	1.8×10^{-3}	-	0.027	-	2.9×10^{-12}	-
Position Sensing Voltage	3.5×10^{-5}	3.5×10^{-5}	0.0038	0.0038	no additional	1.9×10^{-13}
Charge Measurement	9.7×10^{-6}	9.7×10^{-6}	0.0019	0.0019	no additional	no additional
Patch Effect	1×10^{-10}	1×10^{-10}	3×10^{-6}	3×10^{-6}	no additional	no additional
Centrifugal Acceleration ⊥	6.3×10^{-4}	-	0.01	-	1.0×10^{-12}	-

⊥ Perpendicular to the direction of the satellite roll axis

Other Forces Acting on Gyroscopes

Because of the unique environment of the Gravity Probe B gyroscopes other forces acting on the gyroscope are not expected to contribute significantly to the forces or force gradients acting on the gyroscopes. Thermal radiation pressure forces are reduced because of the 2.3 K operating temperature, and the radiometer effects are expected to be negligible because of the operating pressure of less than 10^{-11} Torr. Similarly, damping of the gyroscope motion the residual gas within the gyroscope housing is expected to be small compared with the forces applied by suspension system. The gyroscopes operate within a superconducting magnetic shield where the residual magnetic field is less than 3×10^{-6} gauss, and external magnetic fields are attenuated by twelve orders of magnitude. Because of this environment forces and force gradients due to magnetic forces are expected to be negligible.
Cosmic Rays?

V. EXPECTED SYSTEM PERFORMANCE

The difference in the specific force acting on the gyroscopes is the sum of the contribution from the noise in the position sensing bridge and the contribution from the specific force noise acting on the gyroscopes. Contributions from the noise in the position sensing bridge are proportional to $(ms^2+\beta s+k)n(s)$, so that the specific force at any given frequency depends on the values of the mass, the negative spring constant, and the noise in the position sensing bridge. The table below summarizes the dominant contributions to the specific force noise due to the noise in the position sensing bridge in the frequency band from 2×10^{-3} Hz to 2×10^{-2} Hz.

The contributions to the specific force from various sources are shown in Table 2 below. Estimates are given for both the maximum d.c. force and the spectral density of the specific force acting both the supported and unsupported gyroscopes. Noise in the pointing control system introduces a large specific force noise in the direction perpendicular to the separation between the gyroscopes.

TABLE 2. Dominant Sources of Specific Force Acting on the Gyroscope

	Maximum D.C. Specific Force (m/sec^2)		**Spectral Density of Specific Force at 7×10^{-3} Hz (m/sec$^2\sqrt{Hz}$)**	
Source	**Supported**	**Unsupported**	**Supported**	**Unsupported**
A/D Converter on Control Voltages	-	-	2.6×10^{-10}	-
Thermal Stability of Control Voltages	10^{-11} (roll)	-	$< 10^{-13}$	-
Miscentering and Position Sensing Voltage	6.8×10^{-10}	6.8×10^{-10}	-	-
Back-Reaction of Position Sensing Circuit	-	-	1.2×10^{-14}	1.2×10^{-14}
Charge Rotor and Asymmetric Ground Plane	$< 3 \times 10^{-9}$	$< 3 \times 10^{-9}$	-	-
Patch Effect forces	7×10^{-14}	7×10^{-14}	-	-
Gradient in Earth's Gravitational Field	$<5.6 \times 10^{-7}$	-	-	-
Centrifugal Specific Force ⊥	$< 1 \times 10^{-6}$	-	-	-
Tangential Rotation Acceleration ⊥	2.5×10^{-12}	-	1.25×10^{-11}	-

⊥ Perpendicular to the direction of the satellite roll axis only

These results indicate that the Gravity Probe B gyroscopes may be used as differential accelerometers and are expected to have an overall accuracy of 2.6×10^{-10} m/(sec$^2\sqrt{\text{Hz}}$) in the frequency range from 2×10^{-3} to 2×10^{-2} Hz. The dominant source of noise is the quantization noise in the D/A converter for the control voltages on the supported gyroscopes. The design of the gyroscopes and the electrostatic suspension system have not been optimized to improve the performance as differential accelerometers but, instead, have been optimized to reduce the torques acting on the gyroscopes, provide reliable suspension of the spinning gyroscopes, and not interfere with the operation of the SQUID readout system. Considerable improvement in the sensitivity of the gyroscopes or proof masses as differential accelerometers could be achieved for a system that was specifically designed for that purpose. The specific force noise on the unsupported gyroscope is expected to be considerable smaller, but there is no convenient method of measuring this specific force noise on the satellite.

ACKNOWLEDGEMENTS

The authors would like to thank Stephano Vitale for valuable discussions.

REFERENCES

1. J. P. Turneaure, C. W. F. Everitt, B. W. Parkinson *et al.*, Adv. Space Res. (UK) **9**, 29-38 (1989).
2. W de Sitter, Mon. Not. Royal Astr. Soc. **77**, 155-184 (1916).
3. J. Lense and H. Thirring, Phys. Z. **19**, 156 (1918).
4. B. Mashhoon, F. W. Hehl, and D. S. Theiss, General Relativity and Gravitation **16**, 711-750 (1984).
5. Y. R. Jafry, *Aeronomy Coexperiments on Drag-Free Satellites with Proportional Thrusters: GP-B and STEP*, Ph. D. Thesis, Stanford University, 1992.
6. M. B. Tapley, *A Geodetic Gravitation Gradiometer Coexperiment to Gravity Probe B* , Ph. D. Thesis, Stanford University, 1993.
7. M. Tapley, J. Breakwell, C.W.F. Everitt *et al.*, Adv. in Space Res. **11**, 179-82 (1991).
8. S. Buchman, T. Quinn, G. M. Keiser *et al.*, Rev. Sci. Instr. **66**, 120-9 (1995).
9. W. R. Smythe, *Static and Dynamic Electricity*, New York: Hemisphere Publishing Corp, 1989, pp. 39-40.
10. G. F. Franklin and J. D. Powell, *Digital Control of Dynamic Systems,* Reading, Mass.: Addison-Wesley, 1981.
11. C. C. Speake, Classical and Quantum Gravity **13**, A291-A297 (1996).
12. P. Zhou, S. Buchman, K. Davis *et al.*, Surface and Coatings Technology **77**, 516-20 (1995).
13. N. J. Kasdin and C. Gauthier, J. Astronautical Sciences 44, 129-147 (1996).
14. R. Whelan, Lockheed-Martin Engineering Memorandum Report No. GPB-10742, 1996.
15. J. Kirshcenbaum, Lockheed-Martin Engineering Memorandum Report No. ATCS 210, 1995.

Design Considerations for Drag Free Satellites

Daniel B. DeBra

Department of Aeronautics and Astronautics
Stanford University, Stanford, California 94305

Abstract. Missions dictate the implementation of drag free satellites. A variety of missions are used to illustrate the various technologies associated with the TRIAD flight of 1972 and current and past proposed flights.

DRAG-FREE SATELLITE PRINCIPLE

Objects may be placed in orbit to provide a more benign environment than is available on the ground. While the disturbances are vastly reduced, radiation pressure, atmospheric drag, and other disturbances limit the degree to which a satellite is in free fall to values the order of 10^{-7} m/sec^2. By providing a shield for the satellite it can be protected from the environment. However, the external forces would drive the shield into the satellite if not controlled to stay away. By sensing the relative position of the shield with respect to the satellite, and using thrusters to keep it positioned so that the satellite is at the center of the inside of the shield a drag-free satellite is formed (1). The satellite is now referred to as a proof mass which may be just a reference mass or an instrument and the satellite is the shield (Fig. 1). Depending upon the particular mission there are a variety of requirements both in terms of the degree to which the shield must not disturb the proof mass and the portion of the spectrum in which these disturbances must be brought down to a required level of free fall.

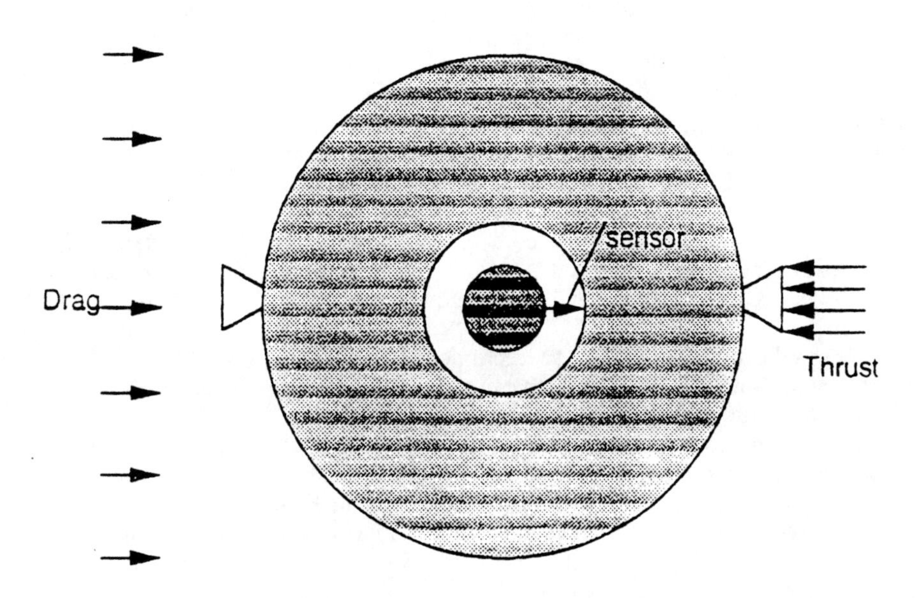

FIGURE 1. The drag-free satellite principle

CP456, *Laser Interferometer Space Antenna*
edited by William M. Folkner
 1-56396-848-7/98/$15.00

Example Missions

The first and to this date the only three axis drag-free satellite flown was TRIAD, officially known as TIP-1 (Transit Improvement Program-1) (2). The motivation for making this navigation satellite drag free was so that its ephemeris could be predicted more accurately. New data would not need to be updated every 12 hours, but rather once a week to once in 10 days. The uncertainty in the residual drag at its polar orbit altitude of 800 to 1000 kilometers was 10^{-8} m/sec^2. To extend the time between updates by a factor of 10, it was desired to reduce the uncertainty in the drag by two orders of magnitude. This could be done best by removing the drag completely and hence the application of the drag-free principle. Since the disturbances come from radiation pressure and other sources as well as atmospheric drag it was preferred to call it a disturbance compensation system or DISCOS.

The flight was sufficiently successful and the performance exceeded what was needed to extend the ephemeris update by a sufficient margin that a follow-on series of satellites TIP-2 and TIP-3 (3) were developed as single axis drag-free satellites. These provided a passive eddy-current suspension of the proof mass (which was a hollow cylinder) in the two axes perpendicular to a conductor that passed through the center of the cylinder. Along the axis of the cylinder, the proof mass was in free fall and provided the reference along track for the satellite to be made drag free.

The Disturbance Compensation System was made up of a housing in which the proof mass was located and to which was attached a caging mechanism. When operated it translated the ball through the housing in the vertical direction with a wiggle fore and aft that permitted verifying that the capacitive pickoff used to detect the position of the satellite with respect to the proof mass was operating normally before lift off. The propellant was cold gas in twin toroidal tanks. The time varying mass of the propellant was thereby centered at the center of the proof mass and its gradient in mass attraction was zero.

The performance goals for the mission were 10^{-11} g and 10^{-11} g/mm which were met. Three days of averaging of the Doppler tracking data were needed for evaluating the performance. For periods of time longer than three days it was impossible to distinguish between proof mass disturbances from the spacecraft and model errors in the earth's gravity field. After some minor corrections in the zonal distribution for the gravity field, the performance was evaluated as having less than 5×10^{-12} g disturbance to the proof mass.

The principal source of disturbance was the mass attraction of parts of the satellite acting on the proof mass. The proof mass was a 22 millimeter diameter sphere made of a gold platinum alloy. It resided in a housing, 40 millimeters in diameter on which electrical plates were located. The capacity between these plates and the spherical proof mass were approximately 0.3 pF. A capacitive detector can be made more stable by having larger active capacity and in the development of the instrument there was a great desire in designing the capacitive pickoff to make the gap between the ball and the wall as small as possible. The amount of null drift grew more rapidly than the gap given that the stray shunt capacity was fixed at about 30 pF and the active capacity varied inversely with the gap. Experimental data confirmed that the null drift grew rapidly and nonlinearly with gap. On the other hand the thermal distortions of the satellite and the need to have relative motion in order to permit the cold gas propulsion system to operate in a pulsed mode, required a larger spacing between the ball and the wall. The trade-off was resolved by choosing the gap that maximized the space remaining between the ball and the wall after the null had drifted. This occurred at a gap of 9 mm.

For many other missions the disturbance of the spacecraft on the proof mass is position dependent and requires much tighter tolerances on the allowable relative motion of the spacecraft with respect to the proof mass, and consequently also of the accuracy of detecting their relative location.

Even in the earliest missions considering the use of drag-free technology it was understood that the larger the ball to wall gap the easier it was to reduce the size of the disturbances of the spacecraft on the proof mass, for example an interplanetary relativity mission developed in France called Sorel in 1973 (4). The design started with a cavity of about 0.5 meters but as the studies developed it grew in size to 0.85 meters with a proof mass diameter of 96 millimeters. Optical shadow sensors were used to obtain an accuracy of readout of 4 micrometers over a range of 4 millimeters.

The need for small relative motion is particularly important for portions of the satellite that are near the proof mass to avoid disturbances from force gradients. A Geodesy mission called GRAVSAT was developed which needed to be at very low altitude in order to sense the higher harmonics of the gravitational field (3). The lower altitude required large thrust for drag-free control. Available propulsion technology could not be modulated adequately and thus had to be operated on-off. The minimum impulse bit was so large that limit cycles causing relative motion of many centimeters were incompatible with the gradient induced disturbances of a spacecraft on the proof mass. This problem was resolved by designing a sensor and housing for the proof mass which would track it the way a drag-free satellite was to track its proof mass. But it was servoed with respect to the satellite rather than using thrusters. Now its position with respect to the spacecraft was the error for the thruster control. The housing then followed the proof mass while providing freedom for the spacecraft to move in response to its limit cycle behavior (Fig. 2).

In Gravity Probe-B (5) drag-free control is used to minimize the support force for gyroscopes that are used to evaluate the interaction of spinning bodies with respect to gravitational fields. The satellite is rotated around a line of site to a star in order to spectrally shift the gyro readouts to a high enough frequency that the 1/f SQUID noise is acceptably low for the mission requirements. This rotation assists in meeting adequate drag-free performance by averaging any body-fixed torques. The only disturbances of concern are now at the rotational frequency with a very narrow bandwidth. Thus the disturbance requirements are easier to meet than they would be for a space fixed experiment.

Similarly for an equivalence principle experiment called STEP (6) the spacecraft is rotated with respect to the principal external disturbances in order to spectrally shift the science information where it can be measured more accurately. In this case the comparison of the free fall of proof masses of different material needs to be done with a precision that requires them to be coaxially mounted. But a number of materials need to be tested. Thus several pairs of differential accelerometers are used in the experiment. Any one of these can be used as the drag-free reference. Furthermore, a virtual reference at locations between differential accelerometers can be used as the drag-free reference by suitably combining the specific force information from the differential accelerometers.

In the baseline design for LISA (7) two proof masses are used in each of three spacecraft forming the corners of the arms of the gravity wave antenna. These faceted proof masses (cubes) are used as the optical reference for the interferometers measuring the arm lengths. A differential measurement is made optically from the front and back of the proof mass to minimize the requirements for how closely the spacecraft must be made drag free. Each of these inertial sensors is constrained in five degrees of freedom and only free in the one along the optical axis for which it is the end mirror. At the corner there are two of these, each drag free along its sensitive direction which are 60 degrees apart.

Thus the implementations that have flown and those that have been proposed for other experiments differ very significantly in the way in which they are implemented depending on the mission. Large separation of the spacecraft cavity walls from the proof mass are desired when such a separation is compatible with the mission. In some cases constraints have to be introduced. For example the support of gyroscopes in GP-B and the four degrees of freedom in the accelerometers in STEP, and five for the faceted end masses for LISA. When the constraints must be introduced, generally the gap must be made small in order to provide the sensing and the actuation for support. This is not always compatible with the desire to minimize the disturbance of the spacecraft acting on the proof mass and leads to compromises.

PROOF MASS DETECTION

The DISCOS that was flown in 1972 employed capacitive detectors as indicated earlier. Subsequently, the design of GRAVSAT required significantly more accurate detection in order to provide a velocity reference with noise less than 10^{-7} m/sec. This required an increase in the active capacity and therefore an increase in the proof mass size. However, the specific force disturbances could be an order of magnitude larger and this permitted the use of a larger proof mass of lower density which as a result could be made hollow. Retaining a 9 millimeter gap but increasing the ball diameter from 22 to 100 millimeters the capacity was increased by an order of magnitude and the noise of the 1970 state-of-the-art electronics was acceptably low for the velocity measurement.

The later single axis DISCOS implementations employed a shadow sensor which allowed the cavity walls to be even further away from the proof mass than they had been in the original DISCOS design. Subsequently, DeHoff (8) developed a sensor with a proof mass that was coated with a phosphor which was excited with ultraviolet light and re-emitted in the infrared. The infrared was then detected by a Schottky barrier diode which acted like a photo potentiometer. With three of these inside the cavity position of the proof mass was measured in three axes.

GRAVSAT

PROBLEM: LARGE THRUSTER--> BIG IMPLUSE BIT--> LARGE LIMIT CYCLE--> LARGE P.O. HOUSING AND DIST FROM MASS ATTRACTION GRADIENT

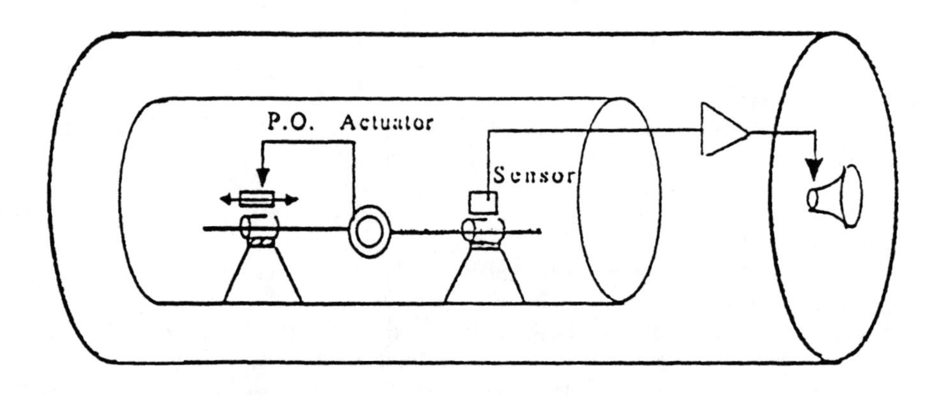

SOLUTION: DUAL CONTROL

- BALL POSITION IN HOUSING ACTIVATES PROPORTIONAL ACTUATION OF P/M HOUSING TO KEEP CENTERED ON BALL
- HOUSING POSITION IN S/C ACTIVATES PROPULSION SYSTEM TO KEEP HOUSING AT CENTER OF S/C

FIGURE 2. Two-stage DISCOS

THE DESIGN TOOLS

There are a number of design options available for meeting the quite varied performance requirements of different missions. These are summarized below.

Surface to Mass Ratio

For a relatively large class of proof mass disturbances the effectis proportional to its surfaceand the response of the proof mass is inversely proportional to its mass. Thus the surface to mass ratio is the most important parameter. For example electrostatic charge, radiation pressure, the radiometer effect, etc. are surfaceeffectswhich are reduced by increasing the size of the proof mass or increasing its density. An example is DISCOS which utilized a gold platinum alloy for the proof mass.

Mass Attraction

Gravitational attraction is attenuated most effectivelyby moving masses as far away from the proof mass as possible. In the DISCOS design the satellite was broken into three parts and 80 percent of the satellite was put on booms which held them the order of 5 meters away from the drag-freecontrol system. This meant that the components in that 80 percent of the satellite did not need to be photographed and their position measured and their mass determined with an accuracy that was inconsistent with normal satellite fabrication. Thus only the drag-free control system parts near by required this type of careful bookkeeping. As the design progressed we found beryllium oxide for the proof mass housing could not be obtained with a density homogeneity of 0.1 percent. At 0.3 percent, density variation in the housing was the dominant source of disturbance on the proof mass.

Magnetic Susceptance

Here gradients in the magnetic field within the spacecraftare minimized but the residual disturbances can still be unacceptable. Gold and platinum are diamagnetic and paramagnetic respectively. The proper alloy achieves a near zero magnetic susceptance. The measured value of the alloy chosen for DISCOS was less than 10^{-8}.

Gap

The gap is the strongest single factor for reducing many of the disturbances including mass attraction for parts that are near the proof mass. We see the very large dimensions associated with Sorel and the trade-offs in the selection of 9 millimeters for DISCOS as examples.

Spectral Shifting

In many cases the engineering or science performance requirements are intrinsically or can be shifted to a specific part of the frequency spectrum. For example if 1/f noise dominates as it does in GP-B, rotating the spacecraft will spectrally shift the sensor information to a part of the spectrum around the spacecraft's roll frequency. This frequency is chosen to be at a sufficiently quiet portion of the detector to make the performance acceptable. Examples are GP-B for the detector of the gyro readouts. For STEP a violation of 10^{-18} would require two satellites in the same period orbit to be only 2 pm different in altitude. The science information is modulated so that zero frequency effects are not a limitation to the readout accuracy.

On the other hand, one might have errors associated with motion of the proof mass (for example rotation of a sphere) and if this contains spectral content near the experiment frequencies, one needs either to shift those to a lower or higher frequency. The baseline design on LISA uses orientation control on the proof mass so that its angular motion doesn't couple into the measurement direction. The heritage for this approach is the development of low level accelerometers in which the free movement in orientation of the proof mass for the low level accelerometers, for example on Cactus (9) were a limitation in its performance.

By contrast one can still employ a spherical proof mass if it is spun. In the GRAVSAT experiment the highest harmonic of the Earth's field that could be measured had a frequency of 0.1 rad/sec. By spinning the proof mass at 10 times that frequency or greater, one spectrally shifts the very small but significant undulations in the lack of roundness of the proof mass to a frequency above the range of science interest.

Two Stage Measurement

A follow-up housing can be used to provide small relative motion for portions of a spacecraft that are near the proof mass and use large separation for the portions of a spacecraft which must have larger motion. An example is the follow-up servo housing for GRAVSAT.

SPACECRAFT CONTROL

We have discussed the need to reduce the disturbances on the proof mass in order to achieve acceptable drag-free performance. This is fundamental but in order to achieve low levels of disturbance and in some cases to provide an environment in which the measurements can be made with sufficient accuracy the feedback control to thrusters on the spacecraft must be of adequate performance. There are requirements for sensing, of actuation (thrusters), and control implementations for particular situations such as the control of a rotating drag-free satellite. For fine control, proportional thrust is necessary. Boil off helium needed as a cryogen is utilized on GP-B and field effect electric propulsion (FEEP) is proposed for LISA.

The degree to which the spacecraft can be made to follow the proof mass is in part determined by how large the external disturbances to the spacecraft are. Thus the selection of an orbit is the most important single design choice in influencing drag-free performance. For example GP-B is maintained at 650 to 700 kilometers to reduce the atmospheric drag and its variations to an acceptable level. Though a lower orbit would slightly favor science sensitivity. Similarly LISA has been chosen to operate in an array considerably away from the earth and moon in order that their gravity disturbances are acceptably low. But this environment is also a benign environment by comparison with near earth for the performance requirements of the thrusters in achieving effective drag-free control.

When the spacecraft is rotating there is the need to provide velocity information to damp the relative motion of spacecraft with respect to the proof mass. When this calculation is done in an inertial non-rotating reference each axis can be implemented independently. When the spacecraft is rotating however, the measurements are made in a rotating coordinate system and thus to get the inertial velocity one needs to apply the Coriolis rule for differentiation in a rotating coordinate frame. Thus some additional terms are needed in a typical implementation for both rate information and also for the integrals used to minimize the errors at low frequencies. In some cases integral control is important to minimize constant disturbances or remove biases that are fixed in the body. This is a conventional integral control term since the disturbance is body fixed. The measurements and the "constant" of integration are both in the computational axes. However, when the disturbance on the spacecraft is fixed, for example with the sun-line for radiation pressure, this will be a rotating disturbance in the body axes. Thus one implements an oscillator for the integral control terms.

In LISA, the drag-free references are broken up into more than one sensor, each provides a single axis measurement of the desired spacecraft error for it to be in free fall. Reciprocal basis control can be employed to decouple the disturbances that correction of one axis might have on the other. Specifically one applies a correction force perpendicular to the second sensor axis rather than along the first sensor axis. The component along the first sensor when its error is being fed back is adequate to correct it which it does without disturbing the second one. The residual is taken up by the constraining forces that support the proof mass.

DRAG-FREE CONTROL WITH AN ACCELEROMETER

There are differences for control using free proof masses versus for accelerometers. In some implementations the suspension of proof mass may have a natural frequency which is comparable to or higher than the bandwidth that is implemented in the drag-free control. In such a case the support force for the proof mass is an acceleration signal. The ratio of the measurement in response to a control force applied to the spacecraft indicates a constant plant. That is, the control produces an acceleration, the acceleration is directly measured so there are no dynamics or integrations between the control actuation and the sensor. In such a case pure integral control is implemented to provide the necessary gain below the control bandwidth and to attenuate feedback which might interact unfavorably with the spacecraft above its bandwidth. By contrast when the proof mass is unsupported and the measurement is made of their relative position, the control force creates an acceleration which is two derivatives away from the measurement. This classic second order control system requires velocity information as well as proportional and sometimes integral feedback for an effective control. If there is a suspension force that is either intentional or unintentional for which the natural frequency is lower than the drag-free control, the control implementation essentially looks the same whether the proof mass is totally free or lightly constrained.

In the first case where pure integral control is used this has a destabilizing effect on the elastic behavior of the proof mass with respect to the spacecraft. So many times a phase stabilizing notch filter is used in addition to the integral control when using accelerometer feedback. This may not be necessary if the accelerometer is itself a feedback device with adequate damping.

A SPHERICAL PROOF MASS FOR LISA?

The baseline design for drag-free reference on LISA employs inertial technology with very good space heritage. The design accommodates the exacting needs of individual interferometer arms that can change their relative orientation by a degree during a year. With a faceted proof mass, separate proof masses are needed for each arm. Each proof mass is constrained in the three degrees of freedom of orientation and in two degrees of freedom of translation. It is free only along the axis of the arm. The sensing and constraint forces associated with the constrained five degrees of freedom increase the interaction forces between the spacecraft and the proof mass, and the possibility of coupling disturbances into the sensitive direction. The design for minimizing these disturbances could be achieved more easily if the gap between the sensor housing and the proof mass were increased from less than 1 millimeter to 10 millimeters or more. Furthermore, if the faceted proof masses could be replaced with a sphere then a single proof mass could be used as the reference for both arms.

If a sphere is employed the sphere doesn't mind whether one is reflecting off of two points on its surface which are changing by an arbitrary angle. Thus the spacecraft could be made drag-free along three translational axes without worrying about the orientation of the proof mass. There would be no constrained degrees of freedom and the difficulty of preventing constraint-force coupling does not exist.

This statement would be true for a perfect sphere. However, the surface undulations of the sphere are very difficult to reduce to levels of 20 nm peak-to-valley. Density inhomogeneity limits how closely one can get the mass center to the center of geometry. These variations are large compared with the performance requirement in the baseline design of LISA of a nanometer. If the proof mass is spun however these variations might be averaged. The apparent relative motion would increase by an order of magnitude but it is highly periodic and the frequency could be well above the science range of frequencies.

GP-B proof masses are manufactured of quartz and of single crystal silicon. Roundness is better than 20 nm peak-to-valley on any great circle. A choice of gold platinum may not be as easy to manufacture to this level of roundness. The requirements on DISCOS were much more modest and only required 2.5 micrometer peak-to-valley variations in roundness which is more relaxed by two orders of magnitude. A more interesting question is whether or not the mass center of the sphere lies sufficiently close to the center of geometry. The single crystal silicon and the specially selected fused silica used for GP-B rotors have extraordinarily uniform densities of less than 10^{-6}. This assures that the rotor has a mass center which is located close enough to the center of geometry that its contribution to the variations in slope of the surface is smaller than the roughness. If the density inhomogeneity of gold platinum is as large as 10^{4}, mass center to center of geometry differences of 1000 nm might be possible. More realistic expectations of 100 nm are still not that promising. However, if the density inhomogeneity is more nearly 10^{-5} then this source of geometric imperfection would be comparable or less than the lack of roundness of the sphere.

These are important questions to be answered if one wants to consider using a spherical proof mass for LISA. On the other hand the opportunity to use a large gap, and no constraints, optical detection and to spin so geometrical imperfections are shifted out of the science measurement band encourages one to think about it as an alternate implementation. With the larger gaps it may be possible to shift some of the contributions in the error budget to allow the proof mass to be made smaller with the consequent reduction in overall mass and size for the physical system. Furthermore, only a single proof mass would be needed for both arms. In addition the larger spacing may reduce the disturbance gradients sufficiently to relax the drag-free performance and permit larger variations in the thrust associated with the FEEP's.

REFERENCES

1. Lange, B., *AIAA Journal*, **Vol. 2, No. 9**, 1590-1606 (1964).

2. TRIAD I: Staff of the Space Dept at Johns Hopkins University Applied Physics Lab, and Staff of the Guidance and Control Lab at Stanford University, *AIAA J. Spacecraft,* **Vol. 11, No. 9,** 637-644 (1974).

3. Ray, J. C., et al, "Attitude and Translation Control of a Low Altitude Gravsat." *Journal of Guidance, Control and Dynamics,* 1982.

4. SOREL, A Space Experiment on Gravitational Theories, Final Report of the Mission Definition Group, ESRO CERS, **Vol. 1**, March 1973.

5. Everitt, C.W.F., et al, *Near Zero: New Frontiers of Physics,* New York: W.H. Freeman and Company, 1987, ch. VI, pp. 587-691.

6. STEP: Satellite Test of the Equivalence Principle, http://Einstein.Stanford.edu/STEP/step2.html.

7. LISA: "Laser Interferometer Space Antenna," Pre-Phase A Report, Second Edition, MPQ 233, July 1998.

8. DeHoff, R.L., "Minimum Thrustors Control of a Spinning Drag-Free Satellite, Including Design of a Large Cavity Sensor," thesis, Stanford University, December 1975.

9. Beaussier, J., et al, "In Orbit Performance of the Cactus Accelerometer (D 5 B Spacecraft)," presented at the International Astronautical Congress, Anaheim, CA, October 10-16, 1976, (IAF paper no. 76-099), ONERA T.P. No 1976-128.

Indium Liquid-Metal Ion Sources as Micronewton Thrusters

M.Fehringer, F.Ruedenauer and W.Steiger

Austrian Research Centre Seibersdorf
A-2444 Seibersdorf

Abstract. The Austrian Research Centre Seibersdorf has been engaged in the development of indium liquid metal ion sources for scientific space applications like spacecraft potential control and mass spectrometry for more than ten years. It holds the unique record of more than 900 hours of operations of this type of ion emitters in space borne instruments. Two operational instruments carrying indium emitters are currently in space and three more are being prepared for launch in the near future. During the last three years the focus of research activities has been targeted onto the development of a micronewton thruster using the already space proven indium emitters. Under a contract with ESA the basic physics of the indium ion sources as they are relevant for thrust generation and characterization have been studied and an engineering thruster model is currently being built.
This paper aims at introducing the indium - FEEPs to the worldwide LISA community for the first time and to give an overview of recent activities and advances.

THE NEEDLE TYPE LIQUID METAL ION SOURCE

When a high electrostatic field is set up between the surface of a conductive liquid and an accelerator electrode, the surface is deformed under the action of electrostatic field stress and surface tension. The resulting equilibrium shape of the liquid is a cone (so called „Taylor cone") with a total apex angle of 98.6° (1). On large free liquid surfaces a 2- or 1- dimensional array of Taylor cones may be observed whereas on the „needle type" - source, which will be discussed here, one single cone is formed on the tip of the needle (Fig.1) If the applied voltage is such that the field at the apex of the cone reaches a few volts per angstrom, the most protruding atoms at the tip are field evaporated, ionized and accelerated towards the accelerator electrode. On liquid tips field evaporated atoms are continuously replenished through liquid flow of atoms towards the cone apex. The result is a dynamic equilibrium between field evaporation and hydrodynamic flow, resulting in the emission of a continuous, space-charge limited ion beam from the cone apex. Depending on charge material and the sharpness of the needle tip, operation of a liquid metal ion source typically sets in at voltages between 4 and 7 kV, while the beam current strongly increases with the applied voltage.

At current onset, a very sharp liquid jet with a diameter of the order of 5 nm forms at the top of the Taylor cone (2). It is there where the actual current emission is confined to. This extremely small emission site is the key to the success of liquid metal ion sources in „focused ion beam" - techniques. Due to this extremely localized emission site already relatively simple ion optical devices are sufficient to produce micro-focused ion beams where beam spot diameters down to 20 nm and current densities up to 10A/cm^2 can be reached. These properties are utilized in a number of innovative "Focused Ion Beam" (FIB) techniques.

The standard Indium Liquid Metal Ion Source (LMIS) developed at the Austrian Research Centre Seibersdorf is of the central needle type, in which a sharpened tungsten needle of tip radius of a few µm is mounted in the center of an indium reservoir (3,4). This reservoir is bonded to a ceramic tube which in turn houses a heater element. During operation this heater element liquefies both the indium in the reservoir and the indium film which covers the tungsten needle (Fig.1). Having the heater at ground potential simplifies the operation of the ion source and

CP456, *Laser Interferometer Space Antenna*
edited by William M. Folkner
 1-56396-848-7/98/$15.00

eliminates the need for power and mass consuming circuitry. According to their application, a number of different sizes of ion emitters are now available, the most basic, because flight proven type is the LMS220. At a total weight of 1.2 g, this emitter has an indium capacity of 220 mg which lasts for about 2500 h of continuous operation at 15 μA. During recent work for high current applications, where several hundred μA are needed, the indium reservoir has been enlarged to hold 1.2 g. A photograph of a space qualified and space proven indium liquid metal ion source is shown in Fig.2.

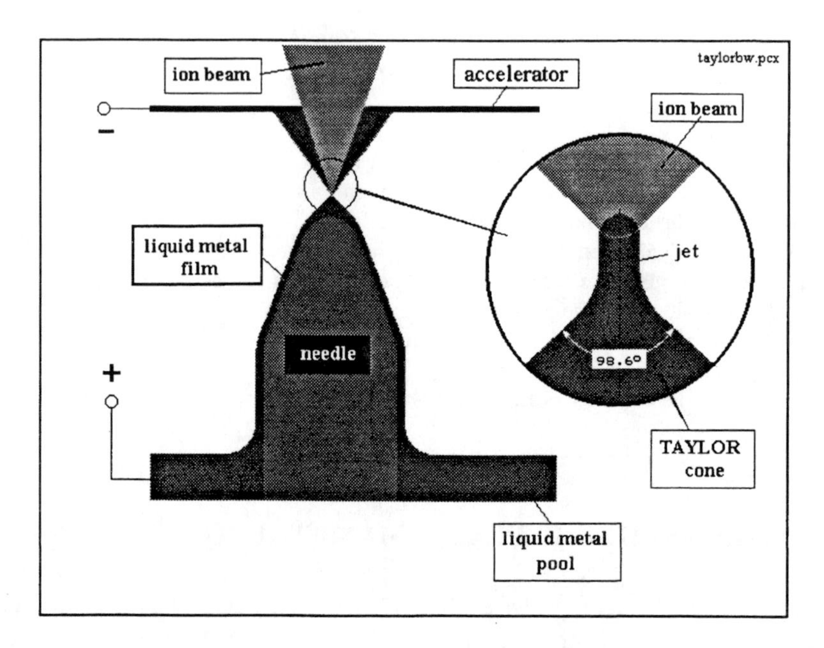

FIGURE 1. Schematic of the operation of a needle type LMIS

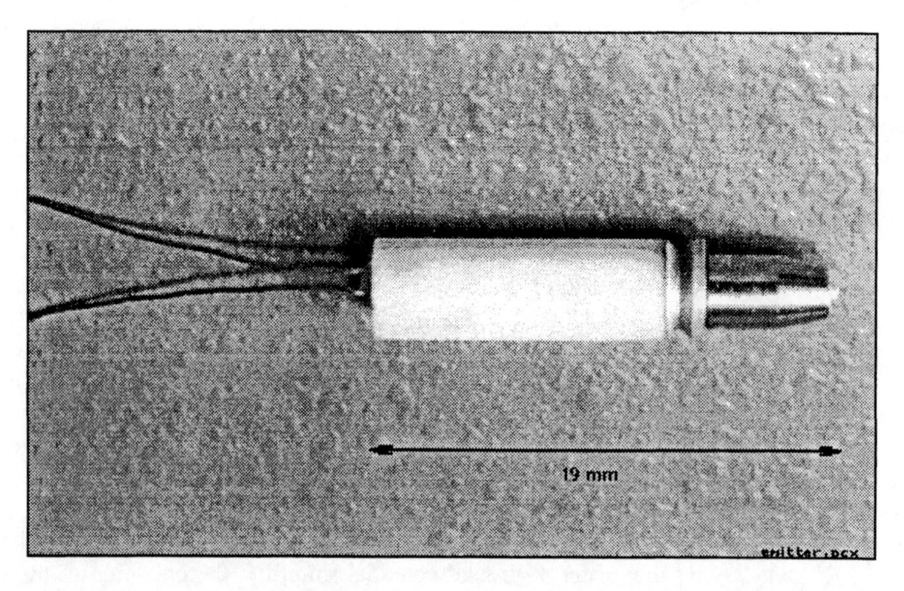

FIGURE 2. Photograph of an indium LMIS developed at Seibersdorf, (length: 19 mm, diameter: 4 mm)

PREVIOUS EXPERIENCE

Originally the indium LMIS has been developed for space based scientific instrumentation like spacecraft charge control devices and mass spectrometers. The first operation in space of an indium LMIS was conducted in 1991 onboard the space station MIR and instruments equipped with indium emitters have been flown on three further missions (MIR, GEOTAIL, EQUATOR-S). A total of more than 900 hours of operation in space has been logged so far. Three more instruments using indium LMIS are currently being built and will be launched within the next few years on the CLUSTER and ROSETTA spacecraft and again on MIR. Except for the inclusion of a neutralizer the S/C potential control instrument ASPOC, that has been developed for CLUSTER and has already been flown on the GEOTAIL and EQUATOR-S missions, already constitutes a low power ion thruster. The left panel of Fig.3 shows an image of ASPOC, which - at a total mass of 1.8 kg and an average power consumption of two Watts - apart of its goal to control charging of the satellite also delivers a couple of μN thrust to the Japanese GEOTAIL satellite. On the right hand side of Fig.3 the mean operational voltage values of all 40 operations of one individual ion emitter onboard GEOTAIL have been plotted together with the integrated emitted charge of that emitter, which gives an indication of the length of the individual operations. As the ion sources on the S/C potential control instruments are run in current controlled mode the operational voltage is set by the control loop in order to drive the requested current. A stable and almost constant value of that self adjusting parameter indicates a very stable operation of the ion emitter during its 127 hours of operation, which were accumulated in 40 individual periods between September 1992 and April 1997.

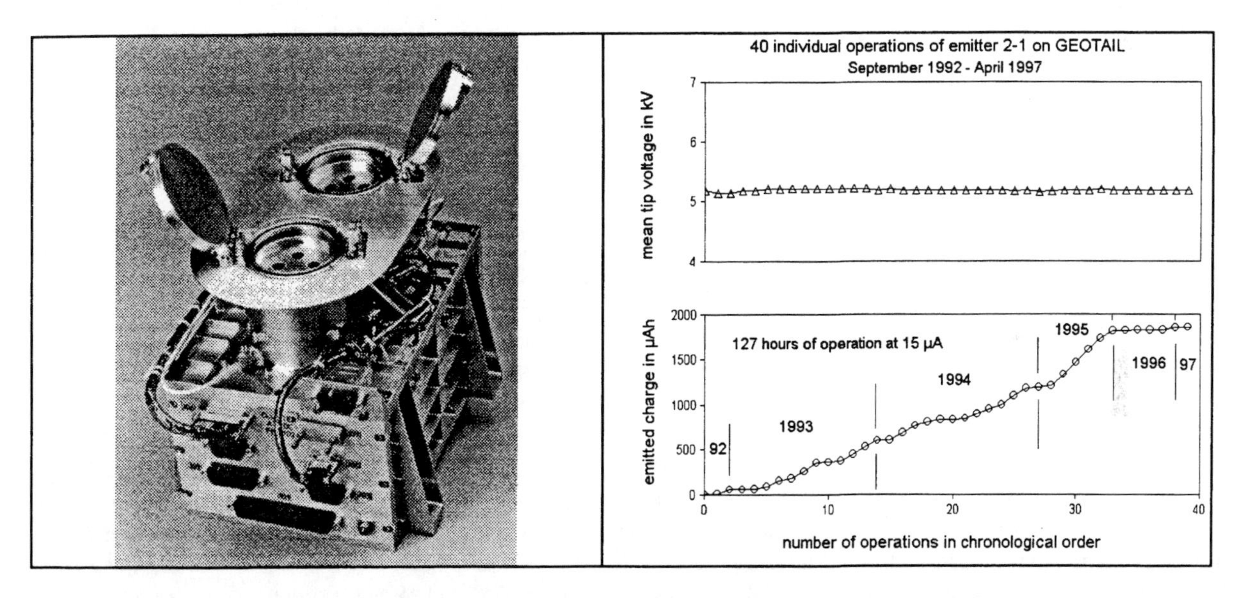

FIGURE 3. Photograph of a flight model of the charge control instrument ASPOC built for the CLUSTER mission (left) and operational data from such a S/C potential control instrument onboard the Japanese GEOTAIL satellite.

THRUSTER RELATED WORK

As a result of the various successful in orbit demonstrations of the indium emitters their possible use in ion thruster applications was considered. A first feasibility study was conducted for ESA in 1996 in order to characterize basic operational parameters with requirements set for a typical LISA mission profile. Table 1 summarizes the findings of that investigation as compared to the requirements given. The specified thrust range T was between 1 and 25 μN. As a result of that study a three years activity to develop a complete prototype thruster package has been started which is funded via ESA´s TRP and GSTP programs.

TABLE 1. Results of first thruster feasibility study carried out in 1996

parameter	requirement	measured
controllability	1%	better than 1 % at T > 3 mN
reproducibility	1%	< 1%
long term stability	-	< 1% over typ. 10 hours
mass efficiency	-	95 % @ 1 μN to 5 % @ 25 μN [a]
electrical efficiency	-	95 % @ 1 μN to 90 % @ 25 μN
pulsed mode	5 Hz, 20 ms	stable operation with 20 ms pulses at 5 Hz demonstrated over typ. 10 hours

[a] has been improved to typically 60 % in the meantime

RECENT ACTIVITIES

Research activities carried out during the past year focused on two equally important and major issues. First, in order to characterize the thrusting behavior of the indium emitters on the basis of measured parameters, a thorough investigation of all detectable beam constituents was performed. Emitted beam constituents were analyzed in respect to their mass, charge state, mass over charge ratio, energy and the direction they were leaving the source. With the resulting database the contribution from each individual beam species to the overall thrust could be calculated and statistical estimates of noise contributions can now be given. A second activity centered on droplet emission and the possible reduction of mass ejected in form of submicrometer sized droplets.

Analysis of Beam Constituents

It has been found earlier that - apart from the well known ionic fraction - ion beams emitted from liquid metal ion sources also contain a fraction of predominantly submicron sized droplets of not well determined but certainly low mass to charge ratios. A wide range of sensitivity in that respect was therefore needed in order to detect all species present in the beam. Magnetic mass spectrometry was chosen for that investigation and three different magnetic mass spectrometers covering a wide range of mass resolution and mass to charge sensitivity have been used. A high abundance sensitivity, double focusing mass spectrometer was employed to measure beam impurities in the mass region around In^+. These beam constituents primarily result from original impurities contained in the procured indium and from elementary species that have been introduced during the emitter production. It turned out that these fractions are so small that they need not be considered as thrust contributors. Those fractions that contribute to the thrust have been investigated with a single focusing, medium mass range spectrometer that allowed the detection of indium clusters up to In_6^+. Respective spectra have been recorded for a current range from 10 μA to 250 μA, corresponding to calculated thrusts between 1μN and 25 μN. A mass spectrum at 10 μA emission current is shown in Fig. 4 (left panel) together with a plot showing the normalized contributions to the calculated thrust for the highest and lowest currents investigated, 10 μA and 250 μA respectively (right panel). It can be seen that In^+ accounts for about 98% of all thrust and that - if at all - only In_2^+ needs to be considered as a thrust contributor.

In order to investigate the charge state of droplets a third mass spectrometer has been employed. An existing 90° deflecting spectrometer was modified to enable a mass to charge resolution of up to 1×10^6 amu/charge. A spectrum taken with this instrument is shown in Fig. 5. No distinct features hinting at a droplet fraction could be found. The two peaks at around 1×10^4 amu/charge are attributed to scattered ions because this mass to charge ratio lies below the theoretical limit where stable droplet existence is possible and which is about 5×10^5 amu/charge /5/. Droplets with lower mass/charge figures would disintegrate due to field evaporation. Furthermore, these peaks have not appeared in spectra taken with our medium mass range spectrometer. It has to be concluded that the droplet population above 5×10^5 amu/charge is either below the detection limit of our instrument or non existent in this range including the possibility of the existence of a neutral droplet fraction.

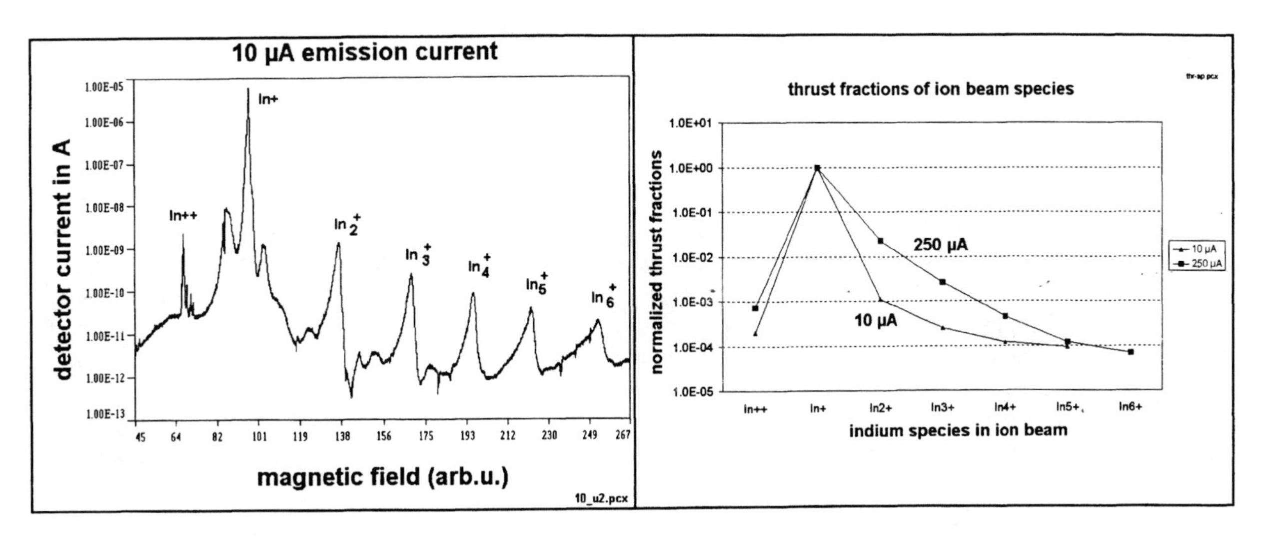

FIGURE 4. Mass spectrum of an indium ion beam at a total emission current of 10 µA (left panel). The right panel shows calculated thrust fractions of the individual species found in the ion beam for emission currents of 10 µA and 250 µA.

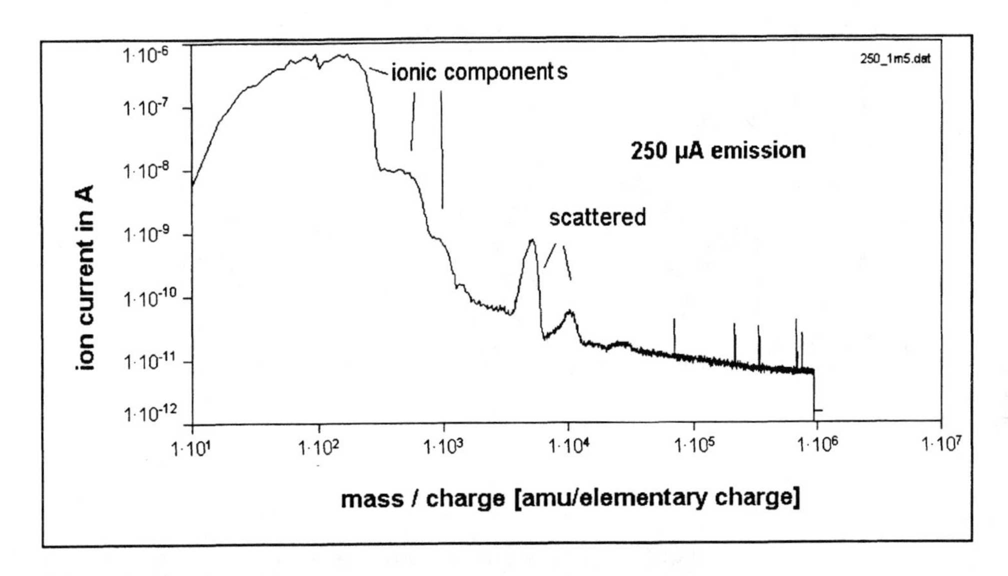

FIGURE 5. Low resolution, high mass range spectrum of thruster beam at 250 µA emission current.

In order to clarify this issue an experimental arrangement combining the high mass range magnetic spectrometer with an electrostatic deflection unit was set up. In the magnetic sector the ion beam was cleared from ionic fractions which are easily deflected out of the measurement path by the magnetic field so that only droplets and neutral components could remain in the beam. A subsequent electrostatic stage was used to discriminate between neutral particles and charged droplets and also served as an energy analyzer. Particle detection was done by placing catcher plates made of highly polished, single crystal silicon surfaces into the beam line and exposing them over adequate periods of time, typically hours. Traces of droplet deposition could easily be evaluated by means of optical and electron microscopy. This method was chosen after all attempts of droplet detection via current measurements - both in continuous and single particle counting modes using electron multipliers - have failed. From these measurements a new figure for the lower limit of the mass to charge ratio of droplets of 6×10^6 could be derived. Furthermore there was no evidence of any neutral droplet fraction in the beam. All droplets carry charge.

Droplet Mass Analysis

The amount of indium ejected from the ion emitter in form of droplets has been derived from topographical information on droplets which were caught on suitable surfaces intercepting the ion beam. Flat polished single crystal silicon chips were exposed to the ion beam for well defined periods of time. Images of these chips were then taken by means of electron and atomic force microscopy. Atomic force micrographs yield very precise and three dimensional topographical information which allow absolute volume- and therefore mass determination of droplets. This extremely precise but slow method served to calibrate a much faster but less precise volume evaluation based on electron micrographs taken from the exposed surfaces. Combining these two methods the total indium mass ejected could be determined and the results agree very well with mass loss measurements performed on a precision scale. Figure 6 shows an example of a micrograph from an exposed silicon surface taken with the atomic force microscope. The extension of the field shown is 2.44 x 1.56 μm².
The droplet population was found to be largely dominated by 100 nm diameter particles. Droplets with diameters up to 1 μm have been found. In general the emission of droplets was found to be more forward peaked than the ion emission.

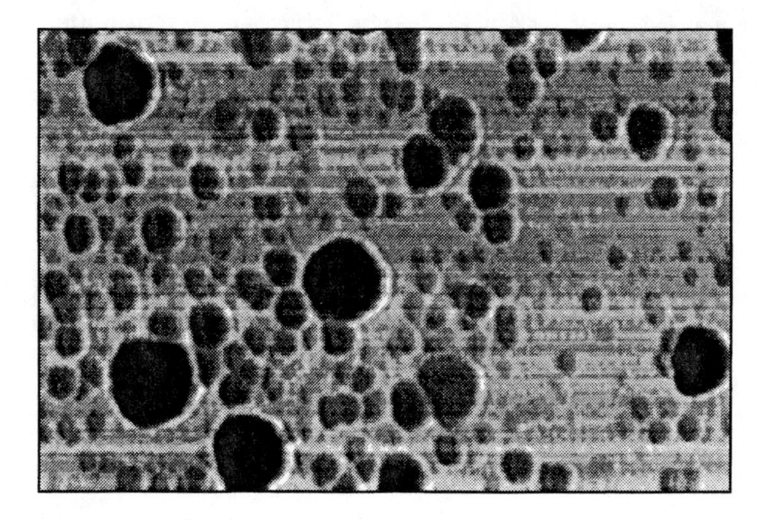

FIGURE 6. Image of a silicon catcher surface exposed to an indium ion beam. The image covering an area of 2.44 x 1.56 μm² was taken with an atomic force microscope.

Mass Efficiency Improvement

Mass efficiency measurements carried out in the past revealed a decrease in mass efficiency with increasing currents. This was attributed to an increased droplet emission at higher currents. Based on theoretical studies on droplet formation inherently present in any liquid metal ion source (6,7) experimental work on reducing this component was performed. Custom tailoring of the indium film covering the needle led to a significant improvement of the measured mass efficiency. Figure 7 shows a comparison of old mass efficiency data with results that were obtained recently with the modified film geometry. Further work towards a better understanding of the improvement is currently under way.

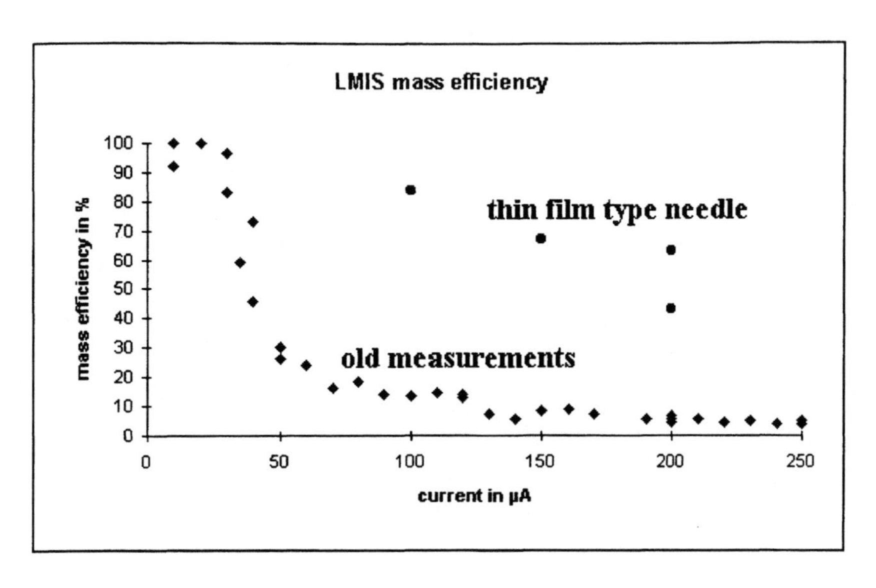

FIGURE 7. Comparison of old and recent mass efficiency measurements.

SUMMARY AND OUTLOOK

A summary of the results of recent work on basic physical properties of the indium ion emitter in respect to its potential use in a low thrust propulsion system is given below, individual items are marked by three asterisks. This work has been carried out under an ESA contract within ESA´s Technology Research Program TRP. A new two years contract aiming at the development of a complete thruster system including neutralizer and electronics has been awarded by ESA, which by mid 2000 should result in the availability of a prototype thruster package. This package should be of a standard that allows in orbit testing.

*** In^+ accounts for about 98% of the total emitted current and calculated thrust

*** In cluster atoms from In_2^+ up to In_6^+ and indium atoms with charge states up to 3 have been measured and their contributions to thrust calculated.

*** the droplet contribution is dominated by 100 nm size particles, their emission is - compared to the ionic emission - forward peaked

*** all droplets carry charge, no neutral component has been found

*** a lower limit for the mass over charge ratio of 6 x 10^6 amu per elementary charge can be given

*** mass efficiency at high emission currents has significantly been increased

REFERENCES

1. Taylor, G. I., *Proc. Roy. Soc.* **A280**, p. 313 (1964)
2. Kingham, D. R. and Swanson, L. W., *Appl. Phys.* **A34**, p. 123 (1984)
3. Ruedenauer, F.G., Steiger, W., Arends, H., Fehringer, M. and Schmidt, R., *J.Phys.* **C6 49**, 161 (1988)
4. Ruedenauer, F.G., Fehringer, M., Schmidt, R. and Arends, H., *ESA-Journal* **17**, p. 147 (1993)
5. Thompson, S. P. and van Engel, A., *J. Phys. D: Appl. Phys.* **15**, (1982)
6. Lord Rayleigh, *Math. Soc.* **10**, p. 4 (1878)
7. Faraday, M., *Phil. Trans. Roy. Soc.* **121**, p. 299 (1831)

corresponding address: Michael.Fehringer@arcs.ac.at

GROUND AND OTHER

GRAVITATIONAL-WAVE DETECTORS

The GEO 600 Ground-Based Interferometer for the Detection of Gravitational Waves

Roland Schilling* for the GEO 600 Team

* *Max-Planck-Institut für Quantenoptik, D – 85748 Garching, Germany*

Abstract. The last few years have brought a great break-through in the quest for earth-bound detection of gravitational waves: on five sites the world over, detectors of armlengths from 0.3 to 4 km are being built. These projects have in common that one prominent noise source, the shot noise, is reduced by the use of *power recycling*. The British-German project GEO 600, although only intermediate in size (600 m), has good chances for a competitive sensitivity by using advanced optical technologies early on. Particularly the use of special narrow-banding schemes, the so-called *signal recycling*, will allow to search for faint sources of only slowly varying frequency (pulsars, close binaries).

We will describe the particular GEO 600 interferometer topology, characterized by the use of a four-pass delay line and *power* as well as *signal recycling*. The current status of the construction of GEO 600 will be outlined (civil engineering, vacuum, optics). Also, recent results from the Garching 30-m prototype interferometer, used as a testbed for GEO 600, will be reported. First operation of GEO 600, and also of other ground-based interferometers, is expected around the year 2000.

THE GEO 600 CONCEPT

GEO 600 is a joint British-German effort to build a medium-size gravitational-wave detector with 600 m armlength at a rather limited budget [1]. The collaborating institutions are University of Glasgow, Max-Planck-Institut für Quantenoptik at Garching and Hannover, Universität Hannover, Laser-Zentrum Hannover, Albert-Einstein-Institut Potsdam, University of Wales College of Cardiff. The detector is being built near Hannover in Lower Saxony, Germany. In order to (at least partly) make up for the relatively short armlength compared to LIGO or VIRGO, advanced technologies like high gain power and signal recycling will be employed. This smaller armlength was not a matter of choice, but one of necessity. The site (on grounds belonging to the University of Hannover) cannot accommodate a larger antenna, and the funds did not allow buying or leasing ground elsewhere. This paper will try to outline how this shortcoming in length is to be compensated by the application of advanced interferometric techniques.

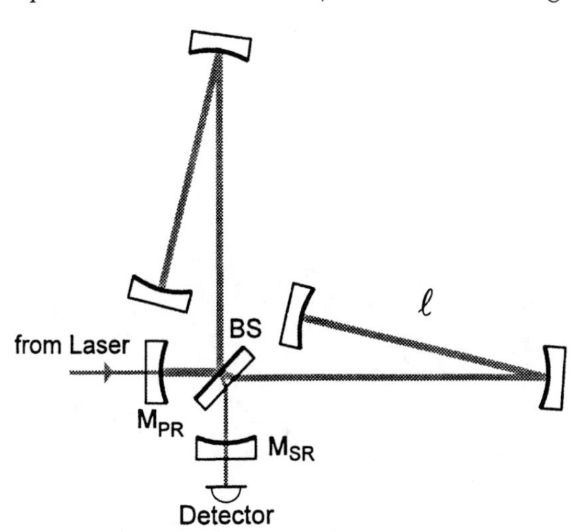

FIGURE 1. DL 4 configuration of the GEO 600 interferometer. The two extra mirrors M_{PR} and M_{SR} are for power and signal recycling, respectively.

In its arms, GEO 600 will employ an optical delay line with only four light transits (DL 4), as shown in Figure 1. After being bounced off the distant mirror the beam is reflected at the near mirror and retraces its path back to the beam-splitter BS. This DL4 scheme differs from the conventional Herriot delay line, which would pose problems in mirror size, in scattered light effects, and in separating incoming and outgoing beam.

A very distinct characteristic of the GEO 600 concept is the use of "dual recycling": In addition to the scheme of power recycling, which is now standard in all the large detectors, a further mirror (M_{SR}) is introduced in the output port, to allow an enhancement of the sidebands that the gravitational wave produces from the carrier beam.

CP456, *Laser Interferometer Space Antenna*
edited by William M. Folkner
 1-56396-848-7/98/$15.00

NOISE CONTRIBUTIONS

To attain the projected sensitivity, extensive experimental work is required to reduce the contributions of noise that would limit the sensitivity of GEO 600.

Mechanical noise. The influence from *external* mechanical vibrations, e.g. *seismic noise*, is strongly reduced by a triple pendulum suspension, which is hung from a cantilever spring isolator. This is supported by a top frame that itself is isolated via a stack of 'rubber' and metal layers. An active seismic control will reduce the motions at very low frequencies. The *internal* mechanical vibrations of the mirrors (the "test masses") can be kept low inside the measurement band by choosing materials of very high mechanical Q, such as pure fused silica, or even silicon. Elaborate methods of attaching the suspending 'wires' are required so as not to compromise the intrinsic Q. Glasgow and Hannover have worked out viable solutions, and Q values of 5×10^6 and better have been achieved.

Laser noise. The development of the laser system for GEO 600 is mainly in the hands of Laser-Zentrum Hannover (LZH). A master/slave system with a highly stabilized Nd:YAG MISER as 'master' and a powerful (up to 20 W) slave are nearing completion. The geometrical noise of the beam (fluctuations in position, orientation, and shape of the beam) as well as amplitude and frequency noise at the modulation frequencies are greatly reduced by the use of two "mode cleaners" in series (Glasgow). Improved stabilization of the laser with respect to frequency and power is the goal of work going on at Hannover.

Shot noise. A very fundamental noise source is the shot noise produced by the 'graininess' of the detected light. Its effect decreases with the root of the available light power. To reach sensitivities that make the detection of events as far out as the Virgo cluster possible, light powers in the order of 10 kW are needed. The cavity composed of the Michelson interferometer and the *power recycling mirror* M_{PR} is designed to enhance the laser power by three orders of magnitude.

THE 30 METER PROTOTYPE

The Garching 30 m prototype (Figure 2) was used for verifying the concepts of power and signal recycling, and for experimentally investigating the demands on the overall control system.

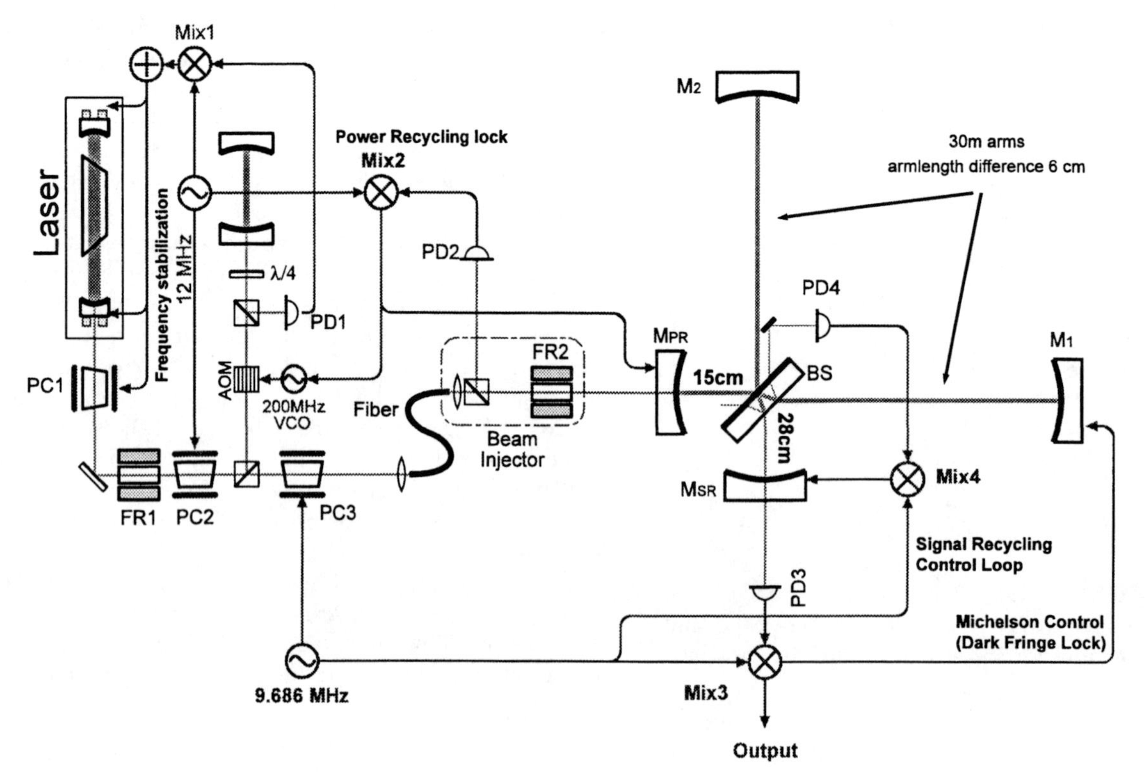

FIGURE 2. Control scheme of the Garching 30 m prototype.

The laser (here still an Argon ion laser) is first stabilised against a reference cavity, and then against the power recycling cavity consisting of the interferometer and the mirror M_{PR}. In an earlier experiment with the 30 m prototype a power-recycling gain of about 300 was achieved [2]. More recently, a fully automatic alignment system for controlling 10 mechanical degrees of freedom was implemented, providing the stability of the settings of all the suspended optical components [3]. The frontal ("Schnupp") modulation [4], applied at the Pockels cell PC3, simplifies the read-out with which the Michelson interferometer is kept in lock, as compared with the previous concept of "external modulation".

Only after this set-up was sufficiently robust the next step, the introduction of the signal recycing mirror M_{SR}, was started. The error signal for its position is obtained with the same Schnupp modulation, using an additional photodetector, PD4. One big problem in such complicated interferometers is the fact that many of the control loops can work properly only after all other loops are locked. The operation of prototypes is essential in finding feasible schemes for this lock acquisition problem. The appropriate arm lengths between beam splitter and the four mirrors were chosen with the help of an elaborate simulation code [5].

With all of this cautious preparatory work done, acquiring lock of the signal recycling loop was then surprisingly easy. The power recycling mirror used in this experiment had a transmission of 7 %, giving a power gain of ~ 50. With a signal recycling mirror of 4 % transmission a signal enhancement of 70 was observed which is due partly to an enhancment of the modulation sidebands and partly to a true signal enhancement [6]. The observed values are in close agreement with that expected from the simulations. The output signal is still dominated by an excessive noise contribution from the laser. Work is continuing to solve this problem.

THE GEO 600 SENSITIVIY

Figure 3 shows the sensitivity, the spectral density of the apparent strain noise, of GEO 600, resulting from the expected contributions of the various noise sources. Making the signal sidebands resonate in the cavity formed by the Michelson interferometer as the one (albeit very complex) mirror, and the signal-recycling mirror M_{SR} as the other, will make the signal response a rather complicated function of GW frequency, which can be adapted to a variety of requirements [7,8].

On the left-hand side, the broadband operation of GEO 600 is shown, having a wide minimum around a frequency of 200 Hz. By appropriate choice of the transmission of the 'signal recycling mirror', a narrowband operation is possible, and the antenna can be tuned by microscopic alignment of the mirror M_{SR} to a given frequency, e.g. to a known or expected GW source.

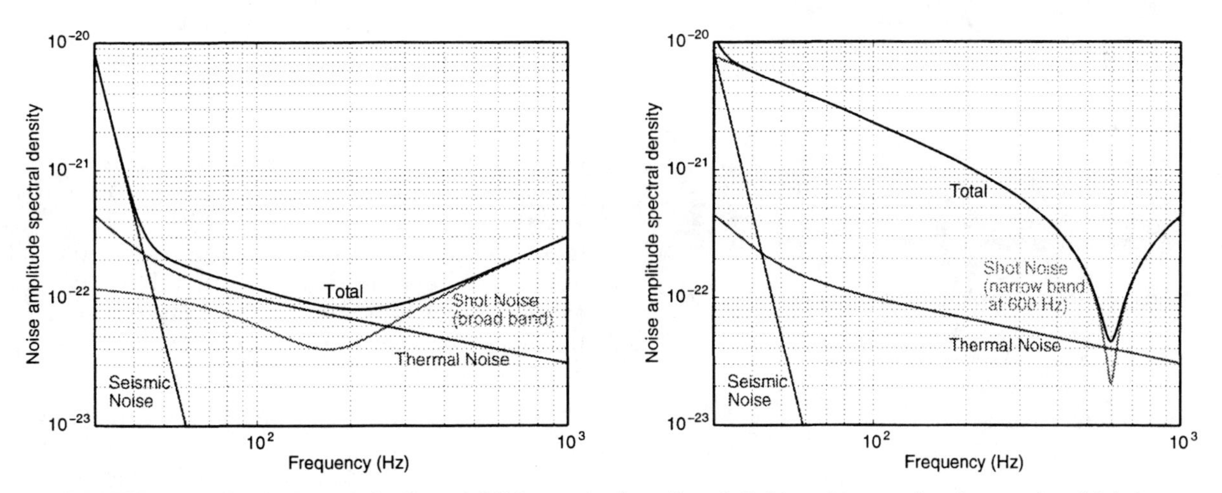

FIGURE 3. Noise spectral density of GEO 600, for broadband (left) and narrowband operation (right).

The limitation due to shot-noise can be reduced if one is willing to "sacrifice" measuring bandwidth. There can be very good reasons to do that: for a nearly continuous-wave signal, as in the early stages of a binary inspiral, the signal frequency changes only very slowly, and narrow-banding at the 'response' level reduces noise much more effectively than narrow-banding at the data-analysis level. All this noise reduction is, however, limited by the other noise contributions, in the case above by thermal noise.

STATUS OF GEO 600

Civil engineering. The construction of the central house and the end houses had been completed in 1997, the connecting trenches that house the vacuum tubes even earlier. The clean-room environment is working very satisfactorily, at class 1000 for the inner area. When it comes to the installation of the final optics this will be further improved to class 100 be means of a "clean tent". The accommodation of electronics and other equipment is scant, but sufficient.

Electronics and data management. Most of the electronics is designed and built by the labs in Glasgow, Garching and Hannover. Communication between the buildings will be exclusively via glass fiber, to avoid crosstalk and ground loops. The data can be sent via direct radio communication to the University of Hannover and from there via fast data links to Potsdam and Cardiff for analysis.

The vacuum pumps. The high vacuum required in the whole apparatus is made possible by powerful turbo-molecular pumps that are situated in the central house and in the end houses. They will run continuously, also during measurements, the magnetic bearings make for a very quiet operation. The total volume of the vacuum system is about $400\,\mathrm{m}^3$, the total surface area about $4000\,\mathrm{m}^2$.

Vacuum tubes. GEO 600 has tried a very cost-effective way of producing the vacuum tubing. The tubes are made of 4.5 m long pieces of 60 cm diameter stainless steel tubes of 0.8 mm thickness. The necessary stiffness against air pressure stems from a bellows-type convolution with a few cm period (see Figure 4). The tube is suspended from crossbars that traverse the trench.

The vacuum achieved. The tubes, wrapped in a 20 cm layer of rock wool, were air-baked at 200 °C for 24 hours, and then baked under vacuum for one week. The vacuum achieved, in the order of 10^{-8} mbar, is just sufficient, even though a few tiny leaks seem to be remaining. The tubes are supposed never to be let up to air again, to be shut off with gate valves when opening the vacuum tanks becomes necessary.

The vacuum tanks. The vacuum tanks (9 in the central house, one each in the end houses) have been leak-tested and baked out, and some of them are already installed. For the modecleaner tanks, the metal-and-rubber stacks, three per tank, are currently being built. They will support the top suspension plates, from which double or triple pendulums will carry the main optical components.

FIGURE 4. The convoluted tubes inside the trench, suspended from cross-bars, before thermal insulation was wrapped around them.

More information about the status and the schedule of GEO 600 can be found in the regularly updated GEO 600 web page `http://www.geo600.uni-hannover.de`.

OUTLOOK

Groups at Cardiff and Potsdam are working on the theory of gravitational wave sources, as well as on problems of data acquisition and analysis.

The aim of GEO 600 is to have first science runs of the interferometer in or around the year 2000. This is the time when also the other large detectors might begin taking data. It will be an exciting time to see gravitational wave astronomy come about.

REFERENCES

1. K. Danzmann *et al.*: in *First Edoardo Amaldi Conference on Gravitational Wave Experiments*, Frascati 1994, (World Scientific, Singapore, 1995) 100–111.
2. D. Schnier *et al.*: Phys. Lett. A **225** (1997) 210–216.
3. G. Heinzel *et al.*: paper in preparation (1998).
4. L. Schnupp: talk at *European Collaboration Meeting on Interferometric Detection of Gravitational Waves* (Sorrento, 1988).
5. J. Mizuno *et al.*: paper in preparation (1998).
6. G. Heinzel *et al.*: paper in preparation (1998).
7. B.J. Meers: Phys. Rev. D **38** (1988) 2317–2326.
8. B.J. Meers: Phys. Lett. A **142** (1989) 465–470.

Status and noise limit of the VIRGO antenna

presented by L. Gammaitoni[1] for the VIRGO collaboration

D. Babusci[2], H. Fang[2], G. Giordano[2], M. Iannarelli[2], G. Matone[2], E. Turri[2], M. Mazzoni[3], R. Stanga[3], E. Calloni[4], S. Cavaliere[4], L. Di Fiore[4], G. Evangelista[4], F. Garifi[4], A. Grado[4], L. Milano[4], S. Solimeno[4], G. Cagnoli[1], C. Cattuto[4], J. Kovalik[1], F. Marchesoni[1], M. Punturo[1], M. Bernardini[5], A. Bozzi[5], S. Braccini[5], C. Bradaschia[5], C. Casciano[5], G. Cella[5], A. Ciampa[5], E. Cuoco[5], G. Curci[5], E. D'Ambrosio[5], V. Dattilo[5], G. De Carolis[5], R. De Salvo[5], A. Di Virgilio[5], D. Enard[5], A. Errico[5], G. Feng[5], I. Ferrante[5], F. Fidecaro[5], F. Frasconi[5], A. Gaddi[5], A. Gennai[5], G. Gennaro[5], A. Giazotto[5], P. La Penna[5], G. Losurdo[5], M. Maggiore[5], S. Mancini[5], F. Palla[5], H. B. Pan[5], F. Paoletti[5], A. Pasqualetti[5], R. Passaquieti[5], D. Passuello[5], R. Poggiani[5], P. Popolizio[5], F. Raffaelli[5], S. Rapisarda[5], R. Taddei[5], A. Vicere[5], Z. Zhang[5], P. Astone[6], F. Bronzini[6], S. Frasca[6], E. Majorana[6], C. Palomba[6], M. Perciballi[6], P. Puppo[6], P. Rapagnani[6] , F. Ricci[6], C. Boccara[7], J. B. Daban[7], M. Leliboux[7], V. Loriette[7], R. Nahoum[7], J-P. Roger[7], P. Ganau[8], B. Lagrange[8], J.M. Mackowski[8], C. Michel[8], N. Morgago[8], L. Pinard[8], A. Remillieux[8], C.Arnault[9], C. Barrand[9], J-L. Beney[9], R. Bilhaut[9], V. Brisson[9], F. Cavalier[9], R. Chiche[9], J-P. Coulon[9], S. Cuzon[9], M. Davier[9], M. Dehamme[9], M. Dialinas[9], C. Eder[9], M. Gaspard[9], P. Hello[9], P. Heusse[9], A. Hrisoho[9], E. Jules[9], J-C. Marrucho[9], M. Mencik[9], P. Marin[9], L. Matone[9], M. Mencik[9], A. Reboux[9], P. Roudier[9], M. Taurigna[9], F. Bellachia[10], M. Bermond[10], D. Boget[10], B. Caron[10], T. Carron[10], D. Castellazzi[10], F. Chollet[10], G. Daguin[10], P-Y. David[10], L. Derome[10], C. Drezen[10], D. Dufournaud[10], R. Flamino[10], L. Giacobone[10], C. Girard[10], X. Grave[10], R. Hermel[10], J-C. Lacotte[10], J-C. Le Marec[10], B. Lieunard[10], F. Marion[10], L. Massonnet[10], C. Mehmel[10], R. Morand[10], B. Mours[10], P. Mugnier[10], V. Sannibale[10], R. Sottile[10], D. Verkindt[10], M. Yvert[10], Y. Acker[11], R. Barillet[11], M. Barsuglia[11], J-P. Berthet[11], A. Brillet[11], J. Cachenaut[11], F. Cleva[11], H. Heitmann[11], J-M. Innocent[11], J-C. Lucenay[11], N. C. Man[11], P-T. Manh[11], J-A. Marck[11], D. Pelat[11], V. Reita[11], J-Y. Vinet[11]

[1] *Dipartimento di Fisica dell'Universita' and INFN Sezione di Perugia I-06100 Perugia (Italy), luca.gammaitoniLapg.infn.it*
[2] *Laboratori Nazionali INFN Frascati*
[3] *Dipartimento di Fisica dell'Universita' and 1NFN Sezione di Firenze*
[4] *Dipartimento di Scienze Fisiche dell'Universita' and INFN Sezione di Napoli*
[5] *Dipartimento di Fisica dell'Universita' and INFN Sezione di Pisa*
[6] *Dipartiniento di Fisica dell'Universita' "La Sapienza" and INFN Sezione di Roma*
[7] *Laboratoire de Spectroscopic en Luniiere Polarisee, Ecole Superieure de Physique et Chimie Industrielle del la Ville de Paris*
[8] *Universite Claude Bernard, IPNL*
[9] *Laboratoire de l'Accelerateur Lineaire, Universite Paris-Sud*
[10] *Laboratoire de Physique del Particules (LAPP)*
[11] *Laser Optics Orsay, Universite Paris-Sud*

CP456, *Laser Interferometer Space Antenna*
edited by William M. Folkner
 1-56396-848-7/98/$15.00

Abstract. The present status of the VIRGO antenna, with special care of the noise limit, is presented. The VIRGO project (a French-Italian collaboration) has been approved in 1993 and it is now in the construction phase. The end of the commissioning phase is foreseen within 2001.

INTRODUCTION

The VIRGO project is a French-Italian collaboration [1]. The laboratories forming the collaboration are: ESPCI-Paris, INFN-Firenze and Urbino, LNF-Frascati, INFN-Napoli, INFN-Perugia, INFN-Pisa, INFN-Roma-1, IPN-Lyon, LAL-Orsay, LAPP-Annecy, LAS-Optics Paris. The VIRGO Project was approved in 1993 by the INFN and CNRS. In 1994 the site for the detector installation was selected and the construction phase started in 1996.

At present, mid 1998, the *central building* and the *mode cleaner building* have been completed on site (in Cascina, few kilometers from the INFN-Pisa Laboratory). The installation of suspension towers is in progress. The end of the commissioning phase is foreseen within 2001.

Pictures of the Cascina site are available on the web at www.pg.infn.it/virgo.

PRINCIPLE OF OPERATION

The VIRGO detector consists in a 3 km arm length-Michelson interferometer, with suspended mirrors (test masses) and laser source. Each arm contains a Fabry-Perot cavity with a F=50 finesse. Due tosmall asymmetries between the two arms a fluctuation in laser amplitude and frequency produce a noise signal which superimpose on the signal generated by the gravitational wave. For this reason special care has been taken in selecting a proper optical scheme: a high power laser ($20W$) injection-locked by an ultra stable low power laser ($1W$). In order to reduce ground induced displacements of test masses a long suspension chain (superattenuator) has been designed. The detector is planned to take data in a large bandwidth ($10Hz - 10kHz$) with a sensitivity as low as $h \approx 10^{-22}\sqrt{Hz}^{-1}$.

THE VIRGO SENSITIVITY

Four major gravitational interferometers are presently under construction: LIGO [2] in the US, TAMA [3] in Japan and GE0600 (a British-German collaboration) in Germany [4] and VIRGO. There is evidence that GW signals in the low frequency range ($1-50Hz$) allow a significant GW source survey [5] and the VIRGO detector, at difference with the other interferometers, has been specifically designed to optimize its sensitivity in this frequency range. Three main noise sources limit the detector sensitivity at low frequency: (a) the *seismic noise*, due to ground vibrations. It is taken care by using a chain of mechanical filters called *superattenuator*; (b) the *newtonian noise* (gravity gradient noise) [6], due to the local ground and atmospheric density fluctuations, is unavoidable in earth-based GW detectors; (c) the *thermal noise*, mainly due to the internal friction [7,8] of the suspension (wires, springs, etc) and optical components (mirrors, beam-splitters, etc.).

The thermal noise limit to the VIRGO sensitivity

Thermal noise poses a severe limit to the sensitivity of interferometric gravitational wave detectors due to the fluctuations in the position of the suspended elements (test masses and optics) of the interferometer and to the internal modes of the mirrors [9–13]. The whole suspending structure can be treated as a multi-stage pendulum whose element positions will fluctuate in time. Such fluctuations combine with the gravitational wave induced displacement, thus setting a lower limit to the antenna sensitivity.

The detector sensitivity limit due to thermal noise can be estimated by measuring the dissipation properties of the suspension structure or, to be more precise, the imaginary part of the response function of the observable of interest $x(t)$ (i.e. the position of the mirror center of mass) to the conjugate force $f(t)$. The response function

$$H(\omega) = \frac{X(\omega)}{F(\omega)} \tag{1}$$

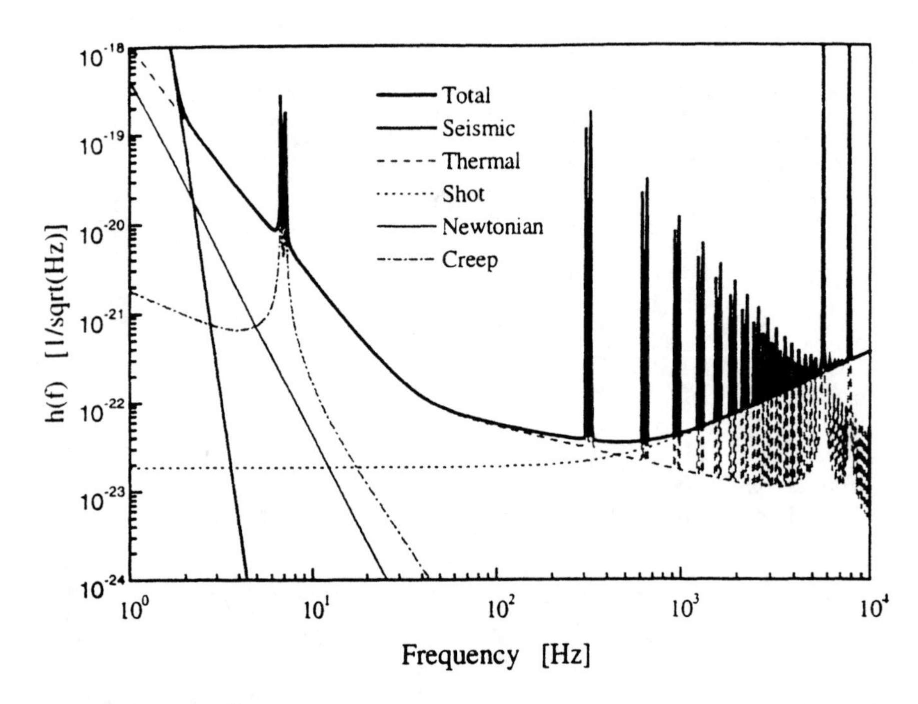

FIGURE 1. The VIRGO sensitivity curve. On the vertical axes we have the expected h (in $(Hz)^{-1/2}$), while on the horizontal axes we have the frequency (in Hz). The major contributions in the low frequency region arise from thermal noise due to the longitudinal *pendulum mode* of the test mass and to the test mass (drum) normal modes (global effect). The low frequency (below $10Hz$) peaks arise from the coupling between the vertical and horizontal motion, due to the geometric curvature of the earth surface.

is obtained by recording the time series of the force and the relevant displacement and, then, taking their Fourier transforms (capital letters). Finally, the fluctuation spectral density (thermal noise) is computed from the $H(\omega)$ imaginary part, $H''(\omega)$ by using the *Fluctuation-Dissipation theorem* (FDT) [14].

$$\langle x(\omega)^2 \rangle = -4kT\frac{H''(\omega)}{\omega} \tag{2}$$

The quantity $H''(\omega)$ can be accessed experimentally and the fluctuation power spectral density $\langle x(\omega)^2 \rangle$ can be obtained accordingly.

In fig. 1 we show the thermal noise limit to the VIRGO sensitivity, as estimated with current mathematical models [11,15], with the help of experimental measurements [15].

A thermal noise related source which can play a role in limiting the interferometer sensitivity is represented by the noise induced by stationary creep in heavily loaded mechanical suspensions (wires, spring blades, etc.). The mechanical losses thus consist of two independent contributions: the conventional internal friction (independent of load), and the mechanical shot noise related to a stationary creep mechanism.

Preliminary estimates based on a simple dynamical model [16] and recent measurements taken in Perugia show that this effect can be significantly reduced by pre-heating the steel suspension wires at a temperature of about 150 C for one week. Under this condition the estimated shot noise is much below the pendulum thermal noise for all the frequencies of interest [16].

PRESENT STATUS AND PERSPECTIVES

An important step in the construction of the VIRGO detector is the commissioning of the so called "test interferometer", a Michelson interferometer situated in the VIRGO central building. This facility will be used to test and debug all the equipment to be installed on the final configuration: superattenuators and suspension systems, injection and detection benches, laser, electronics and software. During the test interferometer commissioning, the arm construction will start together with the vacum tube installation. As a final step the full scale detector commissioning is expected to be completed by 2001.

REFERENCES

1. For more information on the VIRGO Project please consult the VIRGO Central Web site at www.pg.infn.it/virgo, see also *The VIRGO Final Conceptual Design*, VIRGO Collaboration, unpublished (1995) and B. Caron *et al.*, Nuc. Instr. Meth. Phys. Res. **A360** 258 (1995)
2. A. Abramovici *et al.*, Science **256** 325 (1992)
3. see http://taniago.mtk.nao.ac.jp/tama.html
4. K. Danzmann *et al.*, GE0600: Proposal for a 600m Laser Interferometricc Gravitational Wave Antenna, Max-Planck-Institut für Quantenoptik Report 190 (Garching, Germany) (1994)
5. K. S. Thorne, *300 Years of Gravitation*, (CUP, Cambridge, 1987)
6. P. R. Saulson, Phys. Rev. **D30** 732 (1984)
7. A. S. Nowick and B. S. Berry, *Anelastic Relaxation in Crystalline Solids* (Academic, New York, 1972)
8. F. R. N. Nabarro, *Theory of Crystal Dislocations* (Oxford University Press, London, 1967)
9. T. J. Quinn, C. C. Speake, W. Tew, R. S. Davis and L. M. Brown, Phys. Lett. **A197**197 (1995)
10. G. Cagnoli, L. Gammaitoni, J. Kovalik, F. Marchesoni and M. Punturo, Phys. Lett. **A213** 245 (1996)
11. P. R. Saulson, Phys. Rev. **D42** 2437 (1990); P. R. Saulson, R. T. Stebbins, F. D. Dumont and S. E. Mock, Rev. Sci. Instrum. **65** 182 (1994)
12. F. Bondu and J. Y. Vinet, Phys. Lett. **A198** 74 (1995)
13. J. E. Logan, J. Hough and N. A. Robertson, Phys. Lett. **A183** 145 (1993)
14. R. Kubo, M. Toda and N. Hashitsume, *Statistical Phvsics II* (Springer, Berlin, 1985); L.D. Landau, E.M. Lifsits, *Statistical Physics* (Mir, Moscow, 1976)
15. G. Cagnoli, J. Kovalik, L. Gammaitoni, F. Marchesoni, M. Punturo, Phvs. Lett. **A213**, 245 (1996)
16. G. Cagnoli, J. Kovalik, L. Ganiniaitoni, F. Marchesoni, M. Punturo, Phys. Lett. **A237** 21 (1998)

Resonant-mass gravitational-wave detectors in operation

Eugenio Coccia

Physics Department, University of Rome "Tor Vergata" and INFN
Via Ricerca Scientifica 1, 00133 Rome, Italy
e-mail: COCCIA@ROMA2.INFN.IT

Abstract. We report on the status of the five resonant-mass detectors of gravitational waves operating today in Australia, Italy and USA. They are now in the continuous observational mode with burst sensitivity $h \simeq 4 \times 10^{-19}$, or, in spectral units, 2×10^{-22} $\mathrm{Hz}^{-1/2}$ over bandwidth of about 1 Hz. The strongest potential sources of GW bursts in our Galaxy and in the local group are today monitored by such instruments. In parallel with the observations, experimental development work is very active.

Resonant-mass gravitational wave (GW) antennas have been greatly improved over the past 20 years. Cryogenic operation, superconducting electronics, improved vibration isolation and increased acoustic Q-factors have contributed to a 10^4 fold improvement in energy sensitivity over Weber's original antennas [1]. The principle of operation of such detectors is based on the assumption that any vibrational mode of a resonant body that has a mass quadrupole moment, such as the fundamental longitudinal mode of a cylindrical bar, can be excited by a GW with nonzero energy spectral density at the mode eigenfrequency. The mechanical oscillation induced in the antenna by the interaction with the GW is transformed into an electrical signal by a motion or strain transducer and then amplified by a very low noise electrical amplifier. Unavoidably, Brownian motion noise associated with dissipation in the antenna and the transducer and electronic noise from the amplifier limit the sensitivity of the detector. The sum of the contributions due to these noise sources gives the total detector noise. This can be referred to the input of the detector (as if it were a GW spectral density) and is usually indicated as $S_h(f)$ [2]:

$$S_h(f) = \frac{\pi}{8}\frac{kTf_0}{ML^2Qf^4}\{1 + \Gamma[Q^2(1 - \frac{f^2}{f_0^2})^2 + \frac{f^2}{f_0^2}]\} \tag{1}$$

where T is the antenna temperature, M and L are the antenna mass and length, Q is the quality factor of the mode, f_0 the resonant frequency, and

$$\Gamma \simeq \frac{T_n}{2\beta QT} \tag{2}$$

where β is the transducer efficiency and T_n is the noise temperature of the amplifier. Γ gives the ratio of the wide band noise in the resonance bandwidth to the narrow band noise (in practice $\Gamma \ll 1$).

The half height width of $S_h(f)$ gives the bandwidth of a resonant detector:

$$\Delta f = \frac{f_0}{Q}\Gamma^{-1/2} \tag{3}$$

This is much larger than the pure resonance linewidth f_0/Q.

A recent review on the performance of resonant-mass detectors and on the progresses in the field has been given at the GR15 Conference by Cerdonio [3] . Here we briefly summarize the status of the five operating detectors: EXPLORER [4], at CERN (operating since 1990), ALLEGRO [5] at Louisiana State University (1991), NIOBE [6] at the University of Western Australia (1993), NAUTILUS [7] at the INFN Frascati National Labs (1995), and AURIGA [8] at the INFN Legnaro National Labs (1997). Table 1 resumes the main features of the five bar detectors.

CP456, *Laser Interferometer Space Antenna*
edited by William M. Folkner
 1-56396-848-7/98/$15.00

TABLE 1. Main features of the resonant-mass detectors in operation. The detectors are almost parallel oriented.

	EXPLORER	ALLEGRO	NIOBE	NAUTILUS	AURIGA
material	Al5056	Al5056	Nb	Al5056	Al5056
length (m)	3.0	3.0	2.75	3.0	2.9
M (kg)	2270	2296	1500	2260	2230
f_- (Hz)	905	895	694	908	912
f_+ (Hz)	921	920	713	924	930
$Q \times 10^6$	1.5	2	20	0.6	3
T (K)	2.6	4.2	5	0.1	0.25
$S_h(f)^{1/2}$	$6\ 10^{-22}$	$1\ 10^{-21}$	$8\ 10^{-22}$	$2\ 10^{-22}$	$2\ 10^{-22}$
Δf(Hz)	0.2	$\simeq 1$	$\simeq 1$	$\simeq 1$	$\simeq 1$
$T_{eff}(mK)$	10	10	3	3	3
h^{min}	$6\ 10^{-19}$	$6\ 10^{-19}$	$6\ 10^{-19}$	$4\ 10^{-19}$	$4\ 10^{-19}$
duty cycle (%)	75	95	75	75	75
event rate (d^{-1})	100	100	75	100	100
latitude	46°27′00"N	30°27′00"N	31°56′00"S	41°49′26"N	44°21′12"N
longitude	6°12′00"	268°50′0"	115°49′0"	12°40′21"	11°56′54"

All the bars are equipped with a resonant transducer coupled in resonance with the antenna, forming a system of two coupled oscillators. The frequencies of the two resulting normal modes are indicated with f_- and f_+. The vibration of the transducer modulates a dc electric field in the case of the capacitive transducers used by EXPLORER, NAUTILUS and AURIGA and a dc magnetic field in the case of the inductive transducer used by ALLEGRO. All these detectors use a dc SQUID amplifier. An active transducer using a microwave cavity is adopted by the Australian group for the NIOBE detector.

T_{eff} expresses (in Kelvin) the minimum detectable energy innovation and determines the detector sensitivity to short (1 ms) bursts, indicated with h^{min}.

We remark that $h = 6 \times 10^{-19}$ corresponds to the signal of a millisecond GW burst due to the total conversion of about $10^{-4} M_\odot$ in the Galactic center.

A very important feature of a detector is the quality of its noise distribution, represented by the rate of samples that deviate from the statistics. The tails of the distributions determine, in fact, the background for any coincidence experiment. The reported event rate is related with a threshold of about 100 mK. The reported value of the duty cycle give the fraction of time during which the hourly averaged value of T_{eff} is lower than 20 mK.

We can confidently say that today the strongest sources in our Galaxy and in the local group will not pass unnoticed to at least two resonant-mass detectors. This fact is extremely important as the search for GW is based on the technique of coincidences among two or more detectors.

All the resonant-mass detector groups formed the International Gravitational Event Collaboration (IGEC) on July 4th 1997, at CERN, and agreed in a data exchange protocol. Up to now the analysis of the data has been focused in searching for short GW bursts. The data of each detector are filtered with Wiener-Kolmogoroff type filters (matched filters for δ-like bursts). A list of energy innovations above a threshold (choosen a priori) and the corresponding occurrence times is then produced for each detector. The number of coincidences at zero delay in a given (a priori) time window is compared with the number of accidental coincidences.

In addition to the search for short bursts, the data collected with these detectors are used to detect periodic waves over long time periods, for giving new upper limits for the stochastic background of cosmological origin, and for studying possible correlation with gamma ray bursts (for all that, see [1] [11] and references therein).

Towards larger bandwidths

The use of a high-Q resonant-mass seems to imply that this type of detector has a very narrow bandwidth. This is not the case. Both the strain signal and the thermal Brownian motion noise in the antenna exhibit the same resonant response near the mode eigenfrequency. Thus the signal-to-noise ratio is not bandwidth limited by the antenna thermal noise. The significant bandwidth limitation comes from the transducer and amplifier.

Near quantum-limited SQUID amplifiers for GW detectors are under developments by various experts: M. Mueck and J. Gail (Giessen University), F.C. Wellstood [9] and coworkers, and P. Carelli et al. [10] who realized a dc SQUID with coupled energy sensitivity of $5.5\hbar$ at $T = 0.9K$. The work necessary to integrate such sensitive devices into the real detectors without degrading their performance is in progress: it appears possible to reach an effective bandwidth of the order of 50 Hz.

Spherical detectors

Most people consider that the next generation of resonant-mass detectors will be of spherical shape. A single sphere is capable of detecting gravitational waves from all directions and polarizations and is capable of determining the direction information and tensorial character of the incident wave. A sphere will have a larger mass than the present bars (with the same resonant frequency), translating into an increased cross section and improved sensitivity. Omnidirectionality and source direction finding ability makes a spherical detector a unique instrument for gravitational wave astronomy with respect to all present detectors. The measurement of the polarization states and the scalar-tensor discrimination opens new possibilities in the study of gravitational physics. Finally, the different features and technology makes a spherical detector complementary to an interferometer. It emerges that an observatory composed of both a sphere and an interferometer will have unprecedented sensitivity and signal characterization capabilities. Studies and measurements essential to define a project of a large spherical detector, 40 to 100 tons of mass, competitive with large interferometers but with complementary features, have been made in USA, Italy, Netherland and Brasil (see [12] and references therein).

REFERENCES

1. Coccia, E., Pizzella, G., and Veneziano, G (eds.) *Gravitational Waves, Proceedings of the Second Edoardo Amaldi Conference*, CERN, Geneva 1997 , World Sci., Singapore,1998.
2. G.Pizzella *Class. Quantum Grav.* **LISA Proc. Suppl.**, 1481-1485 (1997)
3. Cerdonio, M. et al. *GR 15 Conference Proceedings*, Pune, India 1997, World Sci., in press.
4. Astone, P. et al. *Phys. Rev. D* **47**, 2 (1993).
5. Mauceli, E. et al. *Phys. Rev. D* **54**, 1264 (1996).
6. Blair, D.G.. et al. *Phys. Rev. Lett.* **74**, 1908 (1995).
7. Astone, P. et al.*Astrop. Phys.* **7**, 231 (1997).
8. Prodi, G.A. et al. *Gravitational Waves, Proceedings of the Second Edoardo Amaldi Conference*, CERN, Geneva 1997 , World Sci., Singapore,1998
9. Jin I. et al. *IEEE Trans. Appl. Sup.* **7**, 2742 (1997).
10. Carelli P. et al, *Appl. Phys. Lett.* **72**, 115 (1998).
11. *Gravitational Wave Data Analysis Workshop 2*, Orsay, France 1997, World Sci., in press.
12. Velloso, W.F., Aguiar, O.D., Magalhaes, N.S. *Omnidirectional Gravitational Radiation Observatory* , Sao Jose dos Campos, Brasil 1996, World Sci., Singapore,1997.

Spacecraft Doppler Gravitational Wave Experiments

J. W. Armstrong

Jet Propulsion Laboratory,
California Institute of Technology, Pasadena CA 91109

Abstract. This paper discusses spacecraft Doppler tracking, the current-generation detection technique in the low-frequency gravitational wave band. Unlike other detectors, the ~1-10 AU earth-spacecraft separation makes the Doppler detector large compared with the gravitational wavelength; this has the consequence that a signal is time-resolved into three events in the Doppler time series. The principles of operation, including the transfer functions of the gravitational wave signal and the leading noise processes to the observed Doppler frequency time series, are outlined. Experiments-to-date, and the expected performance of the very-high-sensitivity Cassini experiment (to be done in 2001-2004) are discussed.

Doppler Tracking Technique in the Low-Frequency Band

Observations in the low-frequency (LF, ~0.00001-0.1 Hz) spectral band require space-based detectors. Currently the only broadband technique in the LF band is Doppler tracking of spacecraft. In this method, the earth and a distant spacecraft are free test masses in a "one-armed" interferometer, coherence being maintained through a high-precision frequency standard on the ground (1, 2, 3). The Doppler tracking system of the NASA Deep Space Network (DSN), referenced to an ultra-high-quality frequency standard, monitors a transponded microwave signal from a distant spacecraft. It thus continuously measures the relative dimensionless velocity ($2 \Delta v/c = \Delta f/f_0$, where Δf is the Doppler shift and f_0 is the link center frequency) between the Earth and spacecraft. A gravitational wave of dimensionless strain amplitude h incident on the system causes small perturbations in the tracking record. These perturbations, of order h in $\Delta f/f_0$, are replicated three times in the Doppler data producing a geometry-dependent signature (1), shown schematically in Figure 1. The sum of the Doppler perturbations of the three pulses is zero. Pulses with duration longer than about the one-way light time produce overlapping responses in the tracking record and the net response then begins to cancel. The tracking system thus has a passband to gravitational excitation: the low-frequency band edge depends on wave angle-of-arrival but is approximately set by pulse cancellation to ~(1/two-way light time). Thermal (white phase noise in the receiver system) and frequency standard noise limit the high frequency band edge to ~(1/10 seconds).

Response of Doppler System to Gravitational Wave Excitation

Unlike LISA, LIGO, or bar detectors, the ~1-10 AU earth-spacecraft separation makes the Doppler detector large compared with the GW wavelength for most candidate signals. In this regime, a GW incident on the Earth-spacecraft system resolves into three events in the Doppler time series: buffeting of the earth, buffeting of the spacecraft, and the initial buffeting of the Earth transponded back to the tracking station. Figure 1 shows this situation in cartoon form.

CP456, *Laser Interferometer Space Antenna*
edited by William M. Folkner

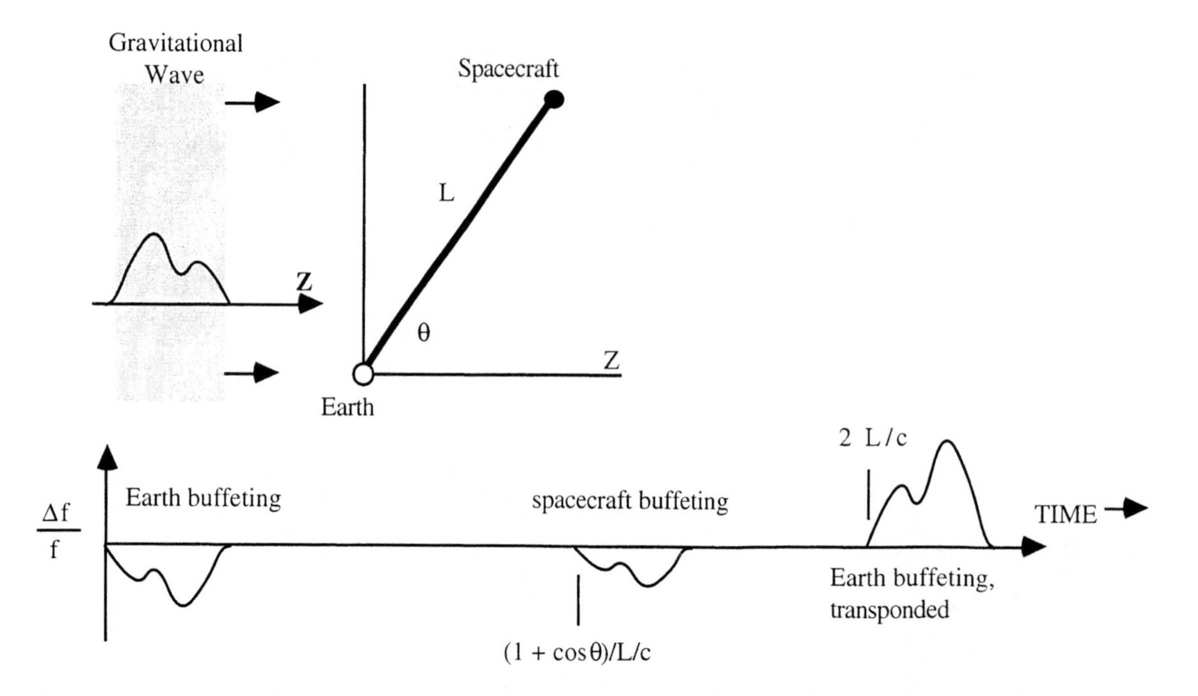

FIGURE 1. Schematic illustration of response of Earth-spacecraft tracking link to gravitational radiation having strain amplitude h. The Doppler perturbations are of order h in $\Delta f/f_0$ and, because the Earth-spacecraft separation is large compared with the GW wavelength, are resolved three times in the Doppler data: once when the wave "buffets" the earth, causing a small change in the difference between the transmitted and received Doppler frequencies, once when the spacecraft is buffeted by the wave, and finally when the initial earth perturbation is transponded back to the earth (1).

GWs compete with other perturbations of the Doppler time series. The noise level is quantified by the statistics of $y(t) = \Delta f/f_0$ for each noise process (4). Table 1 lists the principal noise processes in Doppler experiments and their transfer functions. These noise sources are: propagation noises (solar wind, ionosphere, troposphere--phase scintillation changes the apparent distance between the earth and spacecraft, hence $\Delta v/c$ hence $y(t) = \Delta f/f_0$), frequency standard noise (a fundamental noise, since interferometric techniques cannot be used to cancel it), antenna mechanical noise (unmodeled motion of the phase center as the ground antenna tracks the spacecraft), thermal noise (finite SNR often limits the high frequency sensitivity), and systematic errors. These noises enter the data through transfer functions connecting the noise process to the Doppler observable. Noise transfer functions are different from the transfer function for the GW signal--these differences can be exploited in signal processing for different gravitational signal waveforms. The way this is done depends on the dominant noise sources--their levels and spectral shapes--and on the signal characteristics. The dominant noise sources in turn depend on the technology used in the spacecraft and the ground stations. Previous generation experiments used an S-band (~2.3 GHz) tracking link. At this radio frequency plasma phase scintillation in the solar wind was the dominant noise source. Current generation experiments us X-band (~8.4 GHz) links, and are roughly equally affected by plasma scintillation noise and phase scintillation in the Earth's neutral atmosphere. Next generation experiments will use Ka-band (~32 GHz) and have independent tropospheric calibration and are expected to be limited by residual tropospheric calibration error and antenna mechanical noise. Analyses of the relevant noises have been published (5, 6, 7, 8, 9, 10).

In summary, a Doppler GW experiment residual time series (after removal of known effects) can be written as:

$y(t) = \Delta f/f_0$ = gravity waves + propagation noise + clock noise + receiver thermal noise + systematic errors

$$= \{[(\mu-1)/2]\ \delta(t) - \mu\ \delta[t-(1+\mu)L/c) + [(1+\mu)/2]\ \delta(t - 2L/c)\} * g(t)$$
$$+ \{\delta(t) + \delta[t-2L/c]\} * \text{troposphere}(t)$$
$$+ \{\delta(t) + \delta[t-2(L-x)/c]\} * \text{plasma}(t) + \{\delta(t) - \delta(t-2L/c)\} * \text{clock}(t)$$
$$+ \text{thermal}(t) + \text{systematic errors} \quad (1)$$

where "*" indicates convolution, $g(t) = (1 - \mu^2)^{-1} \{\mathbf{n} \cdot [h_+(t)\ \mathbf{e}_+ + h_\times(t)\ \mathbf{e}_\times] \cdot \mathbf{n}\}$ is the scalar waveform produced by the polarized tensor wave's interaction with the earth-spacecraft system, **n** is a unit vector from Earth to the spacecraft, "troposphere", "plasma", "clock" and "thermal" are noise processes, L is the distance to the spacecraft, and x is the distance to the effective plasma "phase screen" ($x \approx 0.25$ AU for solar wind contribution in experiments done near solar opposition; $x \approx 0$ for the ionosphere).

TABLE 1. Principal Doppler perturbations in a GW experiment and their transfer functions. μ = cosine of the angle between the earth-spacecraft vector and the gravitational wave vector; L = earth-spacecraft distance, x = distance from the earth of equivalent plasma "screen" (x = 0 for the ionosphere; solar wind is an integral over x, but roughly x $\approx$ 0.25 AU for observations at solar opposition).

Doppler perturbation	time-domain transfer function	comment
gravitational waves	$[(\mu-1)/2]\ \delta(t) - \mu\ \delta[t-(1+\mu)L/c) + [(1+\mu)/2]\ \delta(t - 2L/c)$	three pulse response--see Figure 1; see text following equation (1) for polarization coupling
plasma scintillation	$\delta(t) + \delta[t-2(L-x)/c]$	strongly dependent on link radio frequency; dominant noise process for S-band (~2.3 GHz) observations, comparable to raw troposphere for X-band (8.4 GHz), comparable to calibrated troposphere in Ka-band (32 GHz) experiments
tropospheric scintillation	$\delta(t) + \delta[t-2L/c]$	mostly due to water vapor (11); non-dispersive, thus does not improve as one goes to higher-radio frequency observations;
frequency standard ("clock noise")	$\delta(t) - \delta[t-2L/c]$	a fundamental limit; non-dispersive
antenna mechanical	$\delta(t) + \delta[t-2L/c]$	few observations (12); magnitude only approximately known
thermal	$\delta(t)$	level set by SNR on the up- and downlinks; along with frequency standard noise, sets high-frequency band edge
systematic errors and "other" effects	complicated (and in some cases transfer function concept is not applicable)	complicated

In situations where one of the noise processes is dominant, one can immediately see the correlation structure imposed on the Doppler data by its associated transfer function. X-band experiment sensitivity is set by a combination of soar wind plasma and tropospheric scintillation. Although the tropospheric noise is variable on time scales ~tracking pass length (or shorter) it can clearly show its transfer function (10). Figure 2 shows an example of Mars Global Surveyor data with a clear tropospheric noise component. The two-way light time is 512 seconds, and the convolution of the time series with $\delta(t) + \delta[t-2L/c]$ produces an echo in the time series which shows up as positive correlation at the two-way light time. These transfer functions can be exploited in data analysis to improve sensitivity for some waveforms (1, 8, 22).

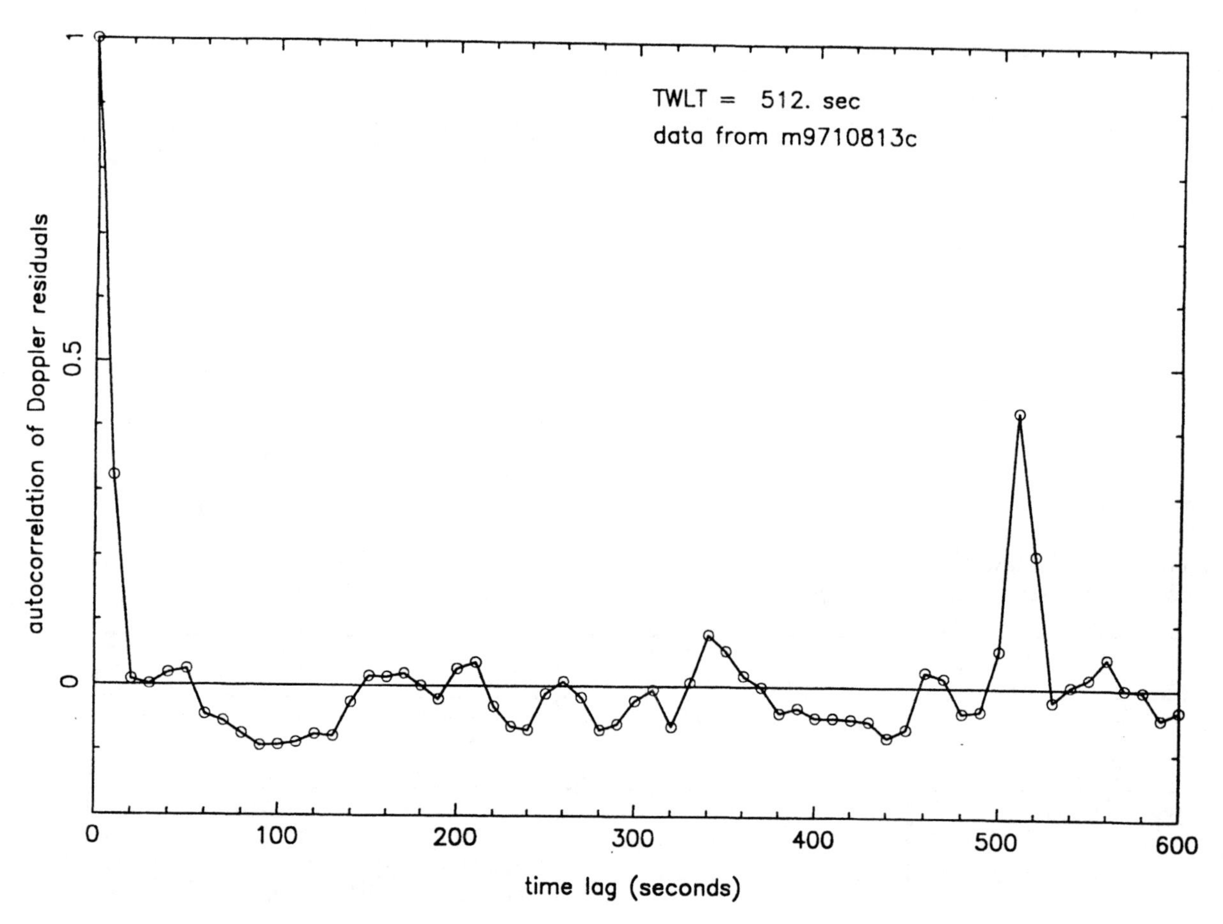

FIGURE 2. Autocorrelation function of X-band Doppler data from Mars Global Surveyor on 1997 DOY 108. The positive correlation at the two-way light time of 512 seconds is due to tropospheric scintillation and its transfer function (see text and Table 1). Differences in the transfer functions for the signal and the noises can be used to improve sensitivity (1, 8, 22)

Doppler Experiments

Spacecraft experiments require a distant spacecraft (low-frequency band edge is set by spacecraft range), a quiet cruising spacecraft (away from gravitational perturbations of a planet or nongravitational perturbations such as spacecraft thrusters), observations near the antisolar direction (to minimize plasma scintillation noise), and a high-stability ground system. Table 2 summarizes past and near-future Doppler observations. The sensitivity achieved depends the radio technology employed, the quality of the ground and spacecraft systems, the gravitational waveform, and the direction-of-arrival of the wave relative to the earth-spacecraft line. Current best sensitivity for, e.g., sinusoidal waveforms is slightly better than 10^{-15} over most of the low-frequency band for the Mars Observer (MO) data (17). The MO noise level is set by a combination of plasma scintillation noise and tropospheric scintillation. The Mars Global Surveyor (MGS) noise level should be somewhat lower than MO's (MGS was taken at larger sun-earth-spacecraft angle thus lower plasma noise) and MGS has better transfer function to, e.g., the galactic center direction.

The Cassini mission promises to improve significantly on X-band experiments. Cassini launched in October 1997 on a gravity assist (Venus-Venus-Earth-Jupiter) trajectory to Saturn. It has an elaborate radio system (20, 21), notably special hardware to allow Ka-band (~32 GHz) up- and down link capability. The NASA Deep Space Network is implementing one ground station (DSS 25, at Goldstone, CA) with a Ka-band uplink capability and an advanced tropospheric calibration system to measure and calibrate most of the atmospheric scintillation. It is expected that these engineering improvements will give about an order of magnitude lowering of noise level relative to X-band experiments, with the level being set by residual uncertainty in the tropospheric calibration and antenna mechanical noise. Cassini's trajectory is such that it has three solar oppositions (nominally December 16, 2001; December 27, 2002, and January 4, 2004) after the Jupiter gravity assist but before insertion into orbit around Saturn. Cassini will be tracked for 40 days centered on each of these opposition opportunities. Cassini's right ascension at the oppositions will be 5h 34m, 6h 21m, and 6h 57m, while its declination will be +22° 39', +21° 51' and +22° 54'. The earth-spacecraft distance at these oppositions will be 5.7, 7.0, and 7.8 AU. The geometry is thus favorable for observations in the directions of most of the members of the Local Group (except, unfortunately, for the galactic center) and also for observations in the direction of Virgo. Additionally, the large two-way light time offers good transfer function to long period waves.

TABLE 2. Doppler GW experiments.

year of observation	spacecraft	comment
1980	Voyager	S-band uplink; a few tracking passes; burst search (13)
1981, 1988	Pioneer 10	S-band; 3 passes (1981); search for periodic waves at targeted frequency and direction; no GW from Geminga (14); 10 days of data (1988); search for chirp and coalescing binaries (15)
1983	Pioneer 11	S-band; 3 days of data; broadband search for periodic waves (16)
1992, 1993	Ulysses	S-band uplink, S/X band down link; 14 days in 1992, 19 days in 1993; part of 1993 coincidence experiment; search for all waveforms (2, 17)
1993, 1994, 1995	Galileo	S-band; 19 days in 1993, 40 days in 1994, 1995; part of 1993 coincidence experiment; search for all waveforms (18)
1993	Mars Observer	first X-band experiment; 19 days; part of 1993 coincidence experiment; search for all waveforms (19)
1997	Mars Global Surveyor	X-band; 21 days
2001, 2002, 2003	Cassini	first Ka-band experiment (advanced spacecraft radio system: X/Ka up and downlinks); 40 days per solar opposition; tropospheric calibration (20, 21)

Summary

Unlike other detectors, the Doppler GW antenna is large compared with the wavelength; thus there is a time-resolved three pulse response for the signal. The system is sensitive in the low frequency band, with geometry-dependent pulse cancellation setting the lowest frequency for full sensitivity to ~1/(two-way light time) (Figure 1) and link SNR (or the requirement to integrate long enough for good stability of the frequency standard) setting the high frequency limit to about 0.1 Hz.

The main noise sources are plasma scintillation noise, frequency standard noise, tropospheric scintillation noise, and antenna mechanical noise; these enter the Doppler observations with transfer functions different than the GW (Table 1). The relative magnitude of these noise sources depends on the tracking technology employed. (For example, plasma phase scintillation in the solar wind dominate S-band (~2.3 GHz) observations.) The differences in the transfer function and in the noise levels can be used in signal processing design.

Current best sensitivity is achieved with X-band radio systems. X-band sensitivity is limited by a combination of tropospheric and plasma scintillation noises. The equivalent sinusoidal sensitivity for, e.g., Mars Observer, is slightly better than 10^{-15} over most of the LF band.

Cassini (Ka-band, 32 GHz radio link) will have strong suppression of plasma scintillation noise and a special tropospheric calibration system. Cassini is expected to be roughly 10 times more sensitive than X-band experiments and will probably be limited by residual ground antenna mechanical motion and residual uncalibrated troposphere.

To do significantly better than Cassini will require (a) moving all the test masses into space, (b) even higher-frequency links, and (c) an interferometric configuration to cancel the frequency standard noise. Cassini is a very important improvement in the LF band and is probably less than a order of magnitude from the ultimate practical sensitivity of the Doppler GW technique. To improve on Cassini will require a LISA-like space-borne observatory.

Acknowledgments

NASA's precision Doppler tracking capability is the result of work by many people. Crucial roles have been and are being played by individuals in the Deep Space Network, the Flight Projects, and in the project radio science and radio science support teams. I thank in particular colleagues who were investigators on the Galileo/Mars Observer/Ulysses coincidence experiment: B. Bertotti, F. Estabrook, L. Iess, and H. Wahlquist. The research described here was carried out at the Jet Propulsion Laboratory, California Institute of Technology, under a contract with NASA.

References

1. Estabrook, F. B. and Wahlquist, H. D. *Gen. Rel. Grav.*, **6**, 439-447 (1975).
2. Bertotti, B., Ambrosini, R., Armstrong, J. W., Asmar, S. W., Comoretto, G., Giampieri, G., Iess, L. Koyama, Y., Messeri, A., Vecchio, A., and Wahlquist, H. D., *Astron. Astrophys.*, **296**, 13 (1995).
3. Iess, L. and Armstrong, J. W. "Spacecraft Doppler Experiments", in *Gravitational Waves: Sources and Detectors.*, I. Ciufolini and F. Fidecaro eds. (World Scientific, 1997), p. 323.
4. Barnes, J. A. *et al.*, Characterization of Frequency Stability, *National Bureau of Standards Technical Note 394*, Boulder CO, 1970.
5. Wahlquist, H. D., Anderson, J. D., Estabrook, F. B., Thorne, K. S., *Atti dei Covegni Lincei,* **34,** 335 (1977).
6. Estabrook, F. B. Gravitational Wave Detection with the Solar Probe. II. The Doppler Tracking Method, in *A Close-Up of the Sun,* edited by M. Neugebauer and R. W. Davies, pp. 441-449, JPL Publication 78-70 (1978).
7. Armstrong, J. W., Woo, R., and Estabrook, F. B., *Ap. J.*, **230**, 570 (1979) (erratum *Ap. J.* **240**, 719, 1980).
8. Armstrong, J. W. "Spacecraft Gravitational Wave Experiments" in *Gravitational Wave Data Analysis*, edited by B. Schutz, Dordrecht: Kluwer, 1989, pp. 153-172.
9. Riley, A. L. *et al.*, "Cassini Ka-Band Precision Doppler and Enhanced Telecommunications System Study", *Joint NASA/JPL/ASI Study on Ka-band*, Jet Propulsion Laboratory, Pasadena CA, 1990.
10. Armstrong, J. W. *Radio Science* (1998, in press).
11. Keihm, S. J., *TDA Progress Report*, **42-122,** 1-11 (1995).
12. Otoshi, T. Y., Franco, M. M., and Lutes, G. F., *Proc. IEEE*, **82**, 788 (1994).
13. Hellings, R. W., Callahan, P. S., and Anderson, J. D., *Phys Rev. D*, **23**, 844 (1981).
14. Anderson, J. D., Armstrong, J. W., Estabrook, F. B., Hellings, R. W., Lau, E. K., and Wahlquist, H. D., *Nature*, **308**, 158 (1984).
15. Anderson, J. D., Armstrong, J. W., and Lau, E. K., *Ap. J.*, **408**, 287 (1993).

16. Armstrong, J. W., Estabrook, F. B., and Wahlquist, H. D., *Ap. J.*, **318**, 536 (1987).
17. Iess, L. and Armstrong, J. W. "Spacecraft Doppler Experiments", in *Gravitational Waves: Sources and Detectors.*, I. Ciufolini and F. Fidecaro eds. (World Scientific, 1997), p. 323.
18. Armstrong, J. W. "The Galileo/Mars Observer/Ulysses Coincidence Experiment", to be published in *Second Amaldi Conference on Gravitational Waves*, E. Coccia *et al.* eds (World Scientific: Singapore), 1998.
19. Iess, L., Armstrong, J. W., Bertotti, B., Wahlquist, H. D., and Estabrook, F. B. "Search for Gravitational Wave Bursts by Simultaneous Doppler Tracking of Three Interplanetary Spacecraft", talk presented at the *15th International Conference on General Relativity and Gravitation*, Pune, India December 16-21, 1997.
20. Comoretto, G., Bertotti, B., Iess, L., and Ambrosini, R., *Nuovo Cimento C*, **15**, 1193 (1992).
21. Bertotti, B., Comoretto, G., and Iess, L.,*Astron. Astrophys.*, **269**, 608 (1993).
22. Tinto, M. and Armstrong, J. W., *Phys. Rev. D*, (1998, in press).

Inertial Sensor for the Gravity Wave Missions

Vincent Josselin, Manuel Rodrigues and Pierre Touboul

Department of Physics, Instrumentation and Sensing, ONERA, BP 72, F-92322 Châtillon Cedex, France

Abstract. Several space missions have been proposed in the present years for the observation of the gravity waves by exploiting laser interferometry between three or six satellites. The laser links between the satellites constitute the arms of the interferometer of a few million kilometres long. Beside the difficulties of the interferometer exhibiting picometer accuracy, the interferometer mirrors are obtained with the proof-masses of inertial sensors that must present outstanding accuracy. The development of such an instrument has been undertaken. It is derived from existing space accelerometers. The mission requirements are presented and the approach to reach the performances is detailed. A sensor prototype, which has been defined on the basis of theoretical analysis and on the experience already acquired with similar accelerometers, is presently produced in view of ground and in orbit design evaluations. This prototype is also described in the paper.

INTRODUCTION

The LISA mission, Laser Interferometer Space Antenna, aims at detect and observe the gravity waves (1) resulting from violent events in the Universe in a frequency range from 10^{-4} Hz to 10^{-1} Hz. On the contrary to the ground-based experiments under development like VIRGO or LIGO which are limited at low frequencies by the seismic noise, LISA will benefit from the very quiet environment in space and from the 5×10^{6} km length of the interferometer arm realised with a cluster of three spacecraft flying in a triangle configuration.

In each spacecraft, two inertial sensors are located at the end of each laser path from/to the two other LISA spacecraft. The proof-masses of the inertial reference sensors reflect the light coming from the YAG laser and define the reference mirrors of the interferometer arms. The same proof-masses are also used as inertial referencesfor the drag-free control of the spacecraft which constitute shields to external forces. The proposed sensor (2) (commonly called CAESAR: Capacitive And Electrostatic Sensitive Accelerometer Reference) can be derived from existing space qualified electrostatic accelerometers already developed for the ESA projects, like the GRADIO accelerometer (3) or the ASTRE sensor delivered to ESTEC for the micro-gravity spacelab survey (4). The last one has flown three times on board the COLUMBIA shuttle in 1996 and 1997.

On the contrary to previous developed sensors, no acceleration measurements are needed but only relative position of the inertial mass. From the analysis of the overall mission performance (5), the requirements can be simply summarised as follows :

- limitation of spurious accelerations applied on the reflecting proof-mass to the following gabarit :

$$3\times10^{-15}\,\mathrm{m\,s^{-2}\,Hz^{-1/2}}\times\left(\frac{10^{-4}\,\mathrm{Hz}}{f}\right)^{1/3}\left[1+\left(\frac{f}{3\times10^{-3}\,\mathrm{Hz}}\right)^{2}\right] \tag{1}$$

- limitation of the relative displacement between each sensor proof-mass and the spacecraft to a few $10^{-9}\,\mathrm{m\,Hz^{-1/2}}$.

In this present paper, the actual operation of the drag compensation of the satellite has not been taken into account. The sensor has been considered as a self consistent instrument for the evaluation of performance without taking into account its dynamics and its coupling with the « drag-free » controller. Such analysis should be addressed during future studies. Furthermore, in the LISA mission now envisaged in cooperation by NASA and ESA (5), two sensors are implemented in the same drag-free spacecraft. The drag-free control strategy is more complex than in the previous

CP456, *Laser Interferometer Space Antenna*
edited by William M. Folkner
 1-56396-848-7/98/$15.00

is more complex than in the previous configuration depicted in the reference (1) with one drag-freesensor per satellite. For what concerns the sensor, this may lead to a less soft acceleration environment with couplings between the motions in translation themselves and with the attitude motions.

INERTIAL SENSOR OVERVIEW

The sensor mechanics is mainly constituted by (see Figure 1) : one quite cubic proof-mass made of a gold-platinum alloy, four gold coated electrode plates made in ULE and constituting the sensor core, one referencesole plate also made in ULE or INVAR that will provide the mechanical interface with the optical bench, eventually one blocking mechanism with gold coated fingers, to keep the proof-mass motionless during the launch and one tight housing made of titanium carrying two windows for the laser beam. Although its high value of thermal expansion coefficient, the gold-platinum alloy is selected because of its high density and its low magnetic susceptibility in order to minimise the effects of all the parasitic forces.

Thanks to an ultrasonic machining specific technique, the electrodes are precisely engraved on the ULE plates. The gaps between the mass and the electrodes result from a compromise between the operation robustness and the possible disturbing effects to the mass motion resulting from the electrode proximity. For the Z sensitive axis, that is the laser beam axis, the limitation of the acceleration disturbances has been privileged and a larger gap of 1.5 mm is selected as a compromise between the capacitive position sensing accuracy, its dynamic range and the spurious acceleration originated in electrical sources.

Derived from the definition of GEOSTEP instrument (2), an electrostatic servo-controlled accelerometers dedicated to space fundamental physics, the proposed design, presented in the Figure 1, is based on the electrostatic suspension of the proof-mass which position and attitude are measured with capacitive sensors. The variations of the capacitances between the mass and the instrument cage depend on the area variations of the electrodes in regard to the mass.

In the case of LISA sensor and on the contrary to standard accelerometers, the electrostatic levitation is only used to centre the mass in orbit and to control it before the operation of the S/C drag compensation system. Then, exploiting the capacitive position sensing, the drag compensation system activates the satellite thrusters in such a way that the satellite follows the mass with an accuracy of 1 nanometer in a bandwidth between DC and 10^{-1} Hz. The electrostatic actuators may also be used along the normal direction to the laser beam or to control the mass attitude because of unavoidable coupling between the different motions.

INERTIAL SENSOR PERFORMANCE EVALUATION

A list of the noise sources (5) which generate disturbing forces on the mass has been established, in order to evaluate the performance of the inertial sensor. They depend on the geometrical configuration of the sensor, the accuracy of the geometry, the quality of the surface and the characteristics of the used materials, but also on the selection of the electrical parameters which define the capacitive sensing and the electrostatic actuator operation. These parameters can be selected with the previous accelerometer development experience. The inertial sensor environment is a major driver of the performances and stringent specifications have to be applied on the satellite like the temperature stability.

For the electrical design, the sine wave detection voltage V_d for the capacitive sensing, the mean difference of potential V_p between proof-mass and electrodes and the contact potential differences dV_p have been characterised by their root mean square and the power spectral density of their fluctuations. The capacitive sensor characteristics, driven by the electrode configuration and by the electronics gains and noise, are also globally modelled by sensitivity and noise. The Table 1 gives a summary of some of the main configuration parameters along the laser beam direction.

As presented in Figure 1, the proposed design appears compatible with the objectives of the gravity wave missions.

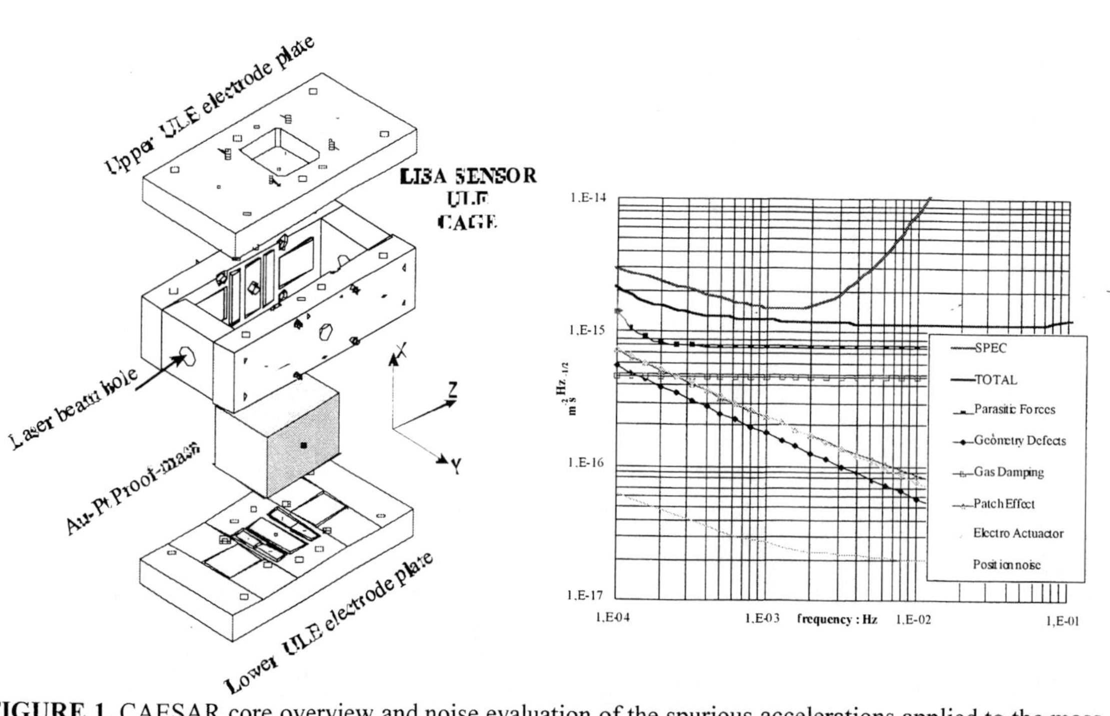

FIGURE 1. CAESAR core overview and noise evaluation of the spurious accelerations applied to the mass.

TABLE 1. Main parameters for the selected sensor configuration along the laser beam direction

Parameters	Values	Variations
Mechanics : Proof-mass mass	1.3 kg	
Mechanics : Considered gap for the Z axis capacitive sensing	1.5 mm	
Electrical : V_p	0.1 V	90 μV Hz^{-}- at 10^{-4} Hz
Electrical : dV_p	15 mV	10 μV Hz^{-}- at 10^{-4} Hz
Electrical : V_d	2 V_{rms}	450 μV Hz^{-}- at 10^{-4} Hz
Electrical : Capacitive sensor gain for the Z axis	100 V pF^{-1}	
Electrical : Capacitive sensor noise for the Z axis		5×10^{-7} pF Hz^{-}- at 10^{-4} Hz
Environment : Mechanics temperature	300 K	10^{-3} K Hz^{-}- at 10^{-4} Hz
Environment : Spatial temperature gradients	0.01 K	10^{-3} K Hz^{-}- at 10^{-4} Hz
Environment : Magnetic moment on board	1.5 Am^2 at 0.35 m	1%

LABORATORY PROTOTYPE DEVELOPMENT

The extraordinary performance necessary for the gravity wave missions cannot be achieved with instruments able to measure or even support 1 g because of the huge range that should be then required. This means that the inertial sensor will be integrated inside the satellite without any ground tests able to verify the final flight configuration of the sensor, its overall operation and performance. Nevertheless, confidence and reliability can be ensured by developing specific experiments with dedicated laboratory testing benches or drop tower falls. With regard to the specification of the ambitious envisaged missions, in-orbit experimentation appears also necessary in particular to verify the operation and the performance of the drag compensation system including the inertial sensor and the ion thrusters. This is what is considered in the ODIE mission (Orbiting Drag-free International Explorer) led by the University of Stanford and the ELITE mission (European LIsa TEchnology demonstration drag-free satellite) studied by ESTEC.

A first laboratory model of the CAESAR sensor has been defined and is being produced in order to be as representative as possible of the flight instrument but taking into account the constraints linked to the ground operation. This model gives the opportunity to implement a blocking mechanism, to verify the resistance to vibration launch and to test a discharge device needed to control the charge of the mass (6). Numerous experimental investigations are envisaged and should be of particular interest for LISA like : the capacitive sensing performancein situ (in electrostatic closed loop), the residual stiffness between the mass and the instrument cage, the damping of the mass motion, the coupling between the different axes, the digital control operation...

This prototype is realised with the same technology and electronics envisaged for the LISA sensor. Only a few parameters have been modified from the configuration depicted on Figure 1, in order to allow ground investigations. The prototype is also compatible with a flight demonstration mission and it is necessary before the production of the LISA model that can be only tested in microgravity conditions during the few second falls in drop tower.

CONCLUSION

The current design of the inertial sensor for gravity wave missions is optimised in one privileged direction in order to minimise the back-actions induced on the proof-mass acceleration by the detection voltages. This optimisation appears compatible with the present space mission objectives. Such a sensor is derived from previous developed space accelerometers.

This model will be tested in 1999 and an up-to-date issue should be integrated in the ODIE mission satellite for a flight by the end of 2000. The ODIE satellite environment and its operation allow in-orbit performance of 10^{-12} m s^{-2} $Hz^{-1/2}$ for the drag-free sensor which is still at an important distance from the gravity wave mission requirements. Nevertheless, such a mission is very fruitful to test the operation and the performance of the drag-free compensation system and to measure the sensitivity to environmental parameters. Moreover the developed instrument will be the first step to realise the sensor needed for more ambitious demonstrators like ELITE, that could be realised in the frame of the NASA Deep Space program.

ACKNOWLEDGEMENTS

The development of the inertial sensor laboratory model is performed under CNES financial support. The authors wish to thank Pr Cruise, Pr Twiggs and their teams for the valuable discussions concerning the electronics architecture compatible with the micro-satellite electrical configuration.

REFERENCES

1. K. Danzmann & LISA study team, *LISA:* Laser Interferometer Space Antenna for graviational wave measurements, *Class. Quantum Grav.* **13**, A247-A250 (1996).

2. P. Touboul, M. Rodrigues, E. Willemenot and A. Bernard, Electrostatic accelerometers for the equivalence principle test in space, Class. Quantum Grav. **13**, A67-A78 (1996).

3. P. Touboul et al, Continuation of the GRADIO accelerometer predevelopment, *ONERA Final Report* **62/6114PY** ESTEC contract (1993).

4. M. Nati, A. Bernard, B. Foulon, P. Touboul, ASTRE-A highly preformant accelerometer for the low frequency range of the microgravity environment, *5th European symposium on Space Environmental Control Systems and 24th International conference on Environmental Systems* (June 20-23, 1994).

5. P. Touboul, M. Rodrigues, Accelerometer Design Optimisation for the LISA mission, *ONERA Final Report* **24/3815DMPH/Y** ESTEC Contract (July 1998)

6. Y. Jafry, T. Sumner and S. Buchman, Electrostatic charging of space-borne test bodies used in precision experiments, *Class. Quantum Grav.* **13**, A97-A106 (1996).

Author Index

A

B

C

D

E

F

R

S

T

V

W

Y

Z